Data Analysis

with
Microsoft® Excel

Duxbury titles of related interest

Data Analysis

with
Microsoft® Excel

Kenneth N. Berk
Illinois State University

Patrick Carey
Carey Associates, Inc.

Duxbury
Thomson Learning™

Australia • Canada • Mexico • Singapore • Spain • United Kingdom• United States

Sponsoring Editor: *Curt Hinrichs*
Marketing Team: *Karin Sandberg, Beth Kroenke, Monica Brown*
Editorial Assistant: *Emily Davidson*
Production Editor: *Mary Vezilich*
Production Service: *GEX Publishing Services*
Permissions Editor: *Mary Kay Hancharick*

Interior Design: *Jeffrey Parks*
Cover Design: *Craig Hanson*
Design Coordinator: *Roy Neuhaus*
Print Buyer: *Vena Dyer*
Typesetting: *GEX Publishing Services*
Printing and Binding: *Webcom Ltd.*

For more information about this or any other Duxbury products, contact:
DUXBURY
511 Forest Lodge Road
Pacific Grove, CA 93950 USA
www.duxbury.com
1-800-423-0563 (Thomson Learning Academic Resource Center)

Printed in Canada

10 9 8 7 6 5 4

Library of Congress Cataloging-in-Publication Data
Berk, Kenneth N., [date-]
    Data analysis with Microsoft Excel/Kenneth N. Berk, Patrick Carey.
        p. cm.
    "Updated for Office 2000."
    Includes bibliographical references and index.
    ISBN 0-534-36278-8
    1. Microsoft Excel (Computer file) 2. Microsoft Office. I. Carey, Patrick, [date-]II.
Title.

HF5548.4.M523 B47 2000
005.369--dc21   00-021372

I thank my wife, Laura, and sons, David and Peter for
their support in a time-consuming effort.
—*Kenneth N. Berk*

Thanks to my wife, Joan, and my sons, John Paul, Thomas, Peter, Michael,
and Stephen for their love and support.
—*Patrick M. Carey*

About the Authors

Kenneth N. Berk

Kenneth N. Berk (Ph.D., University of Minnesota) is a professor of Mathematics at Illinois State University and a Fellow of the American Statistical Association. His recent work has focused on applied statistics and statistical computing. Berk was editor of Software Reviews for the *American Statistician* for six years. He served as chair of the Statistical Computing Section of the American Statistical Association. He has twice co-chaired the annual Symposium on the Interface between Computing Science and Statistics.

Patrick Carey

Patrick Carey received his M.S. in biostatistics from the University of Wisconsin where he worked as a researcher in the General Clinical Research Center designing and analyzing clinical studies. He co-authored his first textbook with Ken Berk on using Excel as a statistical tool. He and his wife Joan founded *Carey Associates, Inc.*, a software textbook development company. He has since authored or co-authored over 20 academic and trade texts for the software industry. Besides books on data analysis, Carey has written on the Windows® operating system, Web page design, database management, the Internet, browsers, and presentation graphics software. Patrick, Joan, and their five young sons live in Madison, Wisconsin.

Brief Contents

Introduction

Data Analysis with Microsoft® Excel Updated for Office 2000 with accompanying software harnesses the power of Excel and transforms it into a tool for learning basic statistical analysis. Students learn statistics in the context of analyzing data. We feel that it is important for students to work with real data, analyzing real world problems so that they understand the subtleties and complexities of analysis that make statistics such an integral part of understanding our world. The data set topics range from business examples to physiological studies on NASA astronauts. Because students work with real data, they can appreciate that in statistics no answers are completely final and that intuition and creativity are as much a part of data analysis as is plugging numbers into a software package. This text can serve as the core text for an introductory statistics course or as a supplemental text. It also allows non-traditional students outside of the classroom setting to teach themselves how to use Excel to analyze sets of real data so they can make informed business forecasts and decisions.

Users of this book need not have any experience with Excel, although previous experience would be helpful. The first three chapters of the book cover basic concepts of mouse and Windows operation, data entry, formulas and functions, charts, editing and saving workbooks. Chapters 4 through 12 emphasize teaching statistics with Excel as the instrument.

Using Excel in a Statistics Course

Spreadsheets have become one of the most popular forms of computer software, second only to word processors. Spreadsheet software allows the user to combine data, mathematical formulas, text, and graphics together in a single report or workbook. For this reason, spreadsheets have been indispensable tools for business, although they have also become popular in scientific research. Excel in particular has won a great deal of acclaim for its ease of use and power.

As spreadsheets have expanded in power and ease-of-use, there has been increased interest in using them in the classroom. There are many advantages to using Excel in an introductory statistics course. An important advantage is that students, particularly business students, are more likely to be familiar with spreadsheets and are more comfortable working with data entered into a spreadsheet. Since spreadsheet software is very common at colleges and universities, a statistics instructor can teach a course without requiring students to purchase an additional software package.

Having identified the strengths of Excel for teaching basic statistics, it would be unfair not to include a few warnings. Spreadsheets are not statistics packages and there are limits to what they can do in replacing a full-featured statistics package. This is why we have included on the accompanying CD our own add-in, StatPlus™. It expands some of Excel's statistical capabilities. (We explain the use of StatPlus where appropriate throughout the text.) Using Excel for anything other than an introductory statistics course would probably not be appropriate due to its limitations. For example, Excel can easily perform balanced two-way analysis of variance, but not unbalanced two-way analysis of variance. Spreadsheets are also limited in handling data with missing values. While we recommend Excel for a basic statistics course, we feel it is not appropriate for more advanced analysis.

For Users of the Previous Editions

Feedback from users of the earlier editions of *Data Analysis with Microsoft Excel* has been very gratifying. Although the success of those editions has been very encouraging, we felt there were several areas of the text and the software that we could improve for students. This new edition contains more data sets and covers several new topics including Stem and Leaf plots and runs tests. The StatPlus™ software provided with this book has been given a new improved interface and now supports several new statistical tools.

System Information

You will need the following hardware and software to use *Data Analysis with Microsoft Excel:*

- An IBM or compatible microcomputer
- Windows 95 or later
- Excel 97 or later. If you are using an earlier edition of Excel, you will have to use an earlier edition of *Data Analysis with Microsoft Excel*. (Go to **www.duxbury.com** for more information.)

The *Data Analysis with Microsoft Excel* package includes:

- The text, which includes 12 chapters, documentation on the data sets, a reference section for Excel's statistical functions, Analysis ToolPak commands, StatPlus Add-In commands, and a bibliography.

- The CD, containing 85 different data sets from real-life situations, 9 interactive Concept Tutorials, and files for StatPlus—our statistical application. The instructor can install these files on a network or on stand-alone workstations to make them available to students. Chapter 1 of the text includes instructions to students for installing the files on their own computers.
- An Instructor's Manual (ISBN 0534-37799-8) with instructions for making the Student files accessible to your students is available to adopting faculty. This ancillary also contains answers and solutions to all the exercises.

The Duxbury Web page has additional resources and information at http://www.duxbury.com.

Excel's Statistical Tools

Excel comes with 81 statistical functions and 59 mathematical functions. There are also functions devoted to business and engineering problems. The statistical functions that basic Excel provides include descriptive statistics such as means, standard deviations and rank statistics. There are also cumulative distribution and probability density functions for a variety of distributions, both continuous and discrete.

The Analysis ToolPak is an add-in that is included with Excel. If you have not loaded the Analysis ToolPak, you will have to install it from your original Excel installation disks, following the instructions in your Excel User's Guide.

The Analysis ToolPak adds the following capabilities to Excel:

- Analysis of variance, including one-way, two-way without replication, and two way balanced with replication
- Correlation and covariance matrices
- Tables of descriptive statistics
- One-parameter exponential smoothing
- Histograms with user-defined bin values
- Moving averages
- Random number generation for a variety of distributions
- Rank and percentile scores
- Multiple linear regression
- Random sampling
- t-tests, including paired and two sample, assuming equal and unequal variances
- z-tests

In this book we make extensive use of the Analysis ToolPak for multiple linear regression problems and analysis of variance.

StatPlus™

Since the Analysis ToolPak does not do everything that an introductory statistics course requires, this textbook comes with an additional add-in called the **StatPlus™ Add-In** that fills in some of the gaps left by basic Excel and the Analysis ToolPak.

Additional commands provided by the StatPlus Add-In give users the ability to:

- Create random sets of data
- Manipulate data columns
- Create random samples from large data sets
- Generate tables of univariate statistics
- Create statistical charts including boxplots, histograms and normal probability plots
- Create quality control charts
- Perform one-sample and two-sample *t*-tests and *z*-tests
- Perform non-parametric analyses
- Perform time series analyses, including exponential and seasonal smoothing
- Manipulate charts by adding data labels and breaking charts down into categories
- Perform non-parametric analyses
- Create and analyze tabular data

A full description of these commands is included in the Reference section of the text and through on-line help on the CD.

Concept Tutorials

Included with the text and CD are several interactive Excel tutorials that provide students a visual and "hands on" approach to learning statistical concepts. These tutorials cover:

- Boxplots
- Probability
- Probability distributions
- Random samples
- Population statistics
- The Central Limit theorem
- Confidence intervals
- Hypothesis tests
- Exponential smoothing

Acknowledgments

We thank Mac Mendelsohn, Managing Editor at Course Technology, for his support and enthusiasm in the first edition of this book. For this edition, our thanks to Curt Hinrichs, Executive Editor at Duxbury Press, for his editorial insight and encouragement, and to Assistant Editor Seema Atwal, Editorial Assistant Emily Davidson, and Marketing Manager Karin Sandberg. We'd also like to thank our copy editor, Connie Day, our cover designer, Craig Hanson, and Karla Russell, our contact at GEX Publishing Services; at Brooks/Cole our thanks to Design Coordinator Roy Neuhaus, and Production Editor Mary Vezilich for their careful and professional attention to all the details of design and production.

Special thanks go to our reviewers, who gave us valuable insights into improving the book in the original CTI edition and in this new Duxbury edition: Robert L. Andrews, Virginia Commonwealth University; David J. Auer, Western Washington University; Sharon Hunter Donnelly, Health Insight; Samuel B. Graves, Boston College, Carroll School of Management; Tom Obremski, University of Denver; Ruth Reingold, University of Illinois, Urbana-Champaign; Earl Rosenbloom, University of Manitoba; Richard D. Spinetto, University of Colorado at Boulder; Wayne L. Winston, Indiana University; and Jack Yurkiewicz, Pace University. Any remaining mistakes or omissions are ours.

We thank Laura Berk, Peter Berk, Robert Beyer, David Booth, Orlyn Edge, Stephen Friedberg, Maria Gillett, Richard Goldstein, Glenn Hart, Lotus Hershberger, Les Montgomery, Joyce Nervades, Diane Warfield, and Kemp Wills for their assistance with the data sets in this book. We especially want to thank Dr. Jeff Steagall, who wrote some of the original material for Chapter 12, Quality Control. If we have missed anyone, please forgive the omission.

Kenneth N. Berk

Patrick M. Carey

Contents

PART 2
FUNDAMENTALS
OF STATISTICS   117

PART I

Excel

Chapter 1
Getting Started with Excel

Chapter 2
Working with Data

Chapter 3
Working with Charts

Getting Started with Excel

Objectives

In this chapter you will learn to:

- Navigate Windows and the Windows Desktop

- Install StatPlus files

- Start Excel and recognize elements of the Excel workspace

- Work with Excel workbooks, worksheets, and chart sheets

- Scroll through the worksheet window

- Work with Excel cell references

- Print a worksheet

- Save a workbook

- Install and remove Excel add-ins

- Work with Excel add-ins

- Learn about the features of StatPlus

I n this chapter you'll learn how to work with Excel in the Windows® operating system. You'll be introduced to basic workbook concepts, including navigating through your worksheets and worksheet cells. This chapter also introduces StatPlus, an Excel add-in supplied with this book and designed to expand Excel's statistical capabilities.

The Windows® Operating System

This book does not require prior Excel experience, but familiarity with basic Microsoft Windows features like dialog boxes, menus, and on-line Help will reduce your start-up time. This section provides a quick overview of the Windows elements you'll use in this book. There are many different versions of Windows. This text assumes that you'll be working with **Windows 2000**, **Windows 98** or **Windows 95**. If you are using an earlier version of Windows, such as Windows 3.1, you should refer to the first edition of *Data Analysis with Microsoft Excel*.

To start Windows:

1   Turn on your computer.

2   On some computer networks you may be required to enter a username and password to access Windows. Talk with your instructor if this is the case.

3   Windows starts, displaying the desktop shown in Figure 1-1.

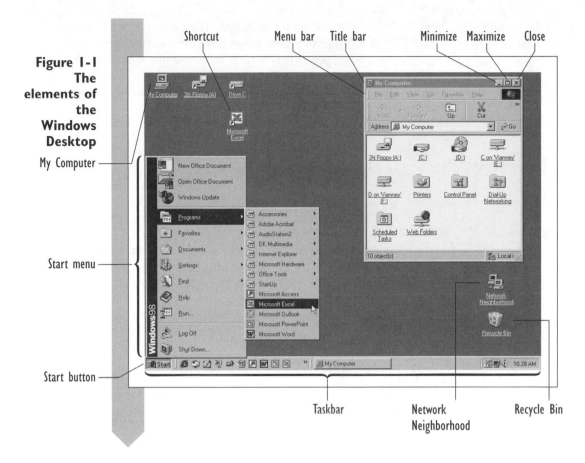

**Figure 1-1
The elements of the Windows Desktop**

Shortcut   Menu bar   Title bar   Minimize   Maximize   Close

My Computer

Start menu

Start button

Taskbar   Network Neighborhood   Recycle Bin

Viewing the Windows Desktop

The Desktop is the home base from which you open program applications, manage files, work with your computer's operating system, and accomplish many other tasks. The Desktop shown in Figure 1-1 is an example of a typical desktop; yours will probably look different. There might be other windows open, with fewer or more elements, depending on how your system has been set up.

Table 1-1 provides a quick identification of the various Windows elements.

Table 1-1 Windows Elements

Windows Element	Image	Purpose
Taskbar		Displays currently active programs and the Start menu
Start menu		Used to launch program and get information about your computer
Title bar		Identifies a window on the Desktop
Menu bar		A list of commands available in the current application

(continued)

Minimize	▭	Minimizes the window to an icon on the Start menu
Maximize	◻	Expands the window to fill the entire screen
Midsize	⊡	Changes a full-screen window to a midsize window
Close	☒	Closes the currently active window
Icon	☒	Small graphic representing applications, windows, or documents
Shortcut	☒	Icon that is used to point to the location of an application, window or document
Start button	🏁Start	A button on the task bar that is used to display a menu of Windows commands and applications
Recycle Bin	🗑	An icon on the Desktop that you drag objects into for deletion
Network Neighborhood	🖧	An icon on the Desktop that contains information about your network
My Computer	🖥	An icon on the Desktop that contains information about your computer and its contents

You work with these various elements with your mouse (you can also use your keyboard, but this book assumes that you'll be using a mouse).

Practicing Mouse Techniques

There are four basic mouse operations used in Windows applications like Excel. These are shown in Table 1-2.

Table 1-2 Mouse Operations

Operation	Description
Clicking	Move your mouse so that the tip of the pointer arrow ⬛ is touching the element you want to work with. Then press and release the left mouse button.
Right-clicking	Same as clicking, except you press and release the right mouse button.
Double-clicking	Point at an element as if you were going to click it, then press and release the left mouse button twice in rapid succession.
Dragging	Point at an element, press and hold down the left mouse, and then, with the button still pressed, move the mouse and the element across the screen. When you have positioned the element where you want it, release the mouse button.

To practice using these mouse techniques:

1 Double-click the **My Computer** icon on the Windows Desktop.

The My Computer window opens on the Desktop, as shown in Figure 1-2.

**Figure 1-2
The My
Computer
window**

Note: If you are using Windows 98 with the Web Desktop, you will only have to single-click the My Computer icon to open the My Computer window.

2 Click the **Minimize** button ▭ to reduce the My Computer window to a button on the Taskbar.

3 Click the **My Computer** icon on the Taskbar to restore the My Computer window to the Desktop.

4 Move the pointer arrow ▷ to the My Computer window title bar and drag the title bar to the center of the Desktop.

5 Click the **Maximize** button ▯.

The My Computer window now fills the entire screen as shown in Figure 1-3.

**Figure 1-3
The My
Computer
window
maximized**

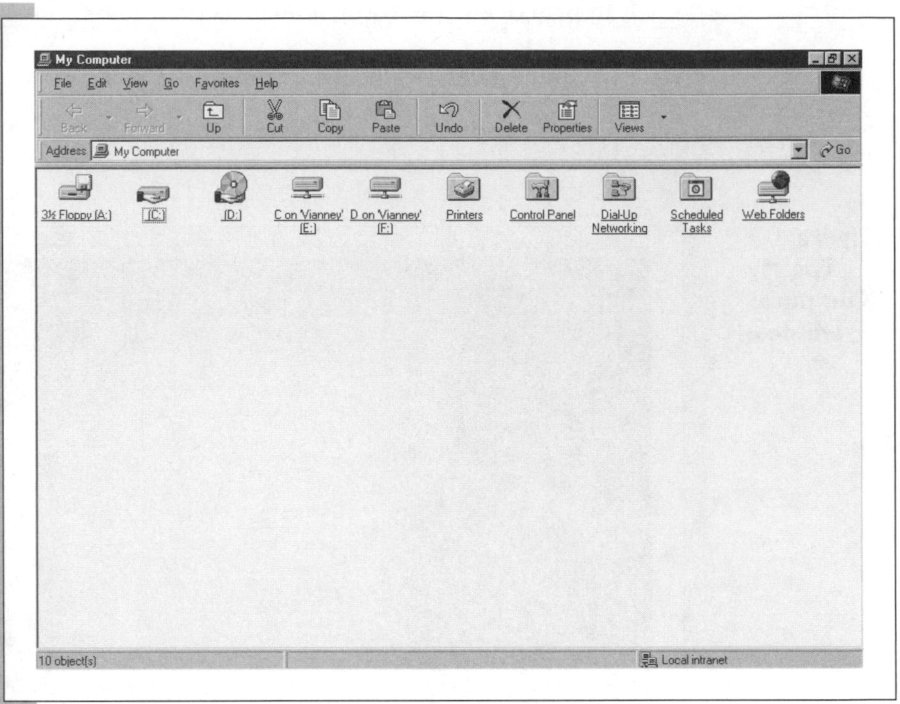

6 Click the **Midsize** button 🗗 to return the My Computer window to its original size.

7 Point to the lower-right corner of the My Computer window so that the pointer turns into a double-headed arrow ⬂, drag the window corner down and to the right, then release the mouse button.

This enlarges the My Computer window. When you are working with more than one window at a time, you will often want to resize windows and arrange them on the Desktop.

8 Click the **Close** button ❌ to close the My Computer window.

Getting Help

The basic mouse operations you just practiced are fundamental to communicating with Windows applications. If you're a new computer user, you may want to spend extra time working with the mouse until you feel comfortable with all of the operations.

If you need more information about Windows, you can also access Windows' on-line Help system. On-line Help provides tips and tutorials to help you master various Windows tasks.

To access on-line Help:

1 Click **Start** 🔲Start on the Taskbar and click **Help**.

2 The Windows Help window opens as displayed in Figure 1-4.

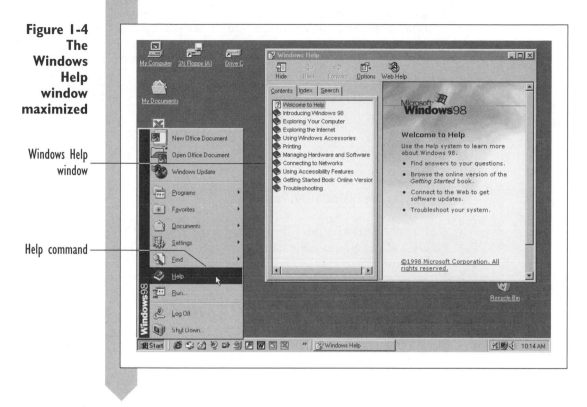

The on-line Help system is extremely useful. It's well worth the time to become familiar with it. Practically every Windows application will provide its own on-line Help windows. Excel is no exception. Much of what you'll learn in this chapter can be found in the on-line Help window for both Windows and Excel.

Special Files for This Book

This book includes additional files to help you learn statistics. There are three types of files you'll work with: StatPlus files, Explore workbooks and Data (or Student) files.

Excel has many statistical functions and commands. However, there are some things that Excel does not do (or does not do easily) that you will need to do in order to perform a statistical analysis. To solve this problem, this book includes **StatPlus**, a software package that provides additional statistical commands accessible from within Excel.

The **Explore workbooks** are self-contained tutorials on various statistical concepts. Each workbook has one or more interactive tools that allow you to see these concepts in action.

The **Data** or **Student files** contain sample data from real-life problems. In each chapter, you'll analyze the data in one or more Data file, employing various statistical techniques along the way. You'll use other Data files in the exercises provided at the end of each chapter.

Installing the Files

The disk comes with an installation program that you can use to install these files on your computer. Install your files now.

To run the installation routine:

1   Place the installation disk in your drive, click the **Start** button ![Start], click **Run** and enter the name of the CD drive, followed by **:/Setup**. For example, if your CD-ROM is drive "d", enter **d:/Setup** in the run box.

2   Click **OK**.

3   Click the **Next** button to start the installation program.

4   Click **Yes** to accept the license agreement for this product.

5   Click the **Next** button after viewing the content of the disk.

6   Enter your name and company (or school) name in the User Information box. Click **Next**.

7   By default, the installation program will install the Data files and Explore workbooks in the Berk-Carey folder of your hard disk. You can click the Browse button to select a different folder. After you've specified a location, click the **Next** button. Please make note of the location you choose because you'll need that information later in this chapter.

    After a few moments, the files are copied to your hard disk.

8   Click **Finish** to complete the installation program.

The files are organized within separate folders on your hard disk. For example, if you specify the Berk-Carey folder in the installation program, the files will be arranged in three separate subfolders as shown in Figure 1-5.

Figure 1-5 Folders on your hard disk

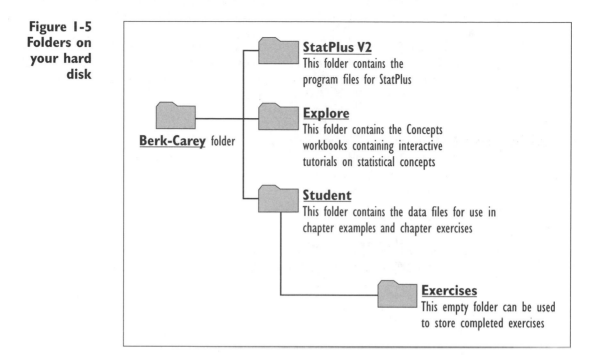

Later in this chapter, you'll learn how to access the StatPlus program from within Excel.

Excel and Spreadsheets

Excel is a software program designed to help you evaluate and present information in a spreadsheet format. **Spreadsheets** are most often used by business for cash-flow analysis, financial reports, and inventory management. Before the era of computers, a spreadsheet was simply a piece of paper with a grid of rows and columns to facilitate entering and displaying information as shown in Figure 1-6.

**Figure 1-6
A sample
Sales
spreadsheet**

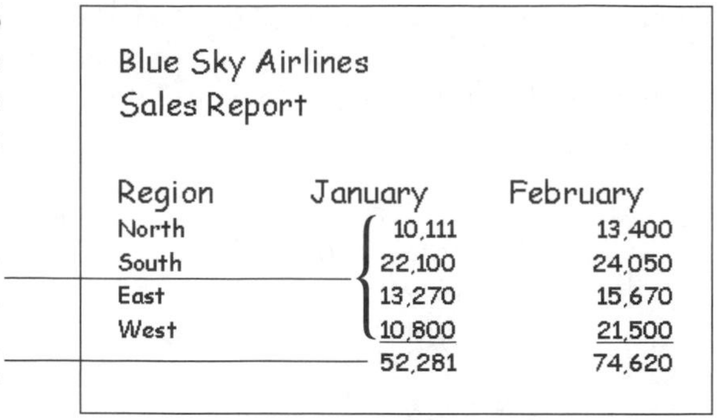

Blue Sky Airlines
Sales Report

Region	January	February
North	10,111	13,400
South	22,100	24,050
East	13,270	15,670
West	10,800	21,500
	52,281	74,620

you add these
numbers

to get this
number

Computer spreadsheet programs use the old hand-drawn spreadsheets as their visual model but add a few new elements, as you can see from the Excel worksheet shown in Figure 1-7.

**Figure 1-7
A sample
spreadsheet
as formatted
within Excel**

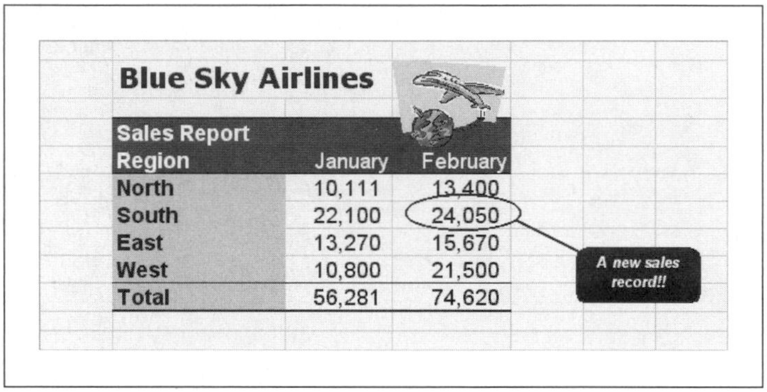

Blue Sky Airlines

Sales Report		
Region	January	February
North	10,111	13,400
South	22,100	24,050
East	13,270	15,670
West	10,800	21,500
Total	56,281	74,620

A new sales record!!

However, Excel is so flexible that its application can extend beyond traditional spreadsheets into the area of data analysis. You can use Excel to enter data, analyze the data with basic statistical tests and charts, and then create reports summarizing your findings.

Launching Excel

When Excel is installed on your computer, the installation program automatically inserts a shortcut icon to Excel in the Programs menu located under the Windows Start button. You can click this icon to launch Excel.

To start Excel:

1 Click the **Start** button 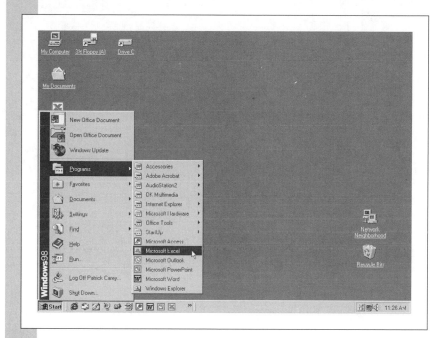 on the Taskbar and point to **Programs**.

2 Click **Microsoft Excel** as shown in Figure 1-8.

Note: Depending on how Windows has been configured on your computer, your Start menu may look different from the one shown in Figure 1-8. Talk to your instructor if you have problems launching Excel.

**Figure 1-8
Starting
Excel from
the
Windows
Start menu**

3 Excel starts up, displaying the window shown in Figure 1-9.

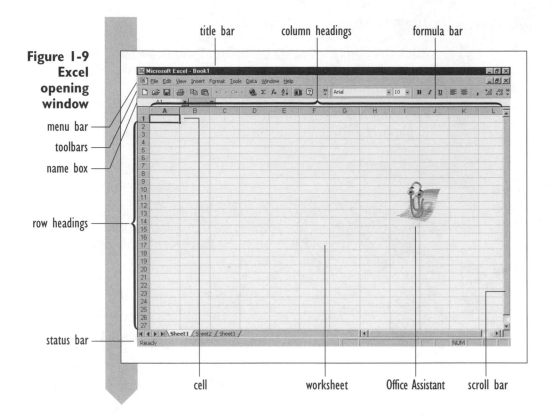

Figure 1-9
Excel opening window

title bar   column headings   formula bar

menu bar

toolbars

name box

row headings

status bar

cell   worksheet   Office Assistant   scroll bar

Viewing the Excel Window

The Excel window shown in Figure 1-9 is the environment in which you'll analyze the data sets used in this textbook. Your window might look different depending on how Excel has been set up on your system. Before proceeding, take time to review the various elements of the Excel window. A quick description of these elements is provided in Table 1-3.

Table 1-3 Excel Elements

Excel Element	Purpose
Title bar	Displays the name of the application and the current Excel document
Menu bar	Contains the names of Excel menus, each of which lists commands for specific Excel tasks
Toolbars	Contain buttons that allow one-click access to various Excel commands and features
Formula bar	Displays the formula or value entered into the currently selected cell
Office Assistant	Provides on-line Help for a variety of Excel tasks
Status bar	Displays messages to user about current Excel operations
Worksheets	Display the contents of an Excel spreadsheet or chart
Cells	Store individual text or numeric entries

(continued)

Column headings	Organize cells into lettered columns
Row headings	Organize cells into numeric rows
Scroll bars	Used to view cell entries that lie outside the main Excel window
Name box	Displays the name or reference of the currently selected object or cell

Running Excel Commands

You can tell Excel what you want to do by using the menus or the toolbars. To run a command from the menu bar, move the mouse pointer on a menu name, click the mouse button to open the menu, and click the command you want to use. Figure 1-10 shows how you would open a file using the Open command available on the File menu. Note that some of the commands have **keyboard shortcuts**—key combinations that run a command or macro. For example, pressing the CTRL and "O" keys simultaneously will also run the Open command.

**Figure 1-10
Opening a
file using
the Excel
menu**

keyboard shortcut

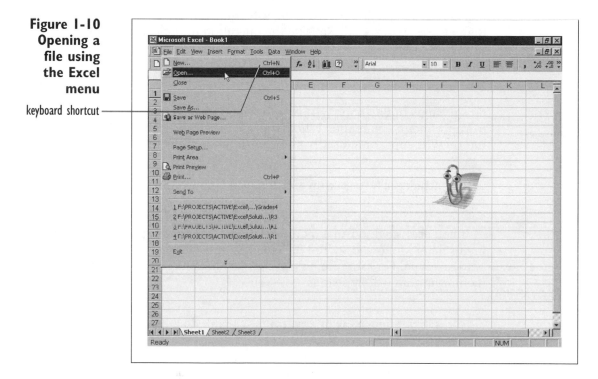

Excel also offers a rich collection of toolbars to allow you one-click access to many of the same commands in the Excel menus. By default, Excel displays two toolbars: the Standard toolbar and the Formatting toolbar. The **Standard toolbar** contains buttons for basic Excel tasks like opening or saving files or printing worksheets. The **Formatting toolbar** contains buttons related to tasks such as formatting your file's appearance. There are other toolbars that may appear to assist you in working with specific Excel tasks such as creating charts or running macros. Table 1-4 shows some of the more commonly used toolbar buttons.

Table 1-4 Toolbar Buttons

Button	Image	Purpose
New		Create a new blank file
Open		Open an existing document
Save		Save the current document
Print		Print the current document
Print Preview		Preview the printout before printing it
Cut		Cut the currently selected text or object and place it in the paste buffer
Copy		Copy the currently selected text or object and place it in the paste buffer
Paste		Insert the contents of the paste buffer into the file
Undo		Undo the last command
Redo		Reverse the operation of the Undo command
Chart		Insert a chart into the document using the Chart wizard

The steps in this book will instruct you either to click a toolbar button or to run a menu command. When you are asked to use a toolbar button, the button will appear pictorially along with the button name in boldface type:

1 Click the **New** button to create a new blank document.

If you are asked to run a menu command, the instruction will include the menu name, followed by the ">" symbol and the name of the command in boldface type. For example, to run the New command located under the File menu, the step will appear as

1 Click **File > New** to create a new blank document.

If you are asked to run a command using a keyboard shortcut, the keyboard combination will be shown in boldface with the keys joined by a plus sign to indicate that you should press these keys simultaneously. For example,

> Press **CTRL+N** to create a new blank document.

In addition to the built-in menus, toolbars, and keyboard shortcuts, you can also customize Excel by creating your own toolbars and commands. StatPlus, for example, adds a new menu called "StatPlus" to the Excel menu bar. Users who have not installed StatPlus will not have this particular menu appear in their document window.

Excel Workbooks and Worksheets

Excel documents are called **workbooks**. Each workbook is made up of individual spreadsheets called **worksheets** and sheets containing charts called **chart sheets**. You'll learn more about worksheets and chart sheets in the next section.

Opening a Workbook

To learn some basic workbook commands, you'll first look at an Excel workbook containing public-use data from Kenai Fjords National Park in Alaska. The data is stored in the file **Park.xls**, located in the Student files folder. Here "Park" is the name of the file, and ".xls" is a letter file extension that identifies the file as an Excel workbook. Open this workbook now.

> **To open the Park workbook:**
>
> Click the **Open** button 📂.
>
> The Open dialog box appears as shown in Figure 1-11. Your dialog box will probably display a different folder and file list.

Figure 1-11
The Open
dialog box

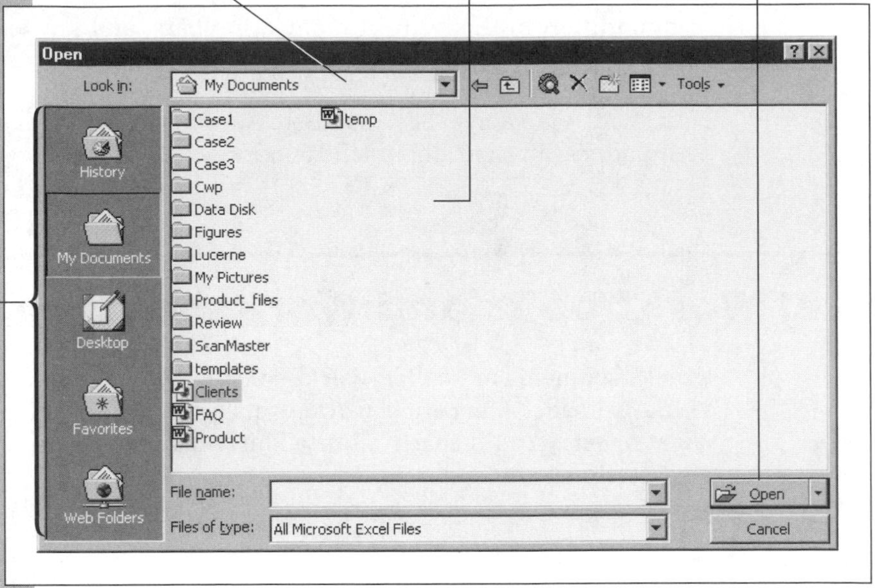

Drop-down
list containing
folder tree

List of files in the
current folder

Click to open the
selected file

Shortcut icons to
specific folders

The Open dialog box

2 Click the **Look In** drop-down arrow to open the tree of folders and objects available on your computer.

3 Locate the folder containing your Student files.

4 Double-click the **Park** workbook.

5 Excel opens the workbook as shown in Figure 1-12.

Figure 1-12
The Park
workbook

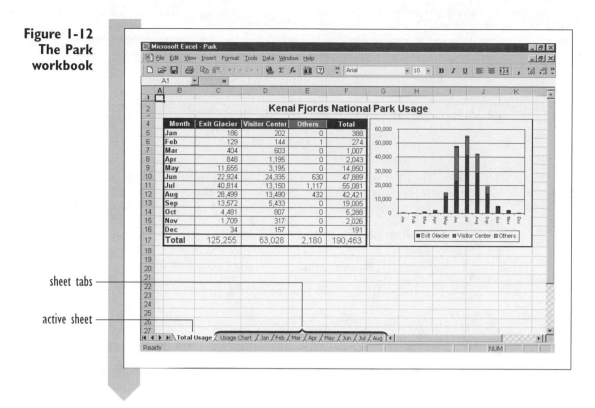

sheet tabs

active sheet

A single workbook can have as many as 255 worksheets. The names of the sheets appears on tabs at the bottom of the workbook window. In the Park workbook, the first sheet is named "Total Usage" and contains information on the number of visitors at each location in the park over the previous year. The sheet shows both a table of visitor counts and a chart with the same information. Note that the chart has been placed within the worksheet. Placing an object like a chart on a worksheet is known as **embedding**. Glancing over the table and chart, we see that the peak usage months were May through September.

The second tab is named "Usage Chart" and contains another chart of park usage. After the first two sheets are worksheets devoted to usage data from each month of the year. Your next task will be to move between the various sheets in the Park workbook.

Scrolling through a Workbook

To move from one sheet to another, you can either click the various sheet tabs in the workbook or use the navigational buttons located at the bottom of the workbook window. Table 1-5 provides a description of these buttons.

Table 1-5 Workbook Navigation Buttons

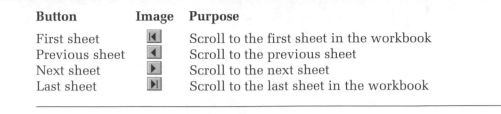

Button	Image	Purpose
First sheet		Scroll to the first sheet in the workbook
Previous sheet		Scroll to the previous sheet
Next sheet		Scroll to the next sheet
Last sheet		Scroll to the last sheet in the workbook

You can also move to a specific sheet by right-clicking one of these navigation buttons and selecting the sheet from the resulting pop-up list of sheet names. Try viewing some of the other sheets in the workbook now.

To view other sheets:

1 Click the **Usage Chart** sheet tab.

2 Excel displays the chart as shown in Figure 1-13.

**Figure 1-13
The Usage
Chart sheet**

Chart toolbar ————

active sheet ————

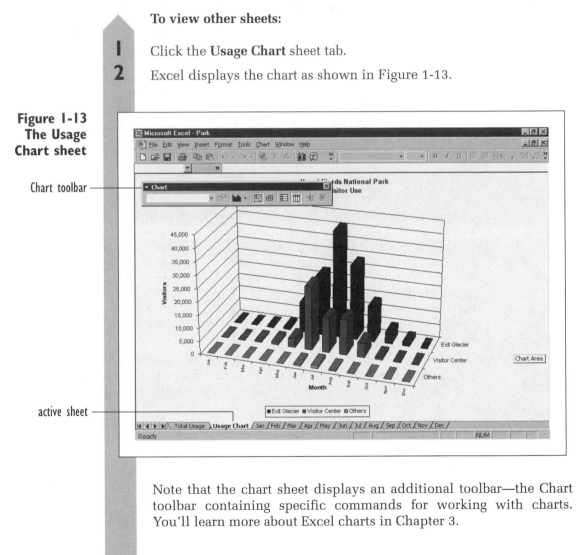

Note that the chart sheet displays an additional toolbar—the Chart toolbar containing specific commands for working with charts. You'll learn more about Excel charts in Chapter 3.

3   Click the **Jan** sheet tab.

4   The worksheet for the month of January is displayed as shown in Figure 1-14.

**Figure 1-14
The January
worksheet**

active sheet

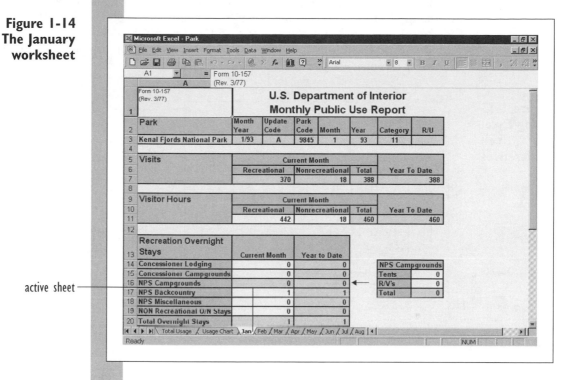

The form that appears in this worksheet resembles the form used by the Kenai Fjords staff to record usage information. It contains information on the park, the number of visits each month, visitor hours, and other important data. Some of this data is hidden beyond the boundary of the worksheet window.

5   Drag the **Vertical scrollbar** down to move the worksheet down and view the rest of the January data.

Clearly, the Park workbook is complex. Its sheets contains many pieces of information, much of it interrelated. This book will not cover all the techniques used to create a workbook like this one, but you should be aware of the formatting possibilities that exist.

Worksheet Cells

Each worksheet can be thought of as a grid of **cells**, where each cell can contain a numeric or text entry. Cells are referenced by their location on the grid. For example, the total number of visitors at the park is shown in cell F17 of the Total Usage worksheet (see Figure 1-15.) As you'll see later in Chapter 2, if you were to use this value in a function or Excel command, you would use the cell reference "F17".

**Figure 1-15
Using Excel
cell
references**

	A	B	C	D	E	F
1						
2				Kenai Fjords Nationa		
4		Month	Exit Glacier	Visitor Center	Others	Total
5		Jan	186	202	0	388
6		Feb	129	144	1	274
7		Mar	404	603	0	1,007
8		Apr	848	1,195	0	2,043
9		May	11,655	3,195	0	14,850
10		Jun	22,924	24,335	630	47,889
11		Jul	40,814	13,150	1,117	55,081
12		Aug	28,499	13,490	432	42,421
13		Sep	13,572	5,433	0	19,005
14		Oct	4,481	807	0	5,288
15		Nov	1,709	317	0	2,026
16		Dec	34	157	0	191
17		Total	125,255	63,028	2,180	190,463

cell F17

Selecting a Cell

When you want to enter data or format a particular value, you must first select the cell containing the data or value. To do this, you simply click on the cell in the worksheet. Try this now with cell F17 in the Total Usage worksheet.

To select a cell from the worksheet:

1 Click the **Total Usage** sheet tab to move back to the front of the workbook.

2 Click **F17** in the worksheet grid.

Cell F17 now has a small box around it, indicating that it is the **active cell** (see Figure 1-16.) Moreover, when you selected cell F17, the Name box displays "F17" indicating that this is the active cell. Also, the formula bar now displayed the formula "=SUM(F5:F16)". This formula calculates the sum of the values in cells F5 through F16. You'll learn more about formulas in Chapter 2.

Formula in the cell sums visitor totals from cells F5 through F16

Figure 1-16 Selecting a cell

cell reference

active cell

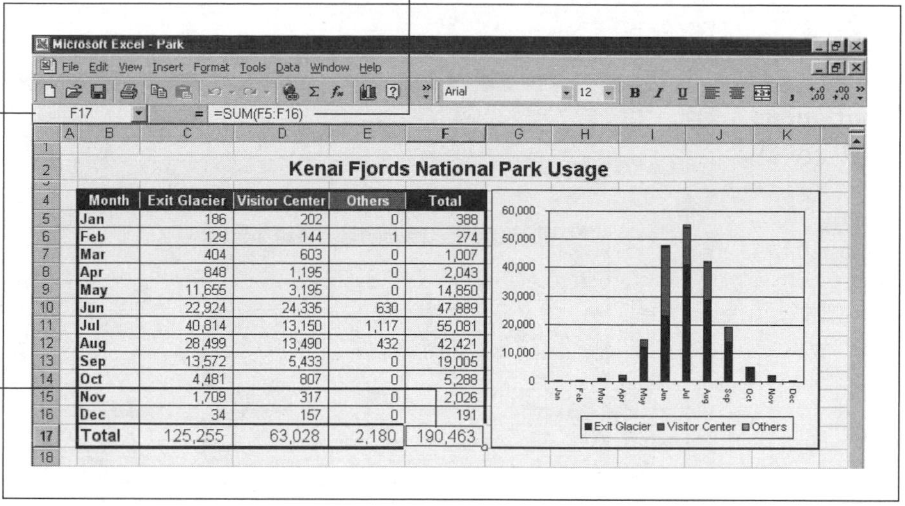

Microsoft Excel - Park
File Edit View Insert Format Tools Data Window Help
Arial
12
F17
=SUM(F5:F16)
Kenai Fjords National Park Usage
Month | Exit Glacier | Visitor Center | Others | Total
Jan 186 202 0 388
Feb 129 144 1 274
Mar 404 603 0 1,007
Apr 848 1,195 0 2,043
May 11,655 3,195 0 14,850
Jun 22,924 24,335 630 47,889
Jul 40,814 13,150 1,117 55,081
Aug 28,499 13,490 432 42,421
Sep 13,572 5,433 0 19,005
Oct 4,481 807 0 5,288
Nov 1,709 317 0 2,026
Dec 34 157 0 191
Total 125,255 63,028 2,180 190,463
60,000
50,000
40,000
30,000
20,000
10,000
0
Exit Glacier
Visitor Center
Others

If you want to select a group of cells, known as a **cell range** or **range**, you must select one corner of the range and then drag the mouse pointer over cells. To see how this works in practice, try selecting the usage table located in the cell range B4:F17 of the Total Usage worksheet.

To select a cell range:

1. Click **B4**.
2. With the mouse button still pressed, drag the mouse pointer over to cell **F17**.
3. Release the mouse button.

Now the range of cells from B4 down to F17 is selected. Observe that a selected cell range is highlighted to differentiate it from unselected cells. A cell range selected in this fashion is always rectangular in shape and contiguous. If you want to select a range that is not rectangular or contiguous, you must use the CTRL key on your keyboard and then select the separate distinct groups that make up the range. For example, if you want to select only the cells in the range B4:B17 and F4:F17, you must use this technique.

To select a noncontiguous range:

1. Select the range **B4:B17**.
2. Press the **CTRL** key on your keyboard.

3   With the CTRL key still pressed, select the range **F4:F17**.

The selected range is shown in Figure 1-17.

Figure 1-17
Non-
contiguous
cell range

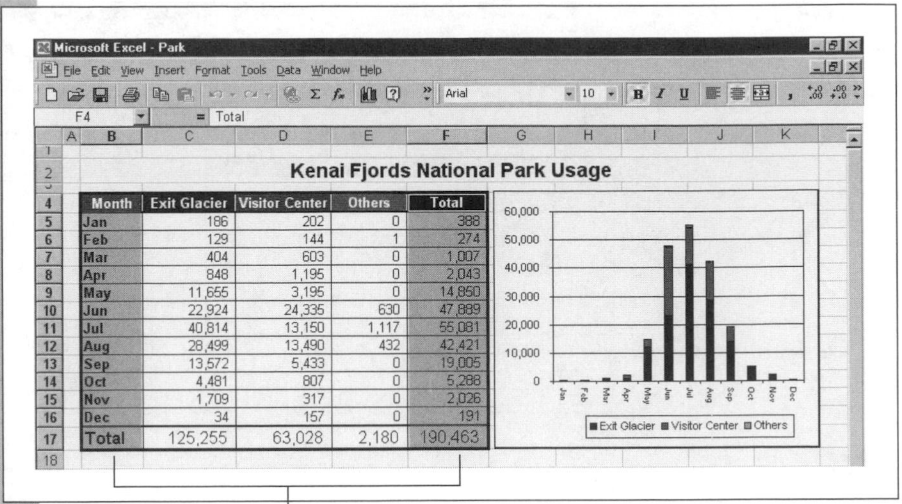

range B4:B17 and
F4:F17 are selected

The cell reference for this group of cells is "B4:B17;F4:F17", where the semicolon indicates a joining of two distinct ranges.

Moving Cells

Excel allows you to move the contents of your cells around without affecting their values. This is a great help in formatting your worksheets. To move a cell or range of cells, simply select the cells and then drag the selection to a new location. Try this now with the table of usage data from the Total Usage worksheet.

To move a range of cells:

1   Select the range **B4:F17**.

2   Move the mouse pointer to the border of the selected area so that the pointer changes from a ✛ to a ↖.

3 Drag the selected area down two cells, so that the new range is now B6:F19, and release the mouse button.

Note that as you moved the selected range, Excel displayed a screen tip with the new location of the range.

4 Click **F19** to deselect the cell range.

When you look at the formula bar for cell F19, note that the formula is now changed from "=SUM(F4:F17)" to "=SUM(F7:F18)". Excel will automatically update the cell references in your formulas to account for the fact that you move the cell range.

You can also use the Cut, Copy, and Paste buttons to move a cell range. These buttons are essential if you want to move a cell range to a new workbook or worksheet (you can't use the drag and drop technique to perform that action). Try using the Cut and Paste method to move the table back to its original location.

To cut and paste a range of cells:

1 Select the range **B6:F19**.

2 Click the **Cut** button ✂ on the toolbar or press **CTRL+x**.

A flashing border appears around the cell range, indicating that it has been cut or copied from the worksheet.

3 Click **B4**.

4 Click the **Paste** button 📋 or press **CTRL+v**.

5 The table now appears back in the cell range, B4:F17.

6 Click cell **A1** to make A1 the active cell again.

If you want to copy a cell range rather than move it, you can use the Copy button 📋 in the above steps, or, if you prefer the drag and drop technique, hold down the CTRL key while dragging the cell range to its new location—this will create a copy of the original cell range at the new location. You can refer to Excel's on-line Help for more information.

Printing from Excel

It would be useful for the chief of interpretation at Kenai Fjords National Park to have a hard copy of some of the worksheets and charts in the Park workbook. To do this, you can print out selected portions of the workbook.

Previewing the Print Job

Before sending a job to the printer, it's usually a good idea to preview the output. With Excel's Print Preview window, you can view your job before it's printed, as well as set up the page margins, orientation, and headers and footers. Try this now with the Total Usage worksheet.

To preview a print job:

1   Verify that Total Usage is still the active worksheet.

2   Click the **Print Preview** button 🔍 or press **File > Print Preview**.

3   The Print Preview opens as displayed in Figure 1-18.

**Figure 1-18
The Print
Preview
window**

Click to zoom in
and out of the
Preview window

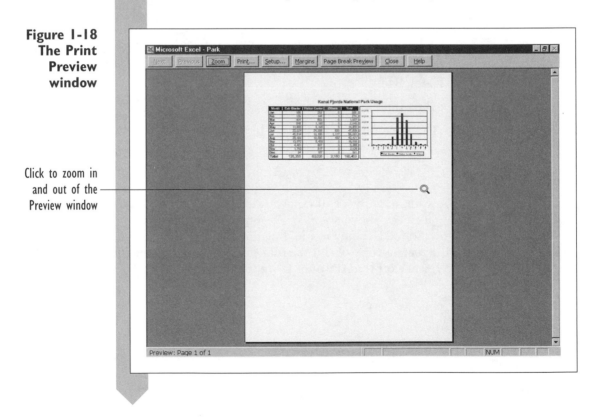

Table 1-6 describes the variety of options available to you from within the Print Preview window.

Table 1-6 Print Preview Options

Button	Description
Previous	View the previous page in the print job
Next	View the next page in the print job
Zoom	Zoom in and out of the Preview window
Print	Print the job displayed in the Preview window
Setup	Select options for the page setup
Margins	Display margins in the Preview window
Page Break Preview	Display the workbook, indicating the location of the page breaks within each worksheet
Close	Close the Preview window
Help	View on-line Help

Setting Up the Page

The Preview window opens with the default print settings for the workbook. You can change these settings for each print job. You may add a header or footer to each page, change the orientation from portrait to landscape, and modify many other features. To see how this works, you'll adjust the settings for the current print job by adding a header and changing the page layout.

To add a header to a print job:

1 Click the **Setup** button in the Preview window.

2 Click the **Header/Footer** dialog sheet tab.

3 Excel provides a list of built-in headers that you can select from the Header drop-down list. You can also write your own—which you'll do now.

4 Click the **Custom Header** button.

5 Type **Yearly Usage Report** in the Center section of the Header dialog box as shown in Figure 1-19.

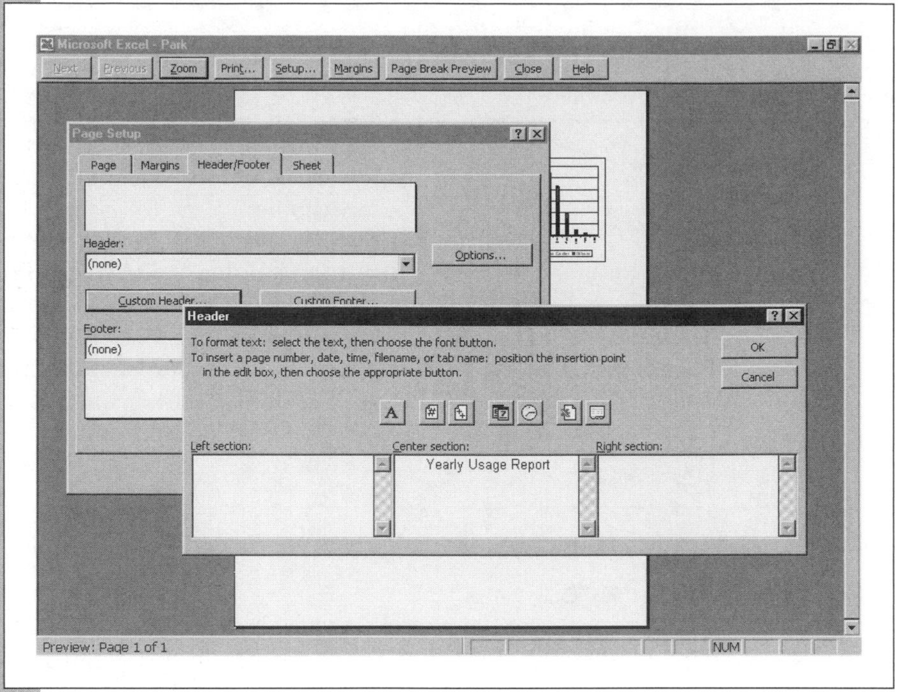

Figure 1-19
Adding a
header to
the print
job

6   Click the **OK** button.

Because the print job is more horizontal than vertical, it would be a good idea to change the orientation from portrait to landscape.

To change the page orientation:

1   Click the **Page** dialog sheet tab within the Page Setup dialog box.

2   Click the **Landscape** option button.

3   Click the **OK** button.

Figure 1-20 shows the new layout of the print job with a header and landscape orientation.

Figure 1-20
Adding a
header and
landscape
orientation
to the print
job

New header ⎯

Landscape
orientation ⎯

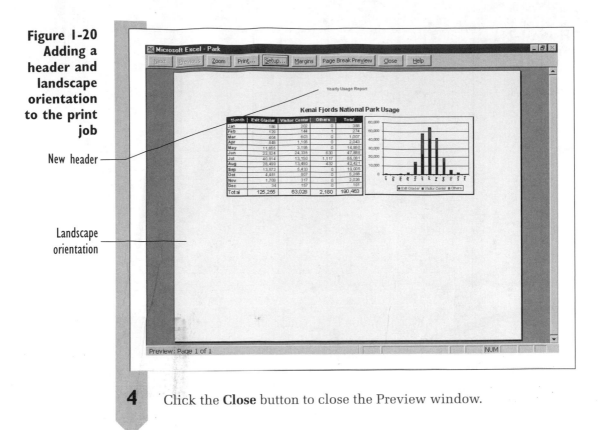

4 Click the **Close** button to close the Preview window.

There are many other printing features available to you in Excel. Check the on-line Help for more information.

Printing the Page

To print your worksheet, you can either click the Print button 🖨 on the toolbar or select the Print command from the File menu. Clicking the Print button is faster because it will bypass the Print dialog box and send output directly to the printer. If you've already set up the printer and the page itself, this is the preferred method. However, if you need to specify which printer to use or which sheets from the workbook to print, you should use the Print command from the File menu. Try printing the Total Usage worksheet now.

To print the Total Usage worksheet:

1 Click **File > Print**.

2 The Print Dialog box appears. See Figure 1-21.

**Figure 1-21
Print dialog
box**

default printer

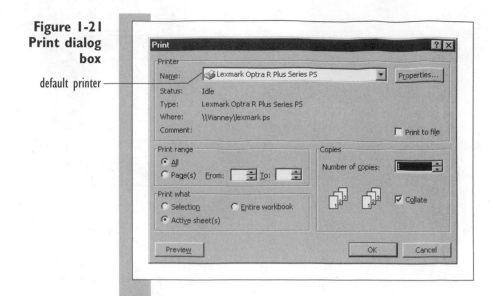

Notice that you can print a selection of the active worksheet (in other words, you can select a cell range and print only that part of the worksheet), the entire active sheet or sheets, or the entire workbook. You can also select the number of copies to print and the range of pages. The other options let you select your printer from a list (if you have access to more than one) and set the properties for that particular printer. You can also click the Preview button to go to the Print Preview window.

3 Click **OK** to start the print job.

Your printer should soon start printing the Total Usage worksheet.

If you were to hand this printout to the chief of interpretation, he or she would be able to use the information contained in it to determine when to hire extra help at the various stations in the park.

Saving Your Work

You should periodically save your work when you make changes to a workbook or when you are entering a lot of data so that you won't lose much work if your computer or Excel crashes. Excel offers two options for saving your work: the Save command, which saves the file; and the Save As command, which allows you to save the file under a new name.

So that you do not change the original files (and can go through the chapters again with unchanged files if necessary), you'll be instructed throughout this book to save your work under new file names, made up of the original file name with the chapter-number suffix. To save the changes you made to the Park workbook, save the file as Park1. If using your own computer, you can save the workbook to your hard drive. If you are using a computer on the school network, you may be asked to save your work to your own floppy disk. This book refers to this disk as your **student disk** and assumes that you'll save your work to such a student disk.

To save the Park workbook as Park1 on your student disk:

1 Click **File > Save As** to open the Save As dialog box.

2 Insert your student disk into drive A (or the drive you use on your computer).

3 Locate your student disk from the Save In list box.

4 Type **Park1** in the File Name box. See Figure 1-22.

**Figure 1-22
Save As
dialog box**

location of saved file

name of saved file

file type

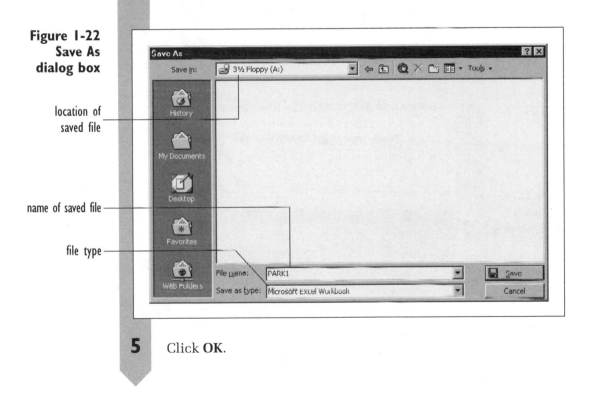

5 Click **OK**.

Excel then saves the workbook under the name Park1.xls on your student disk (it automatically adds the ".xls" extension for you).

Excel Add-Ins

Excel's capabilities can be expanded through the use of special programs called **add-ins**. These add-ins tie into Excel's special features, almost looking like a part of Excel itself. Various add-ins are supplied with Excel that allow you to easily generate reports, explore multiple scenarios, or access databases. To use these add-ins, you have to go through a process of saving the add-in files to a location on your computer, and then telling Excel where to find the add-in file.

Excel comes with an add-in called **Analysis ToolPak** that provides some of the statistical commands you'll need for this book. Another add-in, StatPlus, you have already copied to your hard disk. Now you will install the add-in in Excel.

Loading the StatPlus Add-In

The add-ins on your computer are stored in a list in Excel. From this list, you can activate the add-in, or browse for new ones. First you'll browse for the StatPlus add-in.

To browse and install the StatPlus add-in:

1   Click **Tools** and **Add-Ins** from the Excel menu. The Add-Ins dialog box appears as shown in Figure 1-23.

**Figure 1-23
The Add-Ins
dialog box**

Click a checkbox
to load the
add-in

Click to browse
for a new add-in
to add to the list

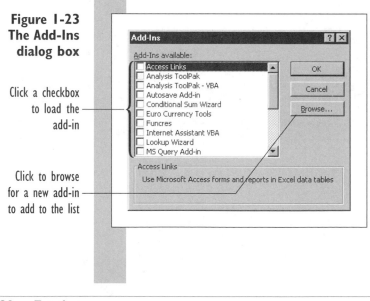

2   Each available add-in is shown in Figure 1-23 along with a checkbox indicating whether that add-in is currently loaded in Excel.

3   Click the **Browse** button.

4   Locate the Berk-Carey folder or the folder you specified earlier when running the StatPlus installation routine.

5   Open the **StatPlusV2** folder.

6   Click **StatPlusV2.xla** and click **OK**.

7   StatPlus Version 2.0 now appears in the Add-Ins dialog box. If it is not checked, click the checkbox. See Figure 1-24.

Figure 1-24
The StatPlus
Add-in

the StatPlus
Add-in is loaded
in Excel

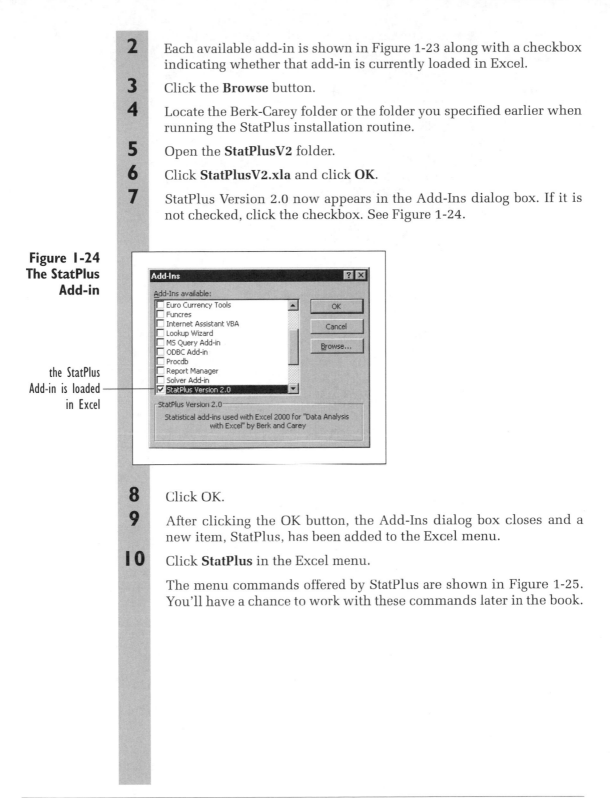

8   Click OK.

9   After clicking the OK button, the Add-Ins dialog box closes and a new item, StatPlus, has been added to the Excel menu.

10   Click **StatPlus** in the Excel menu.

The menu commands offered by StatPlus are shown in Figure 1-25. You'll have a chance to work with these commands later in the book.

**Figure 1-25
Commands
in the
StatPlus
menu**

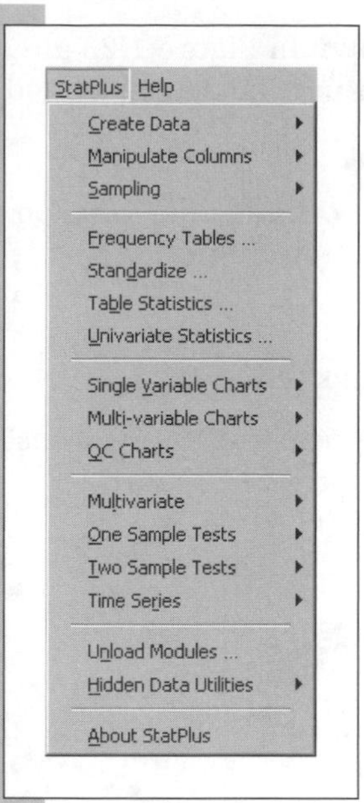

StatPlus   Help

   Create Data ▶

   Manipulate Columns ▶

   Sampling ▶

   Frequency Tables ...

   Standardize ...

   Table Statistics ...

   Univariate Statistics ...

   Single Variable Charts ▶

   Multi-variable Charts ▶

   QC Charts ▶

   Multivariate ▶

   One Sample Tests ▶

   Two Sample Tests ▶

   Time Series ▶

   Unload Modules ...

   Hidden Data Utilities ▶

   About StatPlus

Loading the Data Analysis ToolPak

Now that you've seen how to load the StatPlus add-in, you can load the Data Analysis ToolPak. Since the Data Analysis ToolPak comes with Excel, you may need to have your Excel 2000 or Office 2000 installation disks handy.

To load the Analysis ToolPak:

1 Click **Tools > Add-Ins**.

2 Click the checkbox for the **Analysis ToolPak** and click **OK**.

3 At this point, Excel may prompt you for the installation CD; if so, insert the CD into your CD-ROM drive and follow the installation instructions.

Once the Data Analysis ToolPak is installed, you can access it from the Excel menu. Unlike StatPlus, the Data Analysis ToolPak does not add a new menu item to Excel's menu bar; instead it adds a new command to the Tools menu.

To access commands for the Data Analysis ToolPak:

1 Click **Tools > Data Analysis**.

2 The Data Analysis dialog box appears as shown in Figure 1-26.

Figure 1-26
Commands
in the Data
Analysis
ToolPak

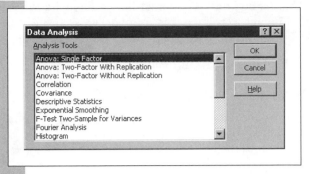

The list of commands available with the Data Analysis ToolPak is shown in the Analysis Tools list box. To run one of these commands, you would select it from the list box and then click the OK button. You'll have an opportunity to use some of the Data Analysis ToolPak's commands later in this book.

3 Click **Cancel** to close the Data Analysis dialog box.

Unloading an Add-in

If at any time, you want to unload the Data Analysis ToolPak or StatPlus, you can do so by clicking **Tools > Add-Ins** from the Excel menu and then deselecting the checkbox for the specific add-in. Unloading an add-in is like closing a workbook; it does not affect the add-in file. If you want to use the add-in again, simply reopen the Add-Ins dialog box and reselect the checkbox. If you exit Excel with an add-in loaded, Excel will assume that you want to run the add-in the next time you run Excel, so it will load it for you automatically.

Features of StatPlus

StatPlus has several special features that you should be aware of. These include modules and hidden data.

Using StatPlus Modules

StatPlus is comprised of a series of add-in files, called **modules**. Each module handles a specific statistical task, such as creating a quality control chart or selecting a random sample of data. StatPlus will load the modules you need upon demand (this way, you do not have to use up more system memory than needed). After using StatPlus for a while, you may have a great many modules loaded. If you want to reduce this number, you can view the list of currently opened modules and unload those you're no longer using.

To view a list of StatPlus modules:

1 Click **StatPlus > Unload Modules**.

StatPlus displays a list of loaded modules. A sample list is shown in Figure 1-27. Yours will be different.

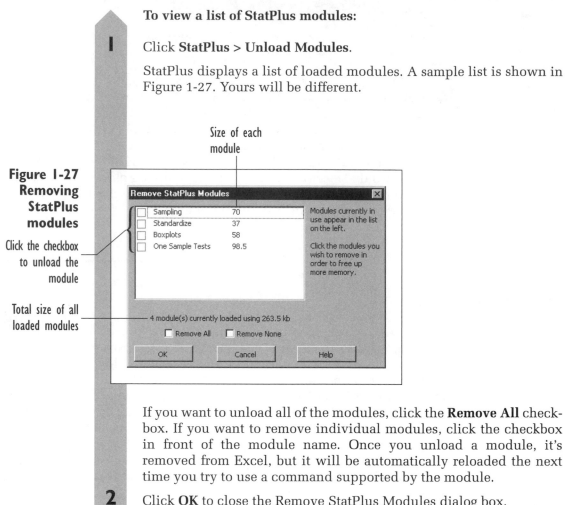

Size of each module

**Figure 1-27
Removing
StatPlus
modules**

Click the checkbox
to unload the
module

Total size of all
loaded modules

Remove StatPlus Modules

	Sampling	70
	Standardize	37
	Boxplots	58
	One Sample Tests	98.5

Modules currently in use appear in the list on the left.

Click the modules you wish to remove in order to free up more memory.

4 module(s) currently loaded using 263.5 kb

☐ Remove All   ☐ Remove None

OK   Cancel   Help

If you want to unload all of the modules, click the **Remove All** checkbox. If you want to remove individual modules, click the checkbox in front of the module name. Once you unload a module, it's removed from Excel, but it will be automatically reloaded the next time you try to use a command supported by the module.

2 Click **OK** to close the Remove StatPlus Modules dialog box.

Hidden Data

Several Excel commands employ hidden worksheets. A **hidden worksheet** is a worksheet in your workbook that is hidden from view. Hidden worksheets are used in creating histograms, boxplots, and normal probability plots (don't worry, you'll learn about these topics in later chapters). You can view these hidden worksheets if you need to troubleshoot a problem with one of these charts. There are three hidden worksheet commands in StatPlus (available in the Hidden Data Utilities submenu):

Table 1-7 Hidden Worksheet Commands

Command	Description
View hidden data	Unhides a hidden StatPlus worksheet
Rehide hidden data	Rehides a StatPlus worksheet
Remove unlinked hidden data	If you delete a chart based on hidden data, that data is no longer needed. Running this command removes that extraneous data from the hidden worksheet.

You can learn more about StatPlus and its features by viewing the on-line Help file. Help buttons are included in every dialog box. You can also open the Help file by clicking **StatPlus > About StatPlus** from the Excel menu.

Exiting Excel

When you are finished with an Excel session, you should exit the program so that all the program-related files are properly closed.

To exit Excel:

1 Click **File > Exit**

If you have unsaved work, Excel asks whether you want to save it before exiting. If you click "No," Excel closes and you lose your work. If you click "Yes," Excel opens the Save As dialog box and allows you to save your work. Once you have closed Excel, you are returned to the Windows desktop or to another active application.

Working with Data

Objectives

In this chapter you will learn to:

- Enter data into Excel from the keyboard

- Work with Excel formulas and functions

- Work with cell references and range names

- Query and sort data using the AutoFilter and Advanced Filter

- Import data from text files and databases

I n this chapter you'll learn how to enter data in Excel through the keyboard and by importing data from text files and databases. You'll learn how to create Excel formulas and functions to perform simple calculations. You'll be introduced to cell references and learn how to refer to cell ranges using range names. Finally, you'll learn how to examine your data through the use of queries and sorting.

Data Entry

Before data analysis comes data entry. Error-free data entry is essential to accurate data analysis. Excel provides several methods for entering your data. Data sets can be entered manually from the keyboard or retrieved from a text file, database, or the Internet. You can also have Excel automatically enter patterns of data for you, saving you the trouble of creating these data values yourself. You'll study all of these techniques in this chapter, but first you'll work on entering data from the keyboard.

Entering Data from the Keyboard

Table 2-1 displays average daily gasoline sales and other (non-gasoline) sales for each of nine service station/convenience franchises in a store chain in a western city. There are three columns in this data set: Station, Gas, and Other. The Station column contains an id number for each of the ten stations. The Gas column displays the gasoline sales for each station. The Other column displays sales for non-gasoline items.

Table 2-1 Service Station Sales

Station	Gas	Other
1	$3,415	$2,211
2	$3,499	$2,500
3	$3,831	$2,899
4	$3,587	$2,488
5	$3,719	$2,111
6	$3,001	$1,281
7	$4,567	$8,712
8	$4,218	$7,056
9	$3,215	$2,508

To practice entering data, you'll insert this information into a blank worksheet. To enter data, you first select the cell corresponding to the upper-left corner of the table, making it the active cell. You then type the value or text you want placed in the cell. To move between cells, you can either press the Tab key to move to the next column in the same row, or press the Enter key to move to the next row in the same column. If you are entering data into several columns, the Enter key will move you to the next row in the first column of the data set.

To enter the first row of the service station data set:

1 Launch Excel as described in Chapter 1.

Excel shows an empty workbook with the name "Book1" in the title bar.

2 Click cell **A1** to make it the active cell.

3 Type **Station** and then press **Tab**.

4 Type **Gas** in cell B1 and press **Tab**.

5 Type **Other** in cell C1 and press **Enter**.

Excel moves you to cell A2, making it the active cell.

6 Using the same technique, type the next two rows of the table, so that data for the first two stations is displayed. Your worksheet should appear as in Figure 2-1.

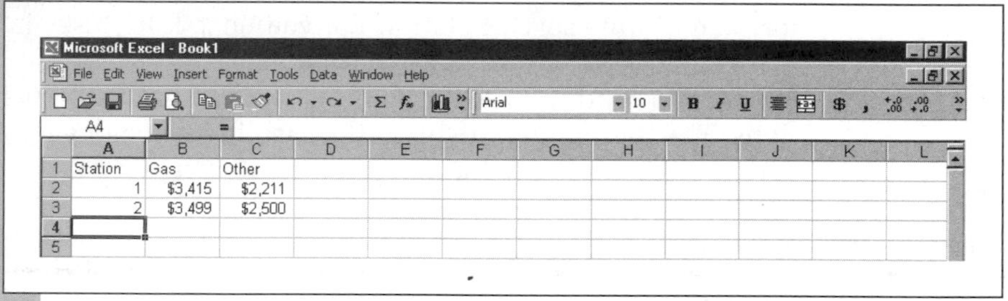

Figure 2-1
The first rows of the service station data set

Entering Data with Autofill

If you're inserting a column or row of values that follow some sequential pattern, you can save yourself time by using Excel's Autofill feature. The **Autofill** feature allows you to fill up a range of values with a series of numbers or dates. You can use Autofill to generate automatically columns containing data values such as:

1, 2, 3, 4, . . . 9, 10
1, 2, 4, 8, ... 128, 256
Jan, Feb, Mar, Apr, ... , Nov, Dec
and so forth.

In the service station data, you have a sequence of numbers, 1–9, that represent the service stations. You could enter this data by hand, but this is also an opportunity to use the Autofill feature.

To use Autofill to fill in the rest of the service station numbers:

1 Select the range **A2:A3**.

Notice the small black box at the lower-right corner of the double border around the selected range. This is called a **fill handle**. To create a simple sequence of numbers, you'll drag this fill handle over a selected range of cells.

2 Move the mouse pointer over the fill handle until the pointer changes from a ✛ to a ✚. Click and hold down the mouse button.

3 Drag the fill handle down to cell **A10** and release the mouse button.

Note that as you drag the fill handle down, a screen is displayed showing the value that will be placed in the active cell if you release the mouse button at that point.

4 Figure 2-2 shows the service station numbers placed in the cell range A2:A10.

Figure 2-2 Using Autofill to insert a sequence of data values

drag the fill handle down to generate a linear sequence of numbers automatically

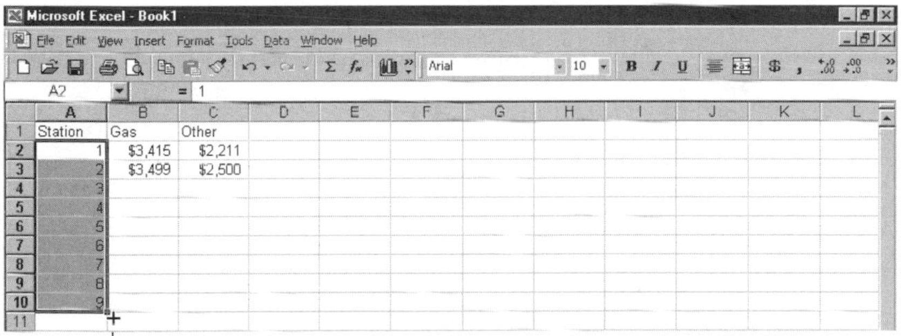

EXCEL TIP

- If you want to create a geometric sequence of numbers, drag the fill handle with your right mouse button and then select "Growth Trend" from the pop-up menu.

- If you want to create a customized sequence of numbers or dates, drag the fill handle with your right mouse button and select "Series…" from the pop-up menu. Fill in details about your customized series in the Series dialog box.

With the service station numbers entered, you can add the rest of the sales figures to complete the data set.

To finish entering data:

1 Select the range **B4:C10**.

2 With B4 the active cell, start typing in the remaining values, using Table 2-1 as your guide.

Note that when you're entering data into a selected range, pressing the Tab key at the end of the range moves you to the next row.

3 Click cell **A1** to remove the selection.

The completed worksheet should appear as shown in Figure 2-3.

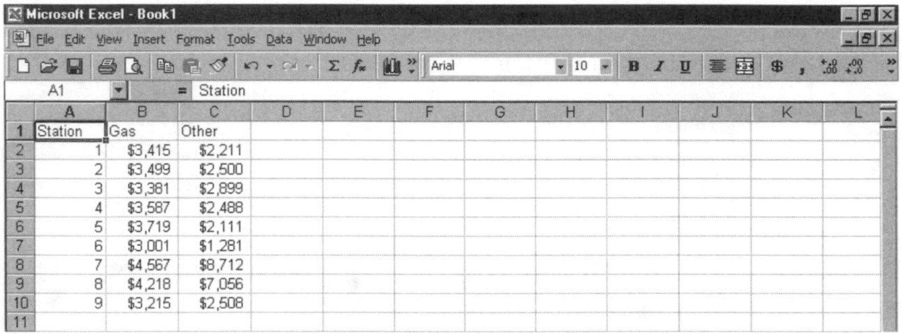

Figure 2-3
The
completed
service
station
data set

Inserting New Data

Sometimes you will want to add new data to your data set. For example, you discover that there is a tenth service station with the following sales data:

Table 2-2 Additional Service Station Sales

Station	Gas	Other
0	$3,995	$1,938

You could simply append this information to the table you've already created, covering the cell range A11:C11. On the other hand, in order to maintain the sequential order of the station numbers, it would be better to place this information in the range A2:C2 and then have the other stations shifted down in the worksheet. You can accomplish this using Excel's Insert command.

To insert new data into your worksheet:

1 Select the cell range **A2:C2**.

2 Right-click the selected range and then click **Insert** from the pop-up menu. See Figure 2-4.

Figure 2-4 Running the Insert command from the pop-up menu

	A	B	C	D
1	Station	Gas	Other	
2	1	$3,415	$2,211	
3	2	$3,499	$2,500	
4	3	$3,381	$2,899	
5	4	$3,587	$2,488	
6	5	$3,719	$2,111	
7	6	$3,001	$1,281	
8	7	$4,567	$8,712	
9	8	$4,218	$7,056	
10	9	$3,215	$2,508	

Pop-up menu: Cut, Copy, Paste, Paste Special..., Insert..., Delete..., Clear Contents, Insert Comment, Format Cells..., Pick From List..., Hyperlink...

3 Verify that the **Shift cells down** option button is selected.

4 Click **OK**.

5 Excel shifts the values in cells A2:C10 down to A3:C11 and inserts a new blank row in the range A2:C2.

6 Enter the data for Station 0 from Table 2-2 in the cell range A2:C2.

7 Click **A1** to make it the active cell.

Data Formats

Now that you've entered your first data set, you're ready to work with data formats. **Data formats** are the fonts and styles that Excel applies to your data's appearance. Formats are applied to either text or numbers. Excel has already applied a currency format to the sales data you've entered. For example, if you click cell B2, note that the value in the formula bar is "3995," but the value displayed in the cell is "$3,995." The extra dollar sign and comma separator are aspects of the currency format. You can modify this format if you wish by inserting additional digits to the value shown in the cell (for example, $3,995.00). You may do this if you want dollars and cents displayed to the user.

For the text displayed in the range A1:C1, Excel has applied a very basic format. The text is left-justified within its cell and displayed in 10-point Arial font (depending on how Excel has been configured on your system, a different font or font size may be used). You can modify this format as well. Try it now, applying a boldface font to the column titles in A1:C1. In addition, center each column title within its cell.

To apply a boldface font and center the column titles:

1 Select the range **A1:C1**.

2 Click the **Bold** button **B** on the toolbar.

3 Click the **Center** button ▤.

4 Click cell **A1** to remove the selection.

5 Your data set should look like Figure 2-5.

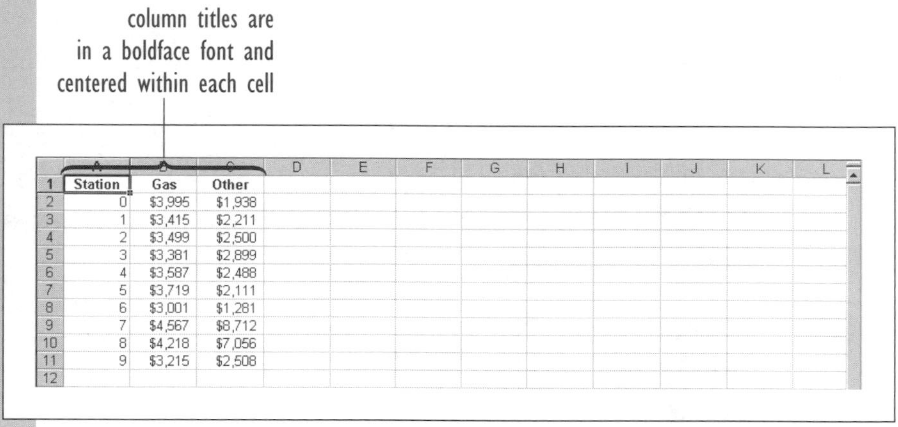

column titles are
in a boldface font and
centered within each cell

**Figure 2-5
Applying a
boldface
font and
centering
the column
titles**

Clicking the Bold **B** and Center ▤ buttons on the toolbar gives you one-click access to two of the more popular formatting commands. Other format buttons are shown in Table 2-3.

Table 2-3 Data Format Buttons

Button	Icon	Purpose
Font	Arial	Apply the font type
Font Size	10	Change the size of the font (in points)
Bold	B	Apply a boldface font
Italic	I	Apply an italic font
Underline	U	Underline the selected text
Align Left	▤	Left-justify the text

(continued)

Align Center	≡	Center the text
Align Right	≡	Right-justify the text
Percent Style	%	Display values as percents (i.e., 0.05 = 5%)
Currency Style	$	Display values as currency (i.e., 5.25 = $5.25)
Comma Style	,	Add comma separators to values (i.e., 43215 = 43,215)
Increase Decimal	.↑0	Increase the number of decimal points (i.e., 4.3 = 4.300)
Decrease Decimal	.↓0	Decrease the number of decimal points (i.e., 4.321 = 4.3)
Fill Color	⬧ ▾	Change the cell's background color
Font Color	A ▾	Change the color of the selected text
Merge and Center	🗗	Merge the selected cells and center the text across the merged cells

You can access all of the possible formatting options for a particular cell by opening the Format Cells dialog box. To see this feature of Excel, you'll use it to continue formatting the column titles, changing the font color to red.

To open the Format Cells dialog box:

1 Select **A1:C1**.

2 Right-click the selection and click **Format Cells** from the pop-up menu.

The Format Cells dialog box contains six dialog sheets labeled **Number**, **Alignment**, **Font**, **Border**, **Patterns**, and **Protection**. Each deals with a specific aspect of the cell's appearance or behavior in the workbook. You'll first change the font color to red. This option is located in the Font dialog sheet.

3 Click the **Font** dialog tab.

4 Click the **Color** drop-down list box and click the **Red** checkbox (located in the first column and third row of the matrix of color options). See Figure 2-6.

**Figure 2-6
Changing
the font
color to red**

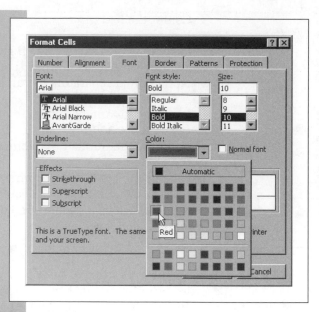

5 Click **OK** to close the Format Cells dialog box.

6 Click **D1** to unselect the cells.

Figure 2-7 displays the final format of the column titles.

text in a
red-colored font

**Figure 2-7
Formatted
column
titles**

	A	B	C	D	E	F	G	H	I	J	K	L
1	Station	Gas	Other									
2	0	$3,995	$1,938									
3	1	$3,415	$2,211									
4	2	$3,499	$2,500									
5	3	$3,381	$2,899									
6	4	$3,587	$2,488									
7	5	$3,719	$2,111									
8	6	$3,001	$1,281									
9	7	$4,567	$8,712									
10	8	$4,218	$7,056									
11	9	$3,215	$2,508									
12												

Formulas and Functions

Another way of entering data is through formulas and functions. Formulas and functions are crucial Excel tools that you enter into cells to perform calculations, displaying the result of the function in the cell. A formula always begins with an equals sign (=) followed by a function name, number, text string, or cell reference. Most functions contain mathematical operators such as + or −. A list of mathematical operators is shown in Table 2-4.

Table 2- 4 Mathematical Operators

Operator	Description
+	Addition
−	Subtraction
/	Division
*	Multiplication
^	Exponentiation

Inserting a Simple Formula

To see how to enter a simple formula, add a new column to your data set displaying the total sales from both gasoline and other sources for each of the ten service stations.

To add a formula:

1 Type **Total** in cell D1 and press **Enter**.

2 Type **=b2+c2** in cell D2 and press **Enter**.

Note that cells b2 and c2 contain the gas and other sales for Station 0. The value displayed in D2 is $5,933—the sum of these two values.

At this point you could enter formulas for the remaining cells in the data set, but it's quicker to use Excel's Autofill capability to add those formulas for you.

3 Click cell **D2** to make it the active cell.

4 Click the Fill Handle and drag it down to cell **D11**. Release the mouse button.

Excel automatically inserts the formulas for the cells in the range D3:D11. Thus, the formula in cell D11 is "**=b11+c11**" to calculate the total sales for Station 9. Note that Excel has also applied the same currency format it used for the values in column B and C to values in column D. See Figure 2-8.

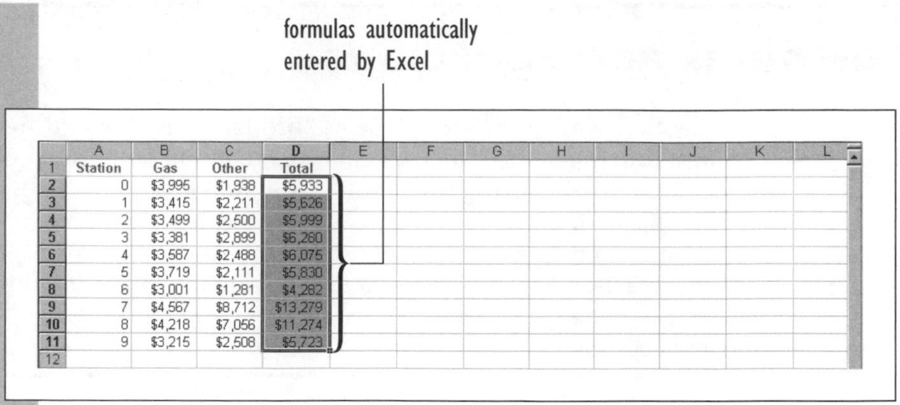

	A	B	C	D
1	Station	Gas	Other	Total
2	0	$3,995	$1,938	$5,933
3	1	$3,415	$2,211	$5,626
4	2	$3,499	$2,500	$5,999
5	3	$3,381	$2,899	$6,280
6	4	$3,587	$2,488	$6,075
7	5	$3,719	$2,111	$5,830
8	6	$3,001	$1,281	$4,282
9	7	$4,567	$8,712	$13,279
10	8	$4,218	$7,056	$11,274
11	9	$3,215	$2,508	$5,723

Figure 2-8 Adding new formulas with Autofill

This example illustrates a simple formula involving the addition of two numbers. What if you wanted to find the total gas and other sales for all ten of the service stations? In that case, you would be better off using one of Excel's built-in functions.

Inserting an Excel Function

Excel has a library containing hundreds of functions covering most financial, statistical, and mathematical needs. Users can also create their own custom functions using Excel's programming language. StatPlus contains its own library of functions, supplementing those offered by Excel. A list of statistics-related functions is included in the Appendix at the end of this book.

A function is composed of the **function name** and a list of **arguments**—values required by the function. To calculate the sum of a set of cells, you would use the SUM function. The general form or syntax of the SUM function is

= SUM(*number1*, *number2*, ...)

where *number1* and *number2* are numbers or cell references. Note that the SUM function allows multiple numbers of cell references. Thus to calculate the sum of the cells in the range B2:B11, you could enter the formula

= SUM(B2:B11)

into a cell in the worksheet.

Although you can type in functions directly, you may find it easier to use Excel's Function Wizard, especially for functions whose syntax you don't know. The **Function Wizard** is a series of dialog boxes that aids you in choosing and constructing your formula. To see the Function Wizard in action, you'll use it now in calculating the total gasoline and other sales over all of the service stations.

To calculate total sales figures for all ten service stations:

1 Type **Total** in cell A12 and press **Tab**.

2 Click the **Paste Function** button 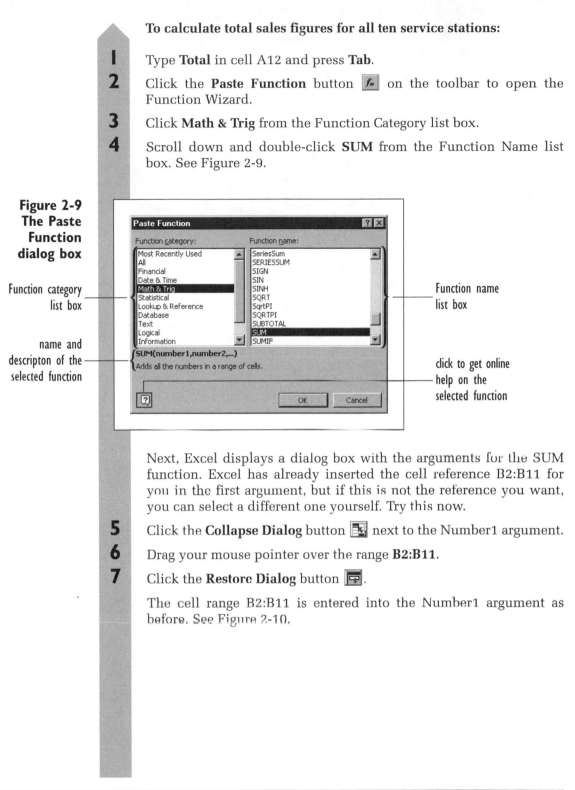 on the toolbar to open the Function Wizard.

3 Click **Math & Trig** from the Function Category list box.

4 Scroll down and double-click **SUM** from the Function Name list box. See Figure 2-9.

**Figure 2-9
The Paste
Function
dialog box**

Function category
list box

name and
descripton of the
selected function

click to get online
help on the
selected function

Function name
list box

Next, Excel displays a dialog box with the arguments for the SUM function. Excel has already inserted the cell reference B2:B11 for you in the first argument, but if this is not the reference you want, you can select a different one yourself. Try this now.

5 Click the **Collapse Dialog** button next to the Number1 argument.

6 Drag your mouse pointer over the range **B2:B11**.

7 Click the **Restore Dialog** button.

The cell range B2:B11 is entered into the Number1 argument as before. See Figure 2-10.

Figure 2-10 Adding arguments to the SUM function

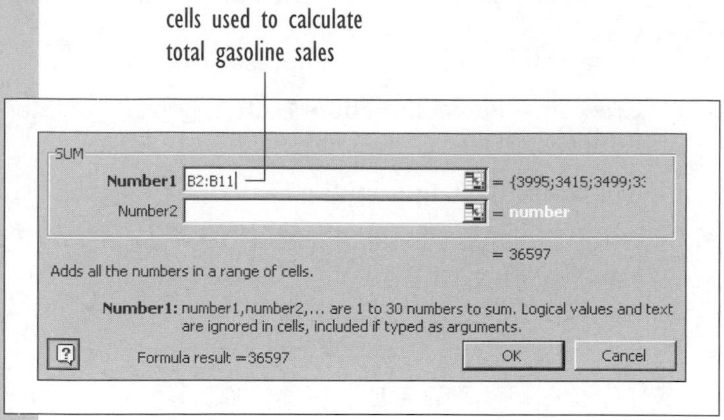

cells used to calculate total gasoline sales

8 Click **OK**.

The total gasoline sales is now displayed in cell B12. You can easily add total sales for other individual products and for all products together using the same Autofill technique used earlier.

To add the remaining total sales calculations:

1 Select **B12**.

2 Click the Fill Handle and drag it to cell **D12**.

3 Release the mouse button.

4 Total sales figures are now shown in the range B12:D12. See Figure 2-11.

Figure 2-11 All sales totals

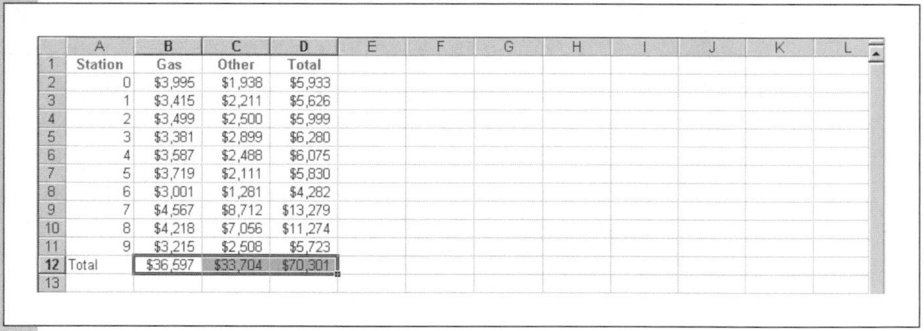

	A	B	C	D
1	Station	Gas	Other	Total
2	0	$3,995	$1,938	$5,933
3	1	$3,415	$2,211	$5,626
4	2	$3,499	$2,500	$5,999
5	3	$3,381	$2,899	$6,280
6	4	$3,587	$2,488	$6,075
7	5	$3,719	$2,111	$5,830
8	6	$3,001	$1,281	$4,282
9	7	$4,567	$8,712	$13,279
10	8	$4,218	$7,056	$11,274
11	9	$3,215	$2,508	$5,723
12	Total	$36,597	$33,704	$70,301
13				

Cell References

When Excel calculated the total sales for column C and column D in your worksheet, it inserted the following formulas into C12 and D12, respectively:

=SUM(C2:C11)

=SUM(D2:D11)

At this point you may wonder how Excel knew to copy everything except the cell reference from cell B12 and, in place of the original B2:B11 reference, to shift the cell reference one and two columns to the right. Excel does this automatically when you use relative references in your formulas. A **relative reference** identifies a cell range based on its position relative to the cell containing the formula. One advantage of using relative references, as you've seen, is that you can fill up a row or column with a formula and the cell references in the new formulas will shift along with the cell.

Now what if you didn't want Excel to shift the cell reference when you copied the formula into other cells? What if you wanted the formula *always* to point to a specific cell in your worksheet? In that case you would need an **absolute reference**. In an absolute reference, the cell reference is prefixed with dollar signs. For example, the formula

=SUM($C$2:$C$11)

is an absolute reference to the range C2:C11. If you copied this formula into other cells, it would still point to C2:C11 and would not be shifted.

You can also create formulas that use **mixed references**, combining both absolute and relative references. For example, the formulas

=SUM($C2:$C11)

and

=SUM(C$2:C$11)

use mixed references. In the first example, the column is absolute but the row is relative, and in the second example, the column is relative but the row is absolute. This means that in the first example, Excel will shift the row references but not the column references, and in the second example, Excel will shift the column references but not the row references. You can learn more about reference types and how to use them in Excel's online Help. In most situations in this book, you'll use relative references, unless otherwise noted.

Range Names

Another way of referencing a cell in your workbook is with a range name. **Range names** are names given to specific cells or cells ranges. For example, you can define the range name "Gas" to refer to cells B2:B11 in your worksheet. To calculate the total gasoline sales, you could use the formula

$$=\text{SUM(B2:B11)}$$

or

$$=\text{SUM(Gas)}$$

Range names have the advantage of making your formulas easier to write and interpret. Without range names you would have to know something about the worksheet before you could determine what the formula =SUM(B2:B11) actually calculates.

Excel provides several tools to create range names, and StatPlus also has a range-naming command. You'll find it easier to perform data analysis on your data set if you've defined range names for all of the columns. The simplest way to create a set of range names is with Excel's Create Names command. Try it now with the service station data.

To create range names for the service station data:

1 Select the range **A1:D11** (*not* the range A1:D12).

2 Click **Insert > Name > Create** to open the Create Names dialog box. See Figure 2-12.

Figure 2-12
The Create Names dialog box

range names will be created based on text values in the top row of the selected cell range

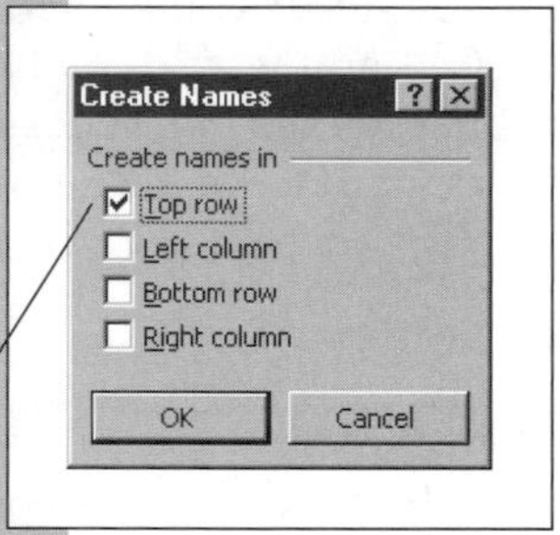

Excel will create range names based on where you have entered the data labels. In this case, you'll use the labels you entered in the top row as the basis for the range names.

3 Verify that the **Top row** checkbox is selected.

4 Click **OK**.

Four range names have been created for you: Station, Gas, Other, and Total. You can use Excel's **Name Box** to select those ranges automatically.

To select the Total range:

1 Click the **Name Box** (the drop-down list box) located directly above and to the left of the worksheet's row and column headers.

2 Click **Total** from the Name Box.

3 The cell range D2:D11 is automatically selected. See Figure 2-13.

**Figure 2-13
Selecting
the Total
range**

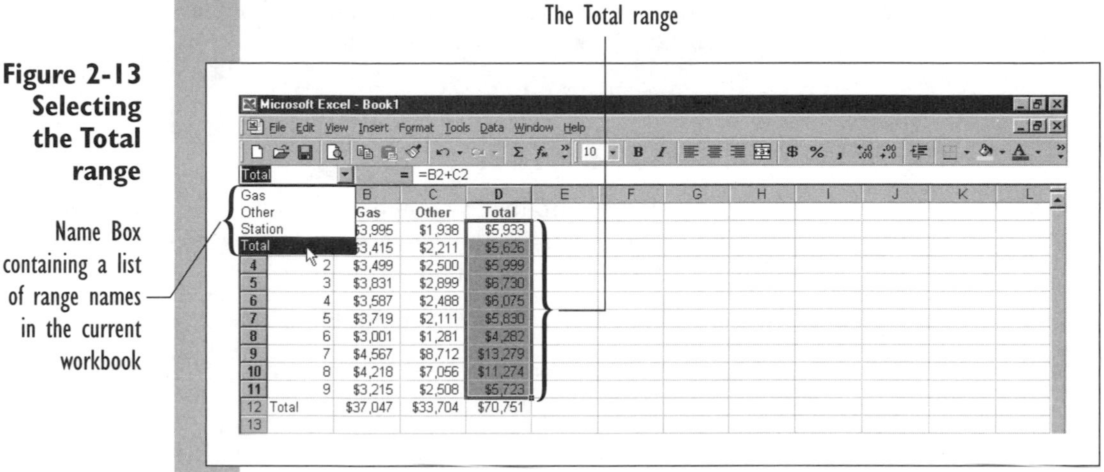

Name Box
containing a list
of range names
in the current
workbook

The Total range

All of the workbooks you'll use in this book contain range names for each of their data columns.

EXCEL TIP

- Another way to create range names is by first selecting the cell range and then typing the range name directly into the Name Box. You can also create range names using the "Insert > Name > Define" command.

- You can duplicate the same range name on different worksheets, but if you wish to reference that name later on, you'll have to specify which worksheet you want to use. For example, you must use the reference 'Sheet 1'!Gas for the cell range "Gas" located on the Sheet 1 worksheet.

- You replace cell references with their range names by first selecting a range of cells containing formulas or functions and then clicking "Insert > Name > Apply" from the Excel menu bar. You will then be prompted for the range names you want to apply to formulas in those selected cells.

Sorting Data

Once you've entered your data into Excel, you're ready to start analyzing it. One of the simplest analyses is to determine the range of the data values. Which values are largest? Which are smallest? To answer questions of this type, you can use Excel to sort the data. For example, you can sort the gas station data in descending order, displaying first the station that has shown the greatest total revenue down through the station that has had the lowest revenue. Try this now with the data you've entered.

To sort the data by Total amount:

1 Select the cell range **A1:D11**.

The range A1:D11 contains the range you want to sort. Note that you do *not* include the cells in the range A12:D12, because these are the column totals and not individual service stations.

2 Click **Data > Sort** from the Excel menu bar.

Excel provides three sorting levels. You can sort each level in an ascending or descending order.

3 Choose **Total** from the drop-down list box for the first sort level and click the **Descending** option button. See Figure 2-14.

**Figure 2-14
The Sort
dialog box**

three levels
of sorting

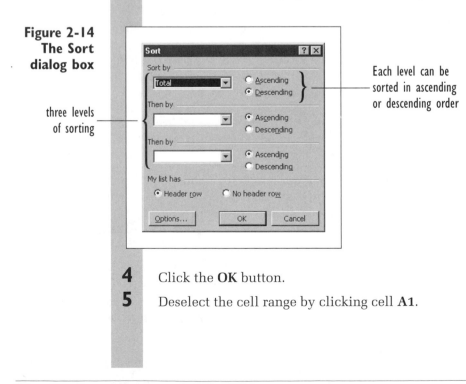

Each level can be
sorted in ascending
or descending order

4 Click the **OK** button.

5 Deselect the cell range by clicking cell **A1**.

The stations are now sorted in order from Station 7, showing the largest total revenue, to Station 6 with the lowest revenue. See Figure 2-15.

**Figure 2-15
The Gas
Station data
sorted in
descending
order of
total
revenue**

	A	B	C	D	E	F	G	H	I	J	K	L
1	Station	Gas	Other	Total								
2	7	$4,567	$8,712	$13,279								
3	8	$4,218	$7,056	$11,274								
4	3	$3,831	$2,899	$6,730								
5	4	$3,587	$2,488	$6,075								
6	2	$3,499	$2,500	$5,999								
7	0	$3,995	$1,938	$5,933								
8	5	$3,719	$2,111	$5,830								
9	9	$3,215	$2,508	$5,723								
10	1	$3,415	$2,211	$5,626								
11	6	$3,001	$1,281	$4,282								
12	Total	$37,047	$33,704	$70,751								
13												

Before going further, this would be a good time to save your work.

To save your work:

1 Click the **Save** button .

2 Save the workbook as **Gas1.xls** in your student folder.

EXCEL TIP

- If you want to sort your list in non-numeric order (in terms of days of the week or months of the year), open the Sort dialog box and click the "Options" button.

Querying Data

Statisticians are often interested in subsets of data rather than in the complete list. For instance, a manufacturing company trying to analyze quality-control data from three work shifts might be interested in looking only at the night shift. An employee firm interested in salary data might want to consider just the subset of those making between $35,000 and $45,000. These subsets fulfill certain criteria, and the process by which Excel selects cells that fulfill those criteria is called **filtering** or **querying**. There are two types of criteria you can use with Excel:

- **Comparison criteria**, which compare variables to specified values or constants. One example might be "Which service stations have gas sales of over $3,500?"
- **Calculated criteria**, which compare variables to calculated values (which are themselves usually calculated from other variables). An example of a calculated criterion might be "Which service stations have gas sales that exceed the average gas sale?"

Excel provides two ways of filtering data. The first method is the **AutoFilter**, which is primarily used for simple queries employing comparison criteria. For more complicated queries and those involving calculated criteria, Excel provides the **Advanced Filter**. You'll have a chance to use both methods.

Using the AutoFilter

Let's say the service station company plans a massive advertising campaign to boost sales for the service stations that are reporting sales of $3,500 or less. You can construct a simple query using comparison criteria to have Excel display only service stations with gas sales < $3,500.

To query the service station list:

1   Click **Data > Filter > AutoFilter**. Notice that Excel adds a drop-down arrow button to the first row of column names.

2   Click the **Gas** drop-down arrow and then click **Custom**. See Figure 2-16.

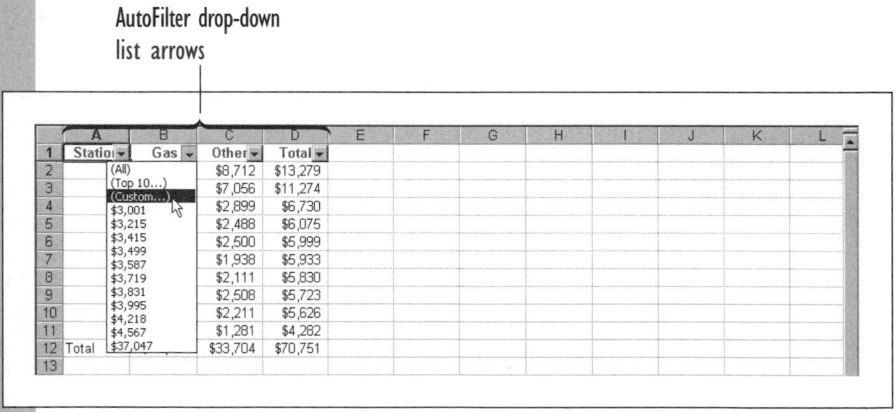

AutoFilter drop-down
list arrows

**Figure 2-16
Selecting
the Custom
AutoFilter
for the
Gas data**

3 In the Custom AutoFilter dialog box, click the **comparison** drop-down list box and click **is less than or equal to**.

4 Press **Tab** and then type **3500** in the text box to the immediate right. See Figure 2-17.

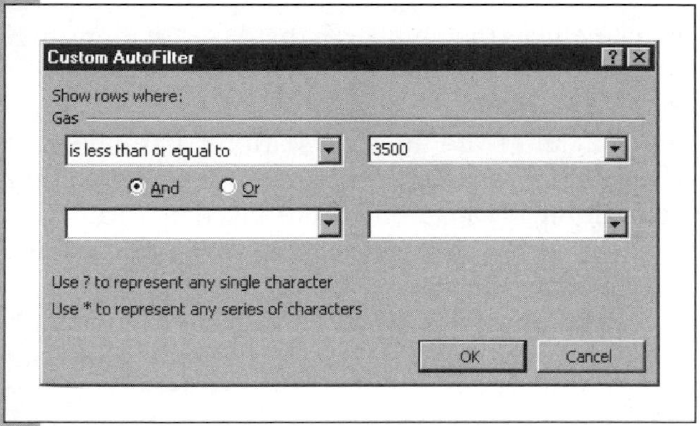

Figure 2-17 Creating a filter for gas sales <= $3500

5 Click **OK**.

6 Excel modifies the list of service stations to show stations 1, 2, 6, and 9. See Figure 2-18. The grand totals are no longer visible, because this row does not fulfill the criteria you specified.

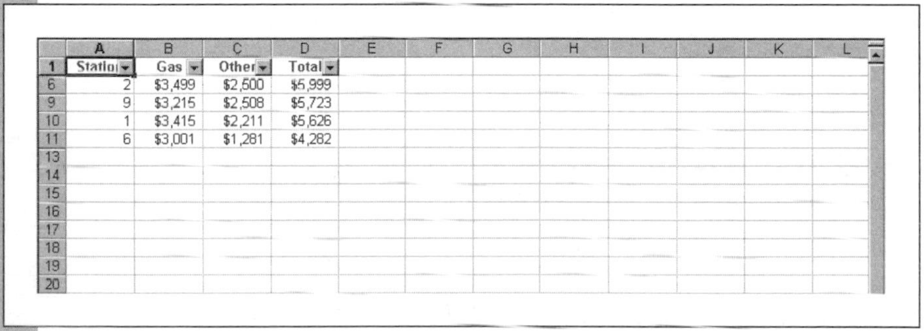

	A	B	C	D
1	Statio	Gas	Other	Total
6	2	$3,499	$2,500	$5,999
9	9	$3,215	$2,508	$5,723
10	1	$3,415	$2,211	$5,626
11	6	$3,001	$1,281	$4,282

Figure 2-18 Stations whose gas sales are <= $3,500

The service station data for the other stations has not been lost, merely hidden. You can retrieve the data by choosing the All option from the Gas drop-down list.

Let's say that you need to add a second filter that also filters out those service stations selling less than $2,500 worth of other products. This filter does not negate the one you just created; it adds to it.

To add a second filter:

1 Click the **Other** drop-down filter arrow and click **Custom**.

2 Enter **is less than or equal to** in the comparison drop-down list box and type **2500** in the accompanying text box.

3 Click **OK**.

Excel reduces the number of displayed stations to Stations 1, 2, and 6.

Stations 1, 2, and 6 are the only stations that have <= $3,500 in gasoline sales *and* <= $2,500 in other sales. Combining filters in this way is known as an **And condition** because only stations that fulfill both criteria are displayed. You can also create filters using **Or conditions** in which only one of the criteria must be true.

To remove the AutoFilter from your data set, you can either stop running the AutoFilter or remove each filter individually. Try both methods now.

To remove the filters:

1 Click the **Other** drop-down filter arrow and click **All**.

The second filter is removed, and now only the results of the first filter are displayed.

2 Click **Data > Filter > AutoFilter** from the Excel menu bar.

Excel stops running AutoFilter altogether and removes the filter drop-down arrows from the worksheet.

Using the Advanced Filter

There might be situations where you want to use more complicated criteria to filter your data. Such situations include criteria that

- Require several *And/Or* conditions
- Involve formulas and functions

Such cases are often beyond the capability of Excel's AutoFilter, but you can still do them using the Advanced Filter. To use the Advanced Filter, you must first enter your selection criteria into cells on the worksheet. Once those criteria are entered, you can use them in the Advanced Filter command. Try this technique by recreating the pair of criteria you just entered— only now you'll use Excel's Advanced Filter.

To create a query for use with the Advanced Filter:

1   Click cell **B14**, type **Advanced Filter Criteria**, and press **Enter**.

2   Type **Gas** in cell B15 and press **Tab**. Type **Other** in cell C15 and press **Enter**.

3   Type **<=3500** in cell B16 and press **Tab**. Type **<=2500** in cell C16 and press **Enter**.

If two criteria occupy the same row in the worksheet, Excel assumes that an And condition exists between them. In the example you just typed in, both criteria were entered into row 16, and Excel assumed that you wanted gas sales <=$3,500 *and* other sales <=$2,500. Thus, these criteria match what you created earlier using the AutoFilter. Now apply these criteria to the service station data. To do this, open the Advanced Filter dialog box and specify both the range of the data you want filtered and the range containing the filter criteria.

To run the Advanced Filter command:

1   Select the cell range **A1:D11**.

2   Click **Data > Filter > Advanced Filter**.

3   Make sure that the "Filter the list, in-place" option button is selected and that "$A$1:$D$11" is displayed in the List Range box.

4   Enter **B15:C16** in the Criteria range box. This is the cell range containing the filter criteria you just typed in. See Figure 2-19.

Figure 2-19 Specifying the cell references for the Advanced Filter

cell range containing filter criteria

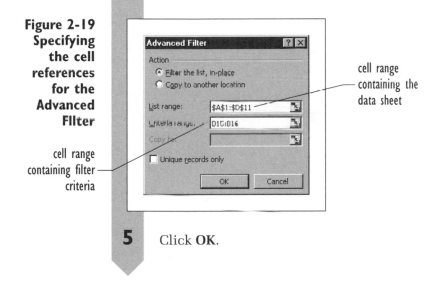

cell range containing the data sheet

5   Click **OK**.

As before, only Stations 1, 2, and 6 are displayed. Note that the column totals displayed in row 12 are not adjusted for the hidden values. You have to be careful when filtering data in Excel because formulas will still be based on the entire data set, including hidden values.

What if you wanted to look at only those service stations with *either* gasoline sales <= $3,500 *or* other sales <= $2,500? Entering an Or condition between two different columns in your data set is not possible with the AutoFilter, but you can do it with the Advanced Filter. You do this by placing the different criteria in different rows in the worksheet.

To create an Or condition with the Advanced Filter:

1 Delete the criteria in cell **C16**.

2 Enter the criterion **<=2500** in cell C17.

3 Click **Data > Filter > Advanced Filter**.

4 Make sure that the cell range "A1:D11" is still entered into the List Range box.

5 Change the cell reference in the Criteria Range box to **B15:C17** to reflect the changes you made to the criteria.

6 Click **OK**.

Excel now displays stations 0, 1, 2, 4, 5, 6, and 9. Each station has gas sales <=$3,500 or sales of other items <=$2,500. See Figure 2-20.

Figure 2-20
Filtering data using an Or condition

filtered data —

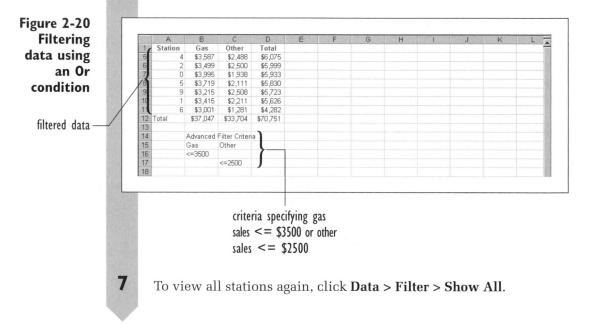

criteria specifying gas
sales <= $3500 or other
sales <= $2500

7 To view all stations again, click **Data > Filter > Show All**.

Using Calculated Values (Optional)

You decide to reward service station managers whose daily gasoline sales were higher than average. How would you determine which service stations qualified? You could calculate average gasoline sales and enter this number explicitly into a filter (either an AutoFilter or an Advanced Filter.) One problem with this approach however, is that every time you update your service station data, you have to recalculate this number and rewrite the query. However, the Advanced Filter allows you to include this information in your query automatically.

To select stations with higher-than-average gas sales:

1 Click cell **E14**, type **HighGas**, and press **Enter**.

2 In cell E15, type **=B2>=AVERAGE($B$2:$B$11)** and press **Enter**.

The formula you just typed tests whether the value in cell B2 is greater than or equal to the average values of the cells in the range B2:B11. When you press the Enter key, the value "TRUE" is displayed in cell E15, indicating that, at least for cell B2, this condition is true. When you use this formula in the Advanced Filter dialog box, it will automatically apply it to the rest of the cells in your data list range.

3 Click **Data > Filter > Advanced Filter**.

4 Enter the range **A1:D11** in the List Range box.

5 Enter the range **E14:E15** in the Criteria Range box and then click **OK**.

Excel displays service stations 0, 3, 5, 7, and 8, indicating that these service stations have higher-than-average daily gasoline sales.

6 View all of the service station data again by clicking **Data > Filter > Show All**.

A few points about creating formulas for calculated criteria:

- A calculated criterion formula must refer to at least one cell in the data list. The reference to the cell must be a relative reference.
- The criterion formula must produce a logical value of TRUE or FALSE. The Advanced Filter will display only those rows whose value (when substituted into the formula) are TRUE.

- Do not use the name of the column label (such as Gas) in a calculated criterion formula because it will produce erroneous results. In the example you created, the label "HighGas" was used. For the same reason, do not use a range name in the formula; instead use an absolute reference. This is why the formula you entered used the cell reference "$B$2:$B$11" rather than the range name "Gas". You could have used a different range name, such as "GasData", as long as it did not match the label given to the column of data values.

You've completed your analysis of the service station data. Save and close the workbook now.

To finish your work:

1 Click the **Save** button 🖫 .

2 Click **File > Close**.

Importing Data from Text Files

Often your data will be created using applications other than Excel. In that case, you'll want to go through a process of bringing that data into Excel called **importing**. Excel provides many tools for importing data. In this chapter you'll explore two of the more common sources of external data: text files and databases.

A **text file** contains only text and numbers, without any of the formulas, graphics, special fonts, or formatted text that you would find in a workbook. Text files are one of the simplest and most widely used methods of storing data, and most software programs can both save and retrieve data in a text file format. Thus, although text files contain only raw, unformatted data, they are very useful in situations where you want to share data with others.

Because a text file doesn't contain formatting codes to give it structure, there must be some other way of making it understandable to a program that will read it. If a text file contains only numbers, how will the importing program know where one column of numbers ends and another begins? When you import or create a text file, you have to know how that data is organized within the file. One way to structure text files is to use a **delimiter**, which is a symbol, usually a space, a comma, or a tab, that separates one column of data from another. The delimiter tells a program that retrieves the text file where columns begin and end. Text that is separated by delimiters is called **delimited text**.

In addition to delimited text, you can also organize data with a **fixed-width file**. In a fixed-width text file, each column will start at the same location in the file. For example, the first column will start at the first space in the file; the second column will start at the tenth space, and so forth.

When Excel starts to open a text file, it automatically starts the **Text Import Wizard** to determine whether the data is in a fixed-width format or a delimited format—and if it's delimited, what delimiter is used. If necessary, you can also intervene and tell it how to interpret the text file.

Having seen some of the issues involved in using a text file, you are ready to try importing data from a text file. In this example, a family-owned bagel shop has gathered data on wheat products that people eat as snacks or for breakfast. The family members intend to compare these products with the products that they sell. The data has been stored in a text file, WHEAT.TXT, shown in Table 2-5. The data was obtained from the nutritional information on the packages of the competing wheat products.

Table 2-5 Wheat Data

Brand	Food	Price	Package oz	Serving oz	Calories	Protein	Carbo-hydrates	Fats
Anderson	Pretzel	$1.55	14	1	110	3	23	1
Uncle B	Bagel	$0.99	15	1.5	120	5	25	0.5
Bays	Eng Muffin	$1.09	12	2	140	5	25	2
Thomas	Eng Muffin	$1.69	12	2	130	4	25	1
Quaker	OHs Cereal	$2.49	12	1	120	1	24	1
Nabisco	Graham Cracker	$2.65	16	0.5	60	1	11	1
Wheaties	Cereal	$3.19	18	1	100	3	23	1
Wonder	Bread	$1.31	16	1	70	3	14	1
Brownberry	Bread	$1.29	16	0.84	60	2	11	1
Pepperidge	Bread	$1.59	16	0.89	80	2	14	2

To start importing WHEAT.TXT into an Excel workbook:

1 Click the **Open** button 📁.

2 Open the folder containing your student files and then click the File of Type drop-down list arrow and select **Text Files**.

3 Double-click **Wheat** (or **Wheat.txt**).

Excel displays the Text Import Wizard to help you select the text to import. See Figure 2-21.

Figure 2-21
Text Import
Wizard –
Step 1 of 3
dialog box

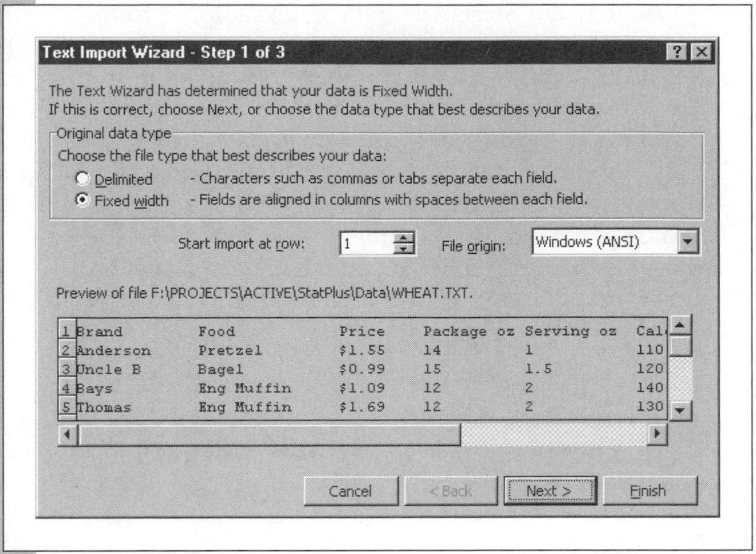

The wizard has automatically determined that the data in the Wheat.txt file is organized as a fixed-width text file. By moving the horizontal and vertical scroll bars, you can see the whole data set.

Once you've started the Text Import Wizard, you can define where various data columns begin and end. You can also have the wizard skip entire columns.

To define the columns you intend to import:

1 Click the **Next** button.

The wizard has already placed borders between the various columns in the text file. You can remove a border by double-clicking it, you can add a border by clicking a blank space in the Data Preview window, or you can move a border by dragging it to a new location. Try moving a border now.

2 Click and drag the right border for Package oz further to the right so that it aligns with the left edge of the Serving oz column. See Figure 2-22.

Figure 2-22
Text Import
Wizard -
Step 2 of 3
dialog box

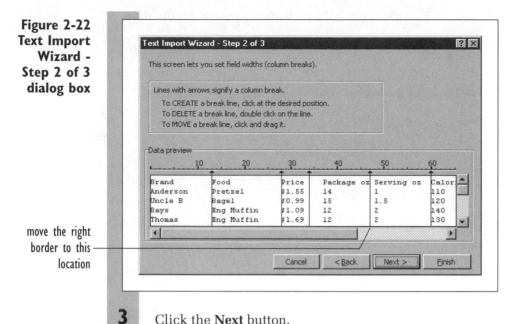

move the right
border to this
location

Figure 2-23
Wheat data
imported
into Excel

3 Click the **Next** button.

The third step of the wizard allows you to define column formats and to exclude specific columns from your import. By default, the wizard applies the General format to your data, which will work in most cases.

4 Click the **Finish** button to close the wizard.

Excel imports the wheat data and places it into a new workbook. See Figure 2-23.

Notice that the data for the first two columns appear to be cut off, but don't worry. When Excel imports a file, it formats the new workbook with a standard column width of about nine characters, regardless of column content. The data are still there but are hidden.

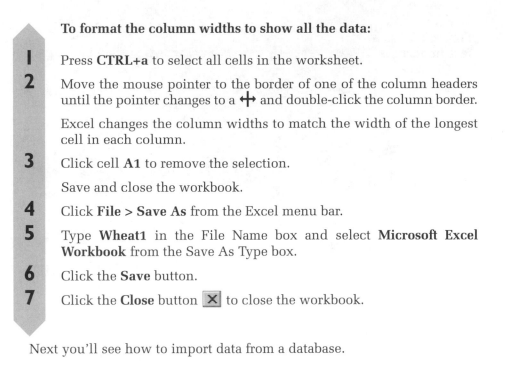

To format the column widths to show all the data:

1 Press **CTRL+a** to select all cells in the worksheet.

2 Move the mouse pointer to the border of one of the column headers until the pointer changes to a ✛ and double-click the column border.

Excel changes the column widths to match the width of the longest cell in each column.

3 Click cell **A1** to remove the selection.

Save and close the workbook.

4 Click **File > Save As** from the Excel menu bar.

5 Type **Wheat1** in the File Name box and select **Microsoft Excel Workbook** from the Save As Type box.

6 Click the **Save** button.

7 Click the **Close** button ☒ to close the workbook.

Next you'll see how to import data from a database.

Importing Data from Databases

A **database** is a program that stores and retrieves large amounts of data and creates reports describing that data. Excel can retrieve data stored in most database programs, including Microsoft® Access, Borland dBASE®, Borland Paradox®, and Microsoft FoxPro®.

Databases store information in **tables**, organized in rows and columns, much like a worksheet. Each column of the table, called a **field**, stores information about a specific characteristic of a person, place, or thing. Each row, called a **record**, displays the collection of characteristics of a particular person, place, or thing. A database can contain several such tables; therefore, you need some way of relating information in one table to information in another. You relate tables to one another by using **common fields**, which are the fields that are the same in each table. When you want to retrieve information from two tables linked by a common field, Excel matches the value of the field in one table with the same value of the field in the second table. Because the field values match, a new table is created containing records from both tables.

A large database can have many tables, and each table can have several fields and thousands of records, so you need a way to choose only the information that you most want to see. When you want to look only at specific information from a database, you create a database query. A **database query** is a question you ask about the data in the database. In response to your query, the database finds the records and fields that meet the requirements of your question and then extracts only that data. When you query a database, you might want to extract only selected records. In this case, your query would contain criteria similar to the criteria you used earlier in selecting data from an Excel workbook.

Using Excel's Database Query Wizard

To see how this works, you'll import the same nutrition data you found in the Wheat text file, only now it will be located in an Access database named "Wheat." The database contains two tables: Product, a table containing descriptive information about each product (the name, manufacturer, serving size, price, and so on), and Nutrition, a table of nutritional information (calories, proteins, etc.). You'll import this data using Excel's Database Query Wizard.

To start the Database Query Wizard:

1 Click the **New** button ⬜ to open a new blank workbook in Excel.

2 Click **Data > Get External Data > New Database Query** from the menu bar.

3 Verify that the Databases dialog sheet tab is selected.

4 At this point, you'll choose a data source. Excel provides several choices from such possible sources as Access, dBase, FoxPro, and other Excel workbooks. You can also create your own customized data source. In this case, you'll use the Access data source because this data comes from an Access database.

5 Click **MS Access Database*** from the list of data sources in the Databases dialog sheet and click the **OK** button.

6 Select the **Wheat.mdb** database file in the folder or disk containing your student files and click the **OK** button.

7 Excel opens the Query Wizard dialog box shown in Figure 2-24.

Figure 2-24
The
Database
Query
Wizard
dialog box

tables in the
Wheat database

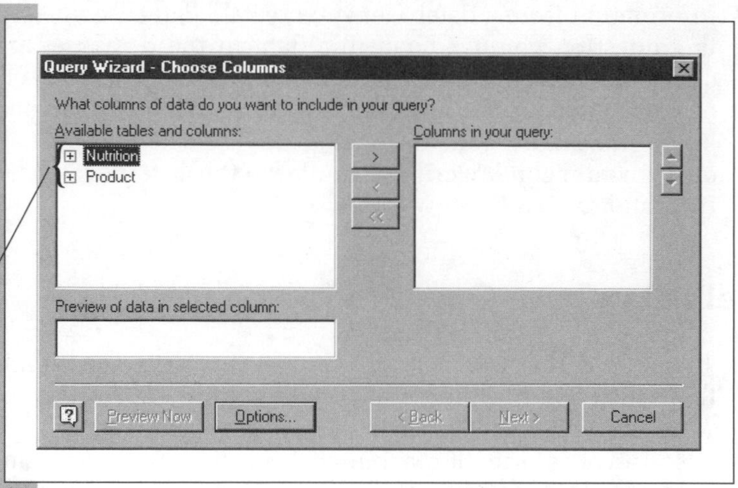

Now that you've started the Query Wizard, you are free to select the various fields that you'll import into Excel. The box on the left of the wizard shown in Figure 2-24 shows the tables in the database. As you expected, there are two: Nutrition and Product. By clicking the "plus" box in front of each table name, you can view and select the specific fields that you'll import into Excel. Try this now by selecting fields from both tables.

To select fields for your query:

1 Click the **plus box [+]** in front of the Product table name.

A space opens beneath the table name displaying the names of each of the fields in the table.

2 Double-click the following names in the list:
Brand
Food
Price
Package oz
Serving oz

As you double-click each field name, the name appears in the box on the right, indicating that they are part of your selection in the query. Note that you do not select the Product ID field. This field is the common field between the two tables and contains a unique id number for each wheat product. You don't have to include this in your query.

3 Click the **plus box [+]** in front of the Nutrition table name and then double-click the following field names:
Calories
Protein
Carbohydrates
Fat

Once again, you do not select the common field, Product ID. Your dialog box should appear as shown in Figure 2-25.

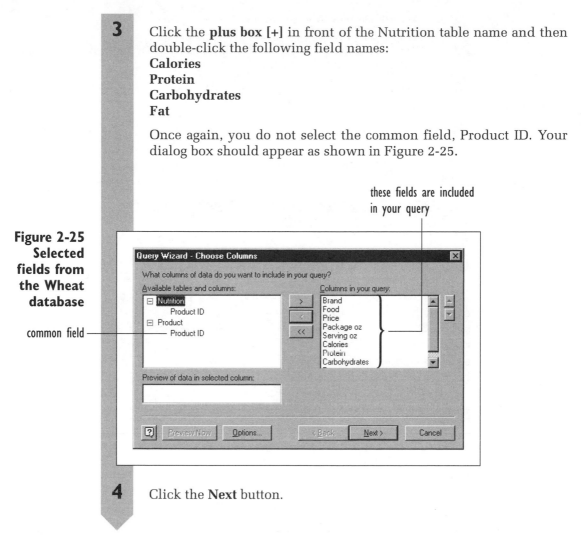

**Figure 2-25
Selected
fields from
the Wheat
database**

these fields are included
in your query

common field

4 Click the **Next** button.

After specifying which fields you'll import, you'll now have the opportunity to control which records to import and how your data will be sorted.

Specifying Criteria and Sorting Data

You can apply criteria to the data you import with the Query Wizard. At this point, your query will import all of the records from the Wheat database, but you can modify that. Say you want to import only those wheat products whose price is $1.25 or greater. You can do that at this point in the wizard. You can specify several levels of And/Or conditions for each of the many fields in your query.

To add criteria to your query:

I Click **Price** from the list of columns to filter in the box at the left of the Query Wizard — Filter Data dialog box.

2 In the highlighted drop-down list box at the right, click **is greater than or equal to**.

3 Type **1.25** in the drop-down list box to the immediate right. See Figure 2-26.

Figure 2-26 Selecting only those records whose price >= $1.25

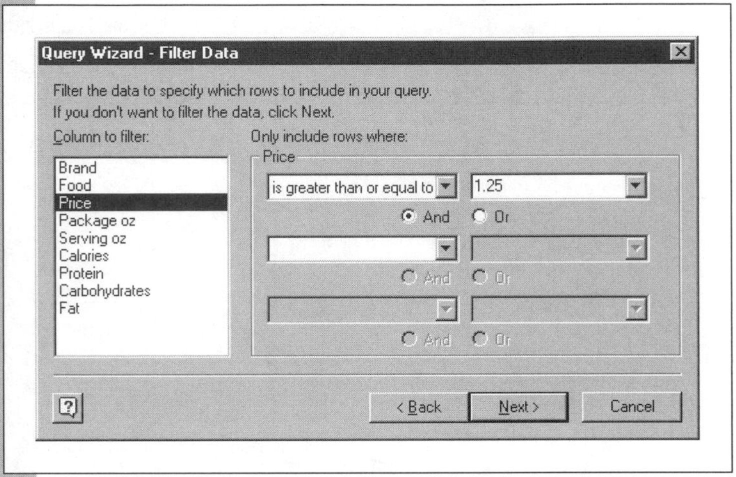

This criterion selects only those records whose price is greater than or equal to $1.25.

4 Click the **Next** button.

The last step in defining your query is to add any sorting options. You can specify up to three different fields to sort by. In this example, you decide to sort the wheat products by the amount of calories they contain, starting with the highest-calorie product first and going down to the lowest.

To specify a sort order:

I Select **Calories** from the Sort by list box.

2 Click the **Descending** option button. See Figure 2-27.

**Figure 2-27
Sorting the
records in
descending
order of
calories**

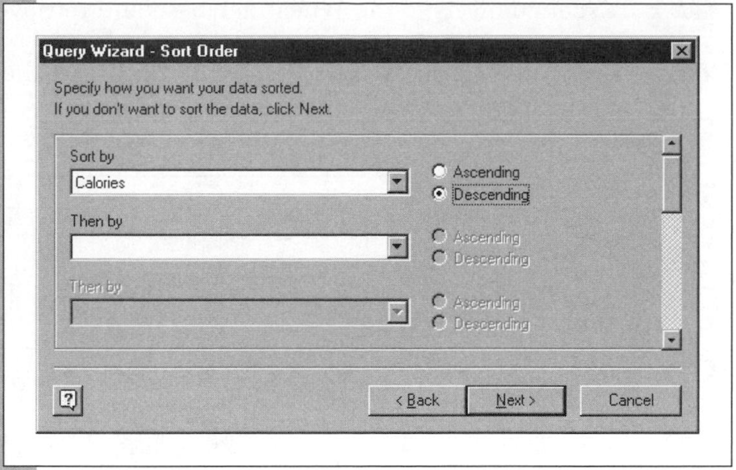

3   Click the **Next** button.

The last step in the Query Wizard is to choose where you want to send the data. You can

1. Import the data into your Excel workbook.
2. Open the results of your query in Microsoft Query. **Microsoft Query** is a program included on your installation disk with several tools that allow you to create even more complex queries.
3. Create an OLAP cube. **OLAP** (On-line Analytical Processing) is a way to organize large business databases. An OLAP cube organizes data so that reports summarizing results of your query are easier to create. An OLAP cube allows you to work with larger data sets than you would otherwise be capable of using in Excel.

You can learn more about these various options in Excel's on-line Help. In this example, you'll simply retrieve the data into your Excel workbook.

To finish retrieving the data:

1   Click the **Return Data to Microsoft Excel** option button.

2   Click the **Finish** button.

You can now specify where the data will be placed. The default will be to place the data in the active cell of the current worksheet. In this case, that is cell A1. Accept this default.

3   Click the **OK** button.

Excel connects to the Wheat database and retrieves the data shown in Figure 2-28. Note that this data includes only those wheat products whose price is $1.25 or greater and that the data is sorted in descending order of calories.

Figure 2-28
Data retrieved from the Wheat database

External Data toolbar

	A	B	C	D	E	F	G	H	I	J	K	L
1	Brand	Food	Price	Package oz	Serving oz	Calories	Protein	Carbohydrates	Fat			
2	Thomas	Eng Muffin	1.69	12	2	130	4	25	1			
3	Quaker	OHs Cereal	2.49	12	1	120	1	24	2			
4	Anderson	Pretzel	1.55	14	1	110	3	23	1			
5	Wheaties	Cereal	3.19	18	1	100	3	23	1			
6	Pepperidge	Bread	1.59	16	0.89	80	2	14	2			
7	Wonder	Bread	1.31	16	1	70	3	14	1			
8	Brownberry	Bread	1.29	16	0.84	60	2	11	1			
9	Nabisco	Grah Cracker	2.65	16	0.5	60	1	11	1			
10												
11												
12												
13												
14												

Unlike importing from a text file, you can have Excel automatically "refresh" the data it imports from a database. Thus, if the source database changes at some point, you can automatically retrieve the new data without recreating the query. Commands like refreshing imported data are available on the External Data toolbar (visible in Figure 2-28). Table 2-6 describes some of these commands.

Table 2-6 External Data Toolbar Commands

Button	Icon	Purpose
Edit Query		Edit the criteria of the query you created to import the data.
Query Properties		Modify the query's properties, allowing you to specify when, where, and how Excel will import the data into your workbook.
Refresh Data		Connect to the database and re-import the data into your workbook.
Refresh All Data		Refresh all data queries located in the current workbook.

Having seen how one would import data from a database into Excel, you are ready to save and close the workbook.

To save and close the Wheat workbook:

1 Click **File > Save As** from the Excel menu bar.

2 Type **Wheat2** in the File Name box and verify that **Microsoft Excel Workbook** is shown in the Save As Type box.

3 Click the **Save** button.

4 Click the **Close** button ✕.

5 Click **File > Exit** to quit Excel.

Exercises

In the exercises that follow, you will be instructed to save your work with the "E" prefix (for Exercise) followed by the chapter number and workbook name.

1. Air quality data has been collected by the EPA (Environmental Protection Agency) and stored in the POLU.XLS workbook. The data shows the number of unhealthful days (heavy levels of pollution) per year for 14 major U.S. cities in the years 1980 and 1985–1989. Open POLU.XLS and

   a. Define range names for each of the columns. Create a new column that calculates the differences between the number of unhealthful days in 1989 and the number of unhealthful days in 1985. Name the new column "Diff8985".

   b. Sort the data in ascending order of the Diff8985 column you just created. Print the results.

   c. Use AutoFilter to view only those cities that showed an increase in the number of healthful days from 1985 to 1989. Print this filtered worksheet.

   d. Remove the filter and create a new column named "Ratio8985" that calculates the ratio of the number of unhealthful days in 1989 to the number of unhealthful days in 1985 (i.e., divide the 1989 values by the 1985 values).

   e. Repeat steps b) and c) for the Ratio8985 column instead of the Diff8985 column.

   f. (Optional) Using the Advanced Filter, show only those cities that had an increase in the number of unhealthful days from 1985 to 1989 *without* using the newly created columns of differences and ratios between the two years. (*Hint*: You'll have to use a calculated criterion.) Print the worksheet.

   g. Write a paragraph summarizing the change from 1985 to 1989. Discuss the difference and also the ratio for the 14 cities.

   h. Save your workbook as E2POLU.XLS.

2. Data on brewers have been saved in a text file named BREWER.TXT, shown in Table 2-7. The file has four variables and nine cases. The first variable is the name of the company; the other three variables are company sales in millions of barrels for the years 1981, 1985, and 1989 (Source: *Beverage Industry Annual Manual 91/92*, page 37).

Table 2-7 Brewer Data

Company	Yr 1981	Yr 1985	Yr 1989
Anheuser-Busch	54.5	68.0	80.7
Miller	40.3	37.1	41.9
Coors	13.3	14.7	17.7
Stroh	23.4	23.4	18.4
Heileman	14.0	16.2	13.0
Pabst	19.2	9.1	6.6
Genesee	3.6	3.0	2.4
Others	10.9	7.0	2.1
Imports	5.2	7.9	8.7

   a. Import the file into an Excel workbook (note that columns are delimited by tabs).

   b. Create range names for each of the four variables in the workbook.

   c. Create two new columns displaying the difference in sales between 1989 and 1981 and the ratio of the 1989 sales to the 1985 sales. Assign range names to these two new columns. Sort the list in descending order of 1981 sales. Print the worksheet.

   d. Is there any relationship between the size of the company and the change in sales? (*Hint:* Are the big companies getting bigger?)

   e. Write a paragraph discussing your conclusions.

   f. Save the workbook in Excel format as E2BREWER.XLS.

3. Table 2-8 shows data for the Big Ten universities. For students entering in 1988-89 and 1989-90 it shows the percentage graduating within six years, the average freshman SAT and ACT scores, and the graduation percentage for athletes entering in 1988–89 and 1989–90 broken down by race and gender. There are some missing values in the table.

Table 2-8 Big Ten Graduation Data

University	Grad 88–90	SAT	ACT	White Males	Black Males	White Females	Black Females
Illinois	80	1146	27	70	80	81	67
Indiana	69	1011	24	59	65	82	100
Iowa	61		24	72	57	76	20
Michigan	85	1189	27	71	59	94	75
Michigan State	70	1003	23	66	35	77	50
Minnesota	50	1026	22	55	21	71	59
Northwestern	89	1250	28	83	60	94	100
Ohio State	61	981	22	63	30	82	70
Penn State	79	1090		81	68	86	67
Purdue	71	1005	24	69	20	88	80
Wisconsin	73	1082	24	64	44	79	30

a. Enter the data from Table 2-8 into a blank workbook.

b. Create two new columns displaying the difference between white male and white female graduation rates and the ratio between white male and white female graduation rates. How do these values compare?

c. Create two more columns calculating the difference and ratio of the white female graduation rate to the overall rate from 1988 to 1990. Summarize your calculations and print the worksheet.

d. Does one university stand out from the others? Use AutoFilter to display only the university with the highest graduation rate of white female athletes to overall graduation rate.

e. Create range names for all of the columns in the worksheet.

f. Save the workbook as E2BIGTEN.XLS.

4. For the E2POLU.XLS data discussed in Exercise 1:

a. Calculate the average number of unhealthful days for each city from 1985 through 1989.

b. Calculate the ratio of this average to the number of unhealthful days in 1980.

c. Sort the list by the largest ratio to the smallest. How is New York ranked? Print your results and write a paragraph describing your observations.

d. Save your changes to E2POLU.XLS.

5. Open the WHEAT.XLS file (this file contains the same data you worked with in this chapter).

a. Add a new column to the worksheet calculating the ratio of calories per unit weight. Do this by dividing the calories by the serving size in ounces. Print the worksheet.

b. Create range names for all of the columns in the worksheet.

c. Use AutoFilter to discover which foods have 100 calories or more per ounce. What is it about these foods that causes them to have more calories? (*Hint*: These foods have only a small amount of a particular ingredient that has weight but no calories.)

d. Save your work as E2WHEAT.XLS.

6. Open the ALUM.XLS file, which contains measurements on eight aluminum chunks from a high school chemistry lab. Both the mass in grams and the volume in cubic centimeters were measured for each chunk.

a. Create a new column in the worksheet, computing the density of each chunk (the ratio of mass to volume.)

b. Sort the data from the chunk with the highest density to that with the lowest.

c. Calculate the average density for all chunks.

d. Is there an outlier (an observation that stands out as being different from the others)? Calculate the average density for all chunks aside from the outlier. Print your results.

e. Which of the two averages gives the best approximation of the density of aluminum? Why?

f. Save the file as E2ALUM.XLS.

7. The ECON.XLS file has seven variables related to the U.S. economy from 1947 to 1962. One of these variables, the Deflator variable, measures the inflation of the dollar and is arbitrarily set to 100 for 1954. The last variable in the list, Total, measures total employment. Open the ECON.XLS file and

a. Create range names for each column in the worksheet.

b. Notice that values in the Population column increase each year. Use the Sort command to find out for which other columns this is true.

c. There is an upward trend to the GNP, although it does not increase each year. Create a new column that calculates the GNP per person for each year. Name this new column GNPPOP and create a range name for the values it contains. Print your results.

d. Save your workbook as E2ECON.XLS.

8. The BASE.XLS file records information related to baseball salaries (in hundreds of thousands of dollars) and batting averages throughout a player's career. Open BASE.XLS and

a. Display only those players whose career batting average is 0.310 or greater. Print your results.

b. Add a new column to the worksheet, displaying the batting average divided by the player's salary and multiplied by 1000.

c. Sort the worksheet in descending order of the new column you created. What are the top ten players in terms of batting average per dollar? Print your results.

d. Examine the number of years played in your sorted list. Where do most of the first-year players lie? What would account for that? (*Hint*: What are some of the other factors besides batting average that may account for a player's high salary?)

e. Save the workbook as E2BASE.XLS.

9. The STATE.XLS workbook shows the 1980 death rates per 100,000 people in each of the 50 states for cardiovascular death and pulmonary death. Open the STATE.XLS workbook and

a. Create a new column calculating the ratio of cardiovascular death rate to the pulmonary death rate in each of the 50 states.

b. Sort the data in descending order of the ratio you created. Print the values.

c. What is the range of values for this ratio? Which state has the largest value? What state has the smallest?

d. Save the workbook as E2STATE.XLS.

10. The CAR database contains data on 392 different car models. The database is in Access format and contains two tables: Car Info and Car Specs. Table 2-9 describes the content of this database:

Table 2-9 CAR Data

Table	Field	Description
Car Info	Model ID	ID number for each car model
	Model	Name of car model
	Year	Year of manufacture
	Origin	Country or region of origin
Car Specs	Model ID	ID number for each car model
	MPG	Miles per gallon
	Cylinders	Number of cylinders in the car engine
	Engine Disp	Engine displacement
	Horsepower	Engine horsepower
	Weight	Car weight
	Accelerate	Car acceleration

Note that Model ID is the common field between the two tables in the database.

a. Create a blank workbook in Excel and, using the Data Query wizard, retrieve all of the fields except the Model ID field from the two tables in the Car database. Have the query sort the data by Year and Origin.

b. Create range names for all of the data columns you retrieved.

Using techniques that you've learned in this chapter, answer the following questions:

c. How many cars come from the model year 1980?

d. Which car has the highest horsepower?

e. Which car has the highest horsepower relative to its weight?

f. Which car has the highest MPG?

g. Which car has the highest MPG relative to its acceleration?

h. If you sort the data by the ratio of MPG to acceleration, where do you find most of the 4-cylinder cars? Where do you find most of the 8-cylinder cars?

i. Save your workbook as E2CARS.XLS.

Chapter 3

Working with Charts

Objectives

In this chapter you will learn to:

- Identify the different types of charts created by Excel

- Create a scatterplot with the Chart Wizard

- Edit the appearance of your chart

- Label points on your scatterplot

- Break a scatterplot down by categories

- Create a bubble plot

- Create a scatterplot containing several data series

I n Chapter 2, you learned how to work with data in an Excel worksheet. In this chapter you'll learn how to display that data through charts. This chapter focuses primarily on two types of charts: scatterplots and bubble charts. Both are important tools in the field of statistics. You'll also learn how to use some features of StatPlus that give you additional tools in working with and interpreting your charts.

Working with Excel Charts

A picture is worth a thousand words. Properly designed and presented, a graph can be worth a thousand words of description. Concepts difficult to describe through a recitation of numbers can be easily displayed in a chart or plot. Charts can quickly show general trends, unusual observations, and important relationships between variables. In Table 3-1, a table of monthly sales values is displayed. How do sales vary during the year? Which month in the table displays an unusual sales result? Can you easily tell?

Table 3-1 Monthly Sales Values

Date	Sales
Jan-2000	16,800
Feb-2000	19,300
Mar-2000	21,100
Apr-2000	21,200
May-2000	20,700
Jun-2000	19,200
Jul-2000	16,100
Aug-2000	14,900
Sep-2000	12,100
Oct-2000	11,900
Nov-2000	12,500
Dec-2000	14,300
Jan-2001	17,500
Feb-2001	19,600
Mar-2001	20,900
Apr-2001	18,200
May-2001	20,600
Jun-2001	18,800
Jul-2001	17,100
Aug-2001	14,100

It's difficult to answer those questions by examining the table. Now let's plot those values. Figure 3-1 displays the result.

**Figure 3-1
Plotted
sales data**

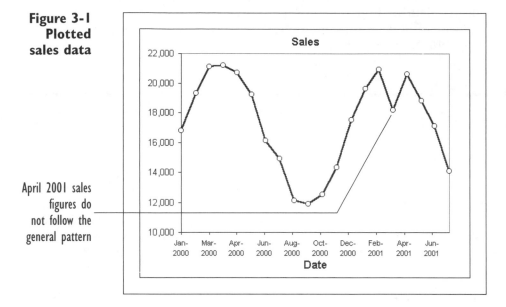

April 2001 sales figures do not follow the general pattern

The chart clarifies things for us. We notice immediately that the sales figures seem to follow a classic seasonal curve with the highest sales occurring during the late winter/early spring months. However, the sales figures for April 2001 seem to be too low. Perhaps something occurred during this time period that should be investigated, or perhaps an erroneous value was entered. In any case, the chart has provided insights that would have been difficult to obtain from a table of values alone.

Excel supports several different chart types for different situations. Table 3-2 shows a partial of list of these.

Table 3-2 Excel Chart Types

Name	Icon	Description
Area		An Area chart displays the magnitude of change over time or between categories. You can also display the sum of group values, showing the relationship of each part to the whole.
Column		A Column chart shows how data changes over time or between categories. Values are displayed vertically, categories horizontally.
Bar		A Bar chart shows how data changes over time or between categories. Values are displayed horizontally, categories vertically.

(continued)

Line		A Line chart shows trends in data, spaced at equal intervals. It can also be used to compare values between groups.
Pie		A Pie chart shows the proportional size of items that make up the whole. The chart is limited to one data series.
Doughnut		The Doughnut chart, like the Pie chart, shows the proportional size of items relative to the whole; it can also display more than one data series at a time.
Stock		The Stock chart is used to display stock market data, including opening, closing, low, and high daily values.
XY(Scatter)		An XY(Scatter) chart displays the relationship between numeric values in several data series. The chart is commonly used for scientific data and is also known as a scatterplot.
Bubble		A Bubble chart is a type of XY(Scatter) chart in which the size of the bubbles is proportional to the value of a third data series.
Radar		A Radar chart shows values from different categories radiating from a center point. Lines connect the values within each data series.
Surface		A Surface chart shows the value of a data series in relation to the combination of the values of two other data series. Surface charts are often used in topographical maps.
Cone, Cylinder, and Pyramid		Cone, Cylinder, and Pyramid charts are similar to Bar and Column charts except that they use cones, cylinders, and pyramids for the markers.

Excel includes variations of each of these chart types. For example, the Column charts can display values across categories or the percentage that each value contributes to the whole across categories. Many of the charts can be displayed in 3D as well.

Most of the charts you'll create in this book will be of the XY(Scatter) type. Other chart types, like Stock charts, are designed for specific types of data (like stock market data) and they are not as useful for general data analysis.

In addition, the StatPlus add-in included with this book gives you the capability of creating other types of charts not currently supported by Excel. You'll learn about these charts as you read through these chapters.

Excel charts are placed in workbooks in one of two ways: **either as embedded chart objects**, which appear as objects within worksheets, or as **chart sheets**, which appear as separate sheets in the workbook. Figure 3-2 shows examples of both ways of displaying a chart.

Figure 3-2
An
embedded
chart object
and a chart
sheet

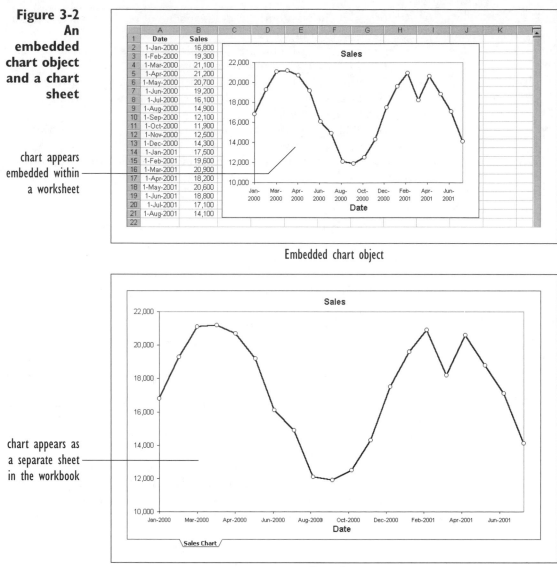

chart appears
embedded within
a worksheet

Embedded chart object

chart appears as
a separate sheet
in the workbook

Chart sheet

Introducing Scatterplots

In this chapter, we'll examine graduation data for a group of universities. The Excel file BIGTEN.XLS contains information on the graduation rates for students who enrolled in Big Ten universities in 1988–89 or 1989–90. Table 3-3 describes the contents of the workbook.

Table 3-3 Big Ten Graduation Rates

Range Name	Range	Description
University	A2:A12	The name of the university
SAT	B2:B12	Average SAT scores of incoming students
ACT	C2:C12	Average ACT scores of incoming students
Graduated	D2:D12	Percentage of students graduating within 6 years of enrolling in 1988–89 or 1989–90
White_Males	E2:E12	Six-year graduation rates for white male athletes
Black_Males	F2:F12	Six-year graduation rates for black male athletes
White_Females	G2:G12	Six-year graduation rates for white female athletes
Black_Females	H2:H12	Six-year graduation rates for black female athletes
Enrollment	I2:I12	The total enrollment at the university
Size	J2:J12	Text indicating whether enrollment is less than or greater than 40,000 students

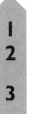

To open the BigTen.xls workbook:

1 Start Excel.

2 Open the **BigTen** workbook from the folder containing your student files.

3 Click **File > Save As** and save the workbook as **BigTen2**.

The workbook opens to a sheet displaying a list of 11 rows and 12 columns. See Figure 3-3. Range names based on the column labels of each column have already been created for you. There are some missing values in the worksheet, such as the SAT value for the University of Iowa in row 4.

Figure 3-3
The BigTen
workbook

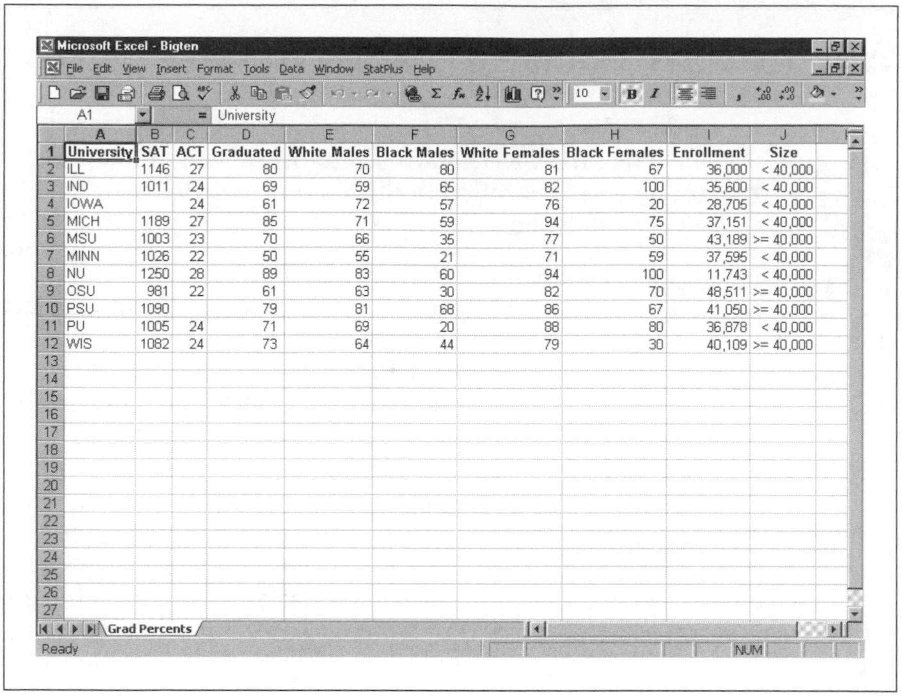

Microsoft Excel - Bigten

File Edit View Insert Format Tools Data Window StatPlus Help

A1 = University

	A	B	C	D	E	F	G	H	I	J
1	University	SAT	ACT	Graduated	White Males	Black Males	White Females	Black Females	Enrollment	Size
2	ILL	1146	27	80	70	80	81	67	36,000	< 40,000
3	IND	1011	24	69	59	65	82	100	35,600	< 40,000
4	IOWA		24	61	72	57	76	20	28,705	< 40,000
5	MICH	1189	27	85	71	59	94	75	37,151	< 40,000
6	MSU	1003	23	70	66	35	77	50	43,189	>= 40,000
7	MINN	1026	22	50	55	21	71	59	37,595	< 40,000
8	NU	1250	28	89	83	60	94	100	11,743	< 40,000
9	OSU	981	22	61	63	30	82	70	48,511	>= 40,000
10	PSU	1090		79	81	68	86	67	41,050	>= 40,000
11	PU	1005	24	71	69	20	88	80	36,878	< 40,000
12	WIS	1082	24	73	64	44	79	30	40,109	>= 40,000
13										

Grad Percents

Ready NUM

We're going to explore the relationship between the average SAT score of incoming freshmen and the percentage of those freshmen who eventually graduate within 6 years of entering college. Do classes with high average SAT scores have higher rates of graduation? We'll get a visual picture of this relationship by producing a scatterplot.

A **scatterplot** is a chart in which observations are represented by **points** on a rectangular coordinate system. Each observation consists of two values: One value is plotted against the vertical or **y-axis**, and the second value is plotted against the horizontal or **x-axis** (see Figure 3-4). By observing the placement of the points on the scatterplot, one can get a general impression of the relationship between the two sets of values. For example, the scatterplot of Figure 3-5 shows that high values of Variable 1 are associated with low values of Variable 2. This is not a perfect association; rather, there is some "scatter" in the points (hence the term "scatterplot").

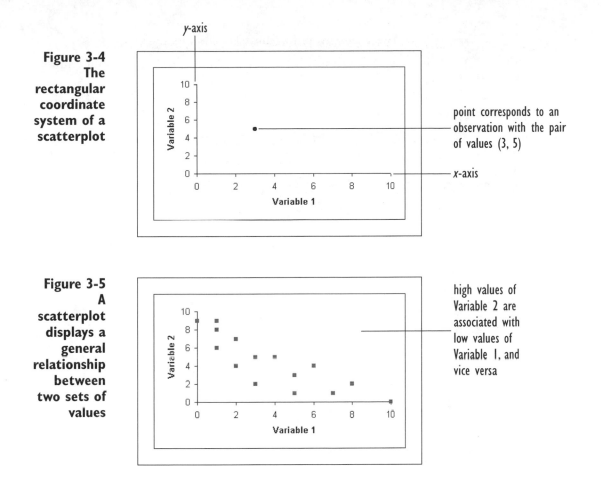

Figure 3-4
The
rectangular
coordinate
system of a
scatterplot

y-axis

point corresponds to an observation with the pair of values (3, 5)

x-axis

Figure 3-5
A
scatterplot
displays a
general
relationship
between
two sets of
values

high values of Variable 2 are associated with low values of Variable 1, and vice versa

Creating Charts with the Chart Wizard

The scatterplot you'll create will have the graduation rate for each university on the *y*-axis and the university's average SAT score on the *x*-axis. To help you create such a chart, Excel supplies the **Chart Wizard**, a series of dialog boxes prompting you to specify the appearance and content of your chart.

To start the Chart Wizard:

Click the **Chart Wizard** button 🔳 on the toolbar.

The first step of four dialog boxes for the Chart Wizard appears. In this step, you choose the chart type and sub-type you intend to create. In this example, you'll create a scatterplot with the points unconnected by any lines.

2   Click **XY(Scatter)** from the Chart Type list box.

3   Click the scatterplot sub-type that displays points without connecting lines. See Figure 3-6.

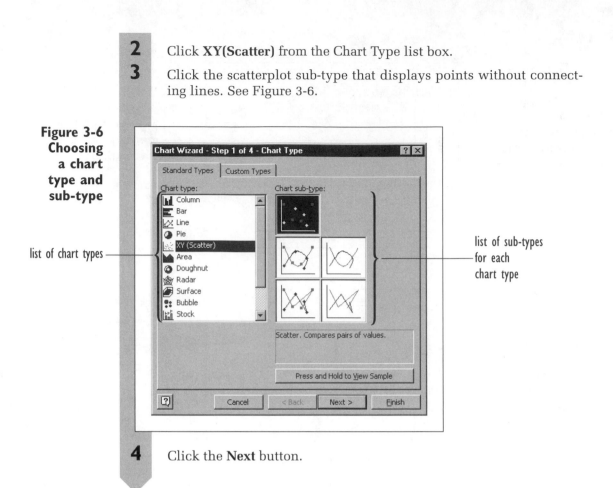

**Figure 3-6
Choosing
a chart
type and
sub-type**

list of chart types

list of sub-types
for each
chart type

4   Click the **Next** button.

In the next series of steps, the wizard will prompt you for the location of the data used in the chart. The set of values is called the **data series**. You will have to enter three pieces of information: the name of the data series, the source of the data to be plotted on the *y*-axis (the graduation rate), and the location of the data to be plotted on the *x*-axis (average SAT score).

To continue the Chart Wizard:

1   Click the **Series** dialog sheet tab.

By default, Excel inserts a data series for you, based on what it assumes you want to plot. You'll have to remove this data series and add one of your own.

2   Click the **Remove** button.

3   Click the **Add** button.

4   Click the **Name** box and then click the **Collapse** button 🔲.

5   Click cell **D1** and then click the **Expand** button 🔲.

6 Click the **X Values** box, click the **Collapse** button ![], select the range **B2:B12**, and click the **Expand** button ![].

7 Using the same technique, insert the range **D2:D12** into the **Y Values** box. The completed dialog sheet appears as shown in Figure 3-7. Note that the dialog sheet shows a preview of your scatterplot.

Figure 3-7 Step 2 of the Chart Wizard

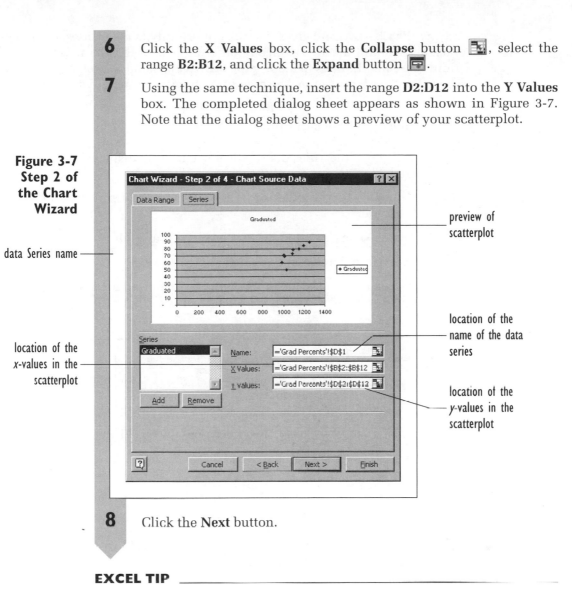

data Series name

preview of scatterplot

location of the name of the data series

location of the x-values in the scatterplot

location of the y-values in the scatterplot

8 Click the **Next** button.

EXCEL TIP

- You can also start the Chart Wizard by clicking "Insert > Chart" from the Excel menu bar.

- Another way of specifying the data series is to select the cells containing the data values first and then start the Chart Wizard. Excel will use your selection in determining the location of the data series. When you use this technique, the columns containing the y-values must lie to the *right* of the column containing the x-values.

- You can add additional data series to your scatterplot by clicking the "Add" button in the second step of the Chart Wizard.

Now the Chart Wizard gives you the opportunity to enter titles, enter legends, and control the appearance of other aspects of the chart. You'll have additional opportunities to modify the chart's appearance later, after it's finished. For now, you can make some basic modifications. First you'll enter the chart titles.

To modify the chart's titles:

1   Verify that the Titles dialog sheet is currently displayed; if not, then click the **Titles** dialog sheet tab.

2   Type **Big Ten Graduation Percentage** in the Chart title box and press **Tab**.

3   Type **SAT** in the Value (X) axis box and press **Tab**.

4   Type **Grad Percent** in the Value (Y) axis box.

As you enter each title, note that the preview window automatically updates to display the text you entered. See Figure 3-8.

**Figure 3-8
Adding
chart titles**

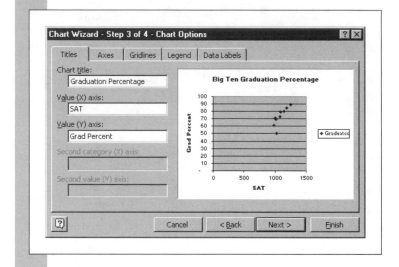

Next you'll remove the legend and gridlines from the plot.

To remove the gridlines and legends:

1   Click the **Gridlines** dialog tab.

2   Deselect the **Major Gridlines** checkbox from the Value (Y) axis to remove the gridlines from the scatterplot.

3 Click the **Legend** dialog tab.

4 Deselect the **Show legend** checkbox to remove the legend from the plot.

Because there is only one set of values being displayed on the chart, you don't need to have a legend. Figure 3-9 shows the current appearance of the chart after your modifications.

Figure 3-9
Step 3 of
the Chart
Wizard,
completed

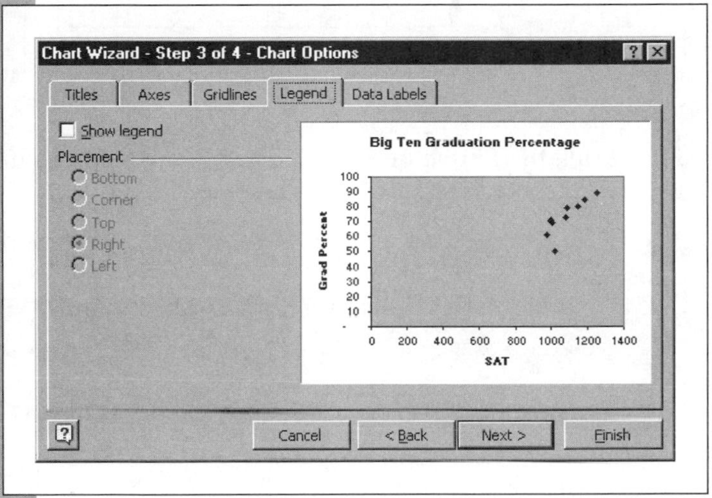

The Data Labels dialog sheet also contains commands to allow you to display values or labels next to each point in the chart. The names are a bit confusing. "Values" are the plot's *y*-values (the graduation rate), and "labels" are the plot's *x*-values (the average SAT scores). In this case, you'll display neither the values nor the labels on the plot.

The final step of the Chart Wizard is to determine whether you want to create an embedded chart located on the currently active worksheet, or a chart sheet containing the chart created by the wizard. You'll create an embedded chart in the Grad Percents worksheet.

To complete the Chart Wizard:

1 Click the **Next** button.

2 Verify that the **As object in** "Grad Percent" option button is selected. See Figure 3-10.

**Figure 3-10
Step 4 of
the Chart
Wizard**

click here to insert
chart into a
Chart sheet

click here to
create an
embedded chart

enter the name
of the Chart
sheet here

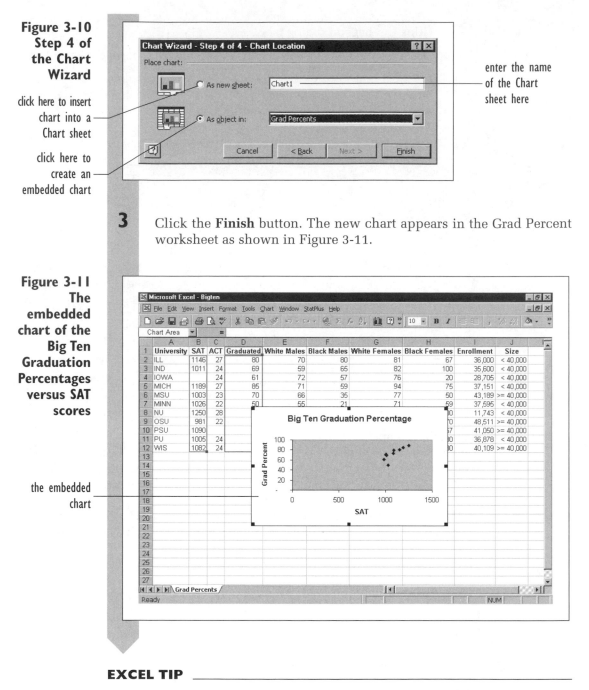

3 Click the **Finish** button. The new chart appears in the Grad Percent worksheet as shown in Figure 3-11.

**Figure 3-11
The
embedded
chart of the
Big Ten
Graduation
Percentages
versus SAT
scores**

the embedded
chart

EXCEL TIP

- You can go backwards and forwards through the Chart Wizard to change your options by clicking the "Next" and "Back" buttons.

- Once the plot is completed, you can return to Step 3 of the Chart Wizard by clicking the chart (to select it) and then clicking "Chart > Chart Options" from the Excel menu bar. In the same way, you can return to Step 4 of the Chart Wizard by clicking the chart and then clicking "Chart > Location."

Now that you've created the initial chart, you'll see how to make further modifications to the chart's appearance.

Editing a Chart

Using Excel's editing tools, you can modify the symbols, fonts, colors, and borders used in your chart. You can change the scale of the horizontal and vertical axes and insert additional data into the chart. To start, you'll edit the size and location of the embedded chart object you just created with the Chart Wizard.

Resizing and Moving an Embedded Chart

Notice that the chart you just created is automatically selected, as you can tell by the eight handles (the small squares on the chart's border, as shown in Figure 3-11). If the chart is not selected, you can select it by clicking any blank space within its border.

When the chart was created, it was placed in a location that covers some of the data in the worksheet. You can move the chart now to a better location.

To move the embedded chart:

1 Click a blank spot within the embedded chart and hold down the mouse button.

Note: If you click the title or other chart element, that element will have a gray selection border around it. If this happens, click elsewhere within the chart, holding down the mouse button. You don't want to select individual chart elements yet.

2 With the mouse button still pressed down, move the chart down so that it covers the range **A13:F26**. As you move the chart, the pointer changes shape to a ⟟.

You use the handles to resize the chart. As you move the pointer arrow over the handles, you'll see the pointer change to a double-headed arrow of various orientations: ↖, ↗, ↕, or ↔. Each pointer allows you to resize the chart in the direction indicated. Try making the chart a little larger. (These instructions assume that you are using a monitor set to 800 × 600 resolution. If your resolution is different, you may have to specify a different size and location for your scatterplot.)

To enlarge the chart:

1   Move your mouse pointer over the handle on the bottom edge of the chart until the pointer changes to a ↕.

2   Drag the pointer down to the bottom of the 28th row and release the mouse button.

The chart now covers the range A13:F28.

Editing the Chart Axes

Now that you've moved and resized the chart, your next task is to edit the individual components of the chart. We'll start with the horizontal and vertical axes. Even though all of the Big Ten graduation rates are 50% or greater in this chart, Excel uses a range of 0 to 100%; and even though the lowest SAT score is 981, Excel uses a lower range of 0 in the chart. The effect of this is that all of the data is clustered in the upper right edge of the chart, leaving a large blank space to the left and below. There are some situations where you want your charts to show the complete range of possible values, and at other times you will want to concentrate on the range of observed values. In this case, you'll rescale the axes so that the scales more closely match the range of the observed values. To do this, you open a Properties dialog box that allows you to change the properties of each scale.

To change the scale of the X-axis:

1   Right-click the X-axis (the axis containing the SAT values) in the scatterplot and click **Format Axis** from the pop-up menu.

2   Click the **Scale** dialog sheet tab.

3   Click the **Minimum** box and type **950**.

4   Click the **Maximum** box, type **1250**, and press **Tab**. The Scale dialog sheet should appear as shown in Figure 3-12.

Figure 3-12
Changing
the scales of
a chart axis

If these checkboxes
are selected, Excel
determines the
scale values
automatically
for you

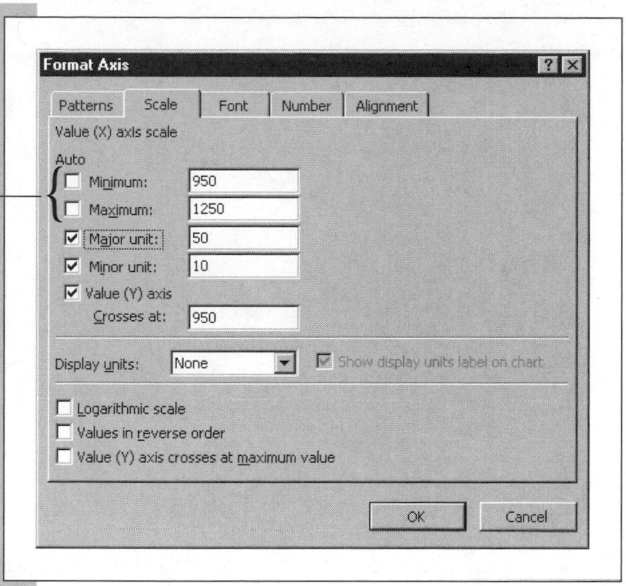

Format Axis

Patterns | Scale | Font | Number | Alignment

Value (X) axis scale

Auto

☐ Minimum: 950
☐ Maximum: 1250
☑ Major unit: 50
☑ Minor unit: 10
☑ Value (Y) axis
Crosses at: 950

Display units: None ▼   ☑ Show display units label on chart

☐ Logarithmic scale
☐ Values in reverse order
☐ Value (Y) axis crosses at maximum value

OK   Cancel

5   Click the **OK** button.

Excel changes the scale of the X-axis in the scatterplot.

You can use the same technique to change the scale of the Y-axis.

To change the Y-axis scale:

1   Right-click the Y-axis and click **Format Axis** from the pop-up menu.

2   Type **45** in the Minimum box in the Scale dialog sheet and then click the **OK** button.

The scales in your scatterplot should resemble those shown in Figure 3-13.

Figure 3-13
Scatterplot
with revised
axes

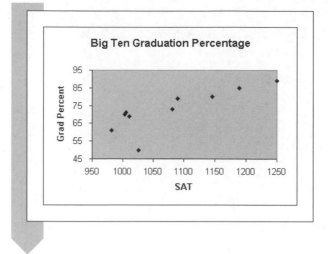

The new scales you've selected show more detail than the default scales used by Excel. Next, you'll see how to modify the plot background and plotting symbols.

EXCEL TIPS

- You can use the Format Axis dialog box to edit other properties of the axis, including the font used for displaying axis values, the format applied to values displayed on the axis, the alignment of axis values, and the line style of the axis line.

- In addition to changing the scale of each axis, you can change the size of the **Major unit** (that is, the space between tick marks on the axis). You can also display values in a logarithmic scale or in reverse order (from highest value down to lowest).

Editing Plot Symbols and Backgrounds

As with each axis, Excel has default options for the display of the plot background and plot symbols. You modify these elements to give your plot a different appearance. Start with the plot background, which Excel has given a light gray color, and change the color to white.

To change the plot background:

I   Right-click an empty spot in the background of the plot and click **Format Plot Area** from the pop-up menu.

2   Click the **White** box located in the last column and fifth row of the matrix of color values. See Figure 3-14.

Figure 3-14
Changing
the plot
area
background
color

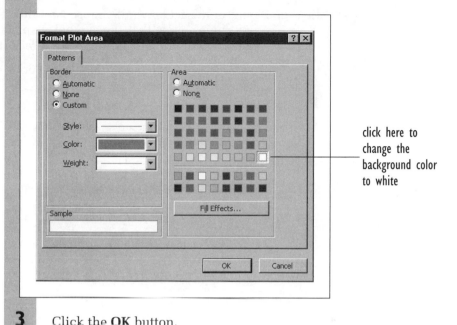

click here to
change the
background color
to white

3   Click the **OK** button.

The background in the plot changes to white. Note that you could have also changed the color and line style of the plot border.

Next, you'll change the type of plot symbol in the chart. By default, Excel uses a blue diamond. You'll change this to an empty circle.

To change the plot symbol:

1   Right-click any plot symbol in the chart and click **Selected Object** from the pop-up menu.

The Format Data Series dialog box appears.

2   Click the **Patterns** dialog sheet tab.

The Patterns dialog sheet is divided into two areas. On the left, you define options for the line that connects one plot symbol with another. In this plot, the symbols are unconnected, which you can detect by noticing that the "None" option button is selected for the type of line. On the right, you define options for the plot symbols. Here is where you'll make your changes.

3   Click the **Style** drop-down list box and choose the Circle symbol.

Symbol colors are divided into foreground and background. The foreground color is essentially the color of the symbol's border, and the background color is the color of the symbol's interior.

4 Click the **Background** list box and choose **No Color** from below the matrix of color values. Note that the symbol displayed in the Style drop-down list box is automatically updated to reflect your choices.

Figure 3-15 shows the completed Format Data Series dialog box.

Figure 3-15
Changing
the plot
symbol

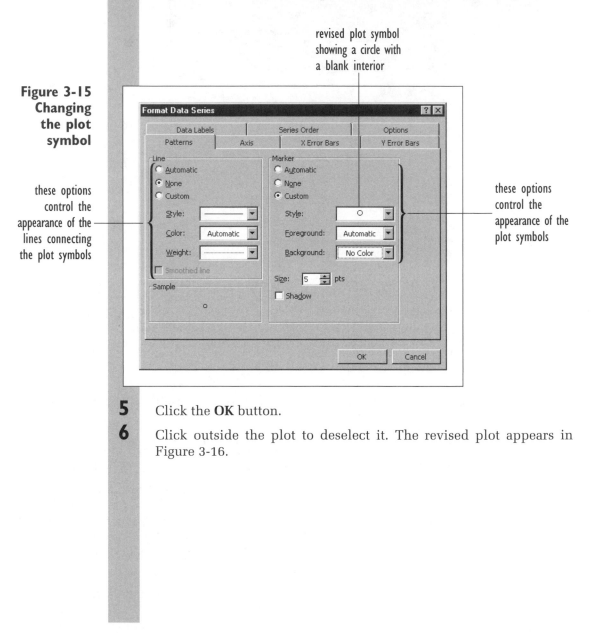

revised plot symbol
showing a circle with
a blank interior

these options
control the
appearance of the
lines connecting
the plot symbols

these options
control the
appearance of the
plot symbols

5 Click the **OK** button.

6 Click outside the plot to deselect it. The revised plot appears in Figure 3-16.

Figure 3-16
Final version
of the Big
Ten
scatterplot

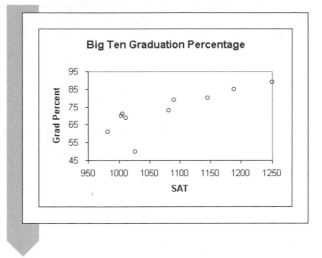

EXCEL TIPS

- You can increase the size of the plot symbols by opening the Format Data Series dialog box, going to the Patterns dialog sheet, and increasing the value entered into the Size box. The default symbol size is 5 points. In the same dialog sheet, you can select the Shadow checkbox to add a drop shadow to your symbol.

- To change the symbol for a single data point in a chart, first click the data point to select the entire data series and then click the data point again to select only that point (wait a brief interval between clicks; you don't want to double-click the data point). Once the single data point is selected, right-click the symbol and choose "Selected Object" from the pop-up menu. Any changes you make to the symbol will apply only to that data point, not to the entire data series.

Let's interpret the scatterplot we've created. Recall that one of the questions we were asking is "What is the relationship (if any) between the average SAT score of a freshman class and its eventual graduation rate?" We can now put forward one hypothesis. Higher average SAT scores seem to be associated with higher graduation rates. There are a couple of exceptions to that relationship. For example, a freshman class at one university showed an average SAT score of 1026 with a graduation percentage of about 50%—much lower than would be expected on the basis of the graduation rates for the other universities with similar average SAT scores. Which university is it?

Identifying Data Points

When you plot data, you will often want to be able to identify individual points. This is particularly important for values that seem unusual. In those cases, you might want to go back to the source of data and check to see whether there were any anomalies in how the data was collected and entered. You may have already noticed that if you pass your mouse cursor over the data points in the BigTen scatterplot, a pop-up label appears to identify the data series name as well as the pair of values used in plotting the point (see Figure 3-17).

Figure 3-17
Getting
information
about a
data point

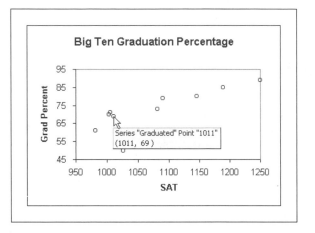

Although this information is interesting and potentially helpful, it doesn't really tell you more about the source of the data point. For example, which university supplied this particular combination of SAT score and graduation percentage? One way to find out is to compare the values given in the pop-up label with the values in the worksheet. For example, you could scroll up the worksheet to see that the point identified in Figure 3-17 is from the University of Indiana (IND), whose freshman class had an SAT average of 1011 and an eventual graduation rate of 69%. In this fashion you could continue to compare values between the chart and the worksheet, finding out which university is associated with which data point. Of course, this is time-consuming and impractical—especially for larger data sets. Excel doesn't provide any other method of identifying specific points, but the StatPlus add-in that comes with this book does provide some additional commands for this purpose (if you haven't installed StatPlus, please read the material in Chapter 1 about StatPlus and installing add-ins).

Selecting a Data Row

One of the StatPlus commands you can use to identify a particular row is the "Select Row" command. This command works only if your data values are organized into columns. To use this command, you select a single point from the chart and then click "Select Row" from the pop-up menu. Try this now and identify the university that had the highest graduation percentage in the Big Ten.

To select a data row:

1 Click a data point in the scatterplot in order to select the entire data series.

2 Pause, and then click the highest data point. This point alone should now be selected.

3 Right-click the data point and click **Select Row** from the pop-up menu.

The eighth row should now be highlighted, indicating that Northwestern University (NU) is the university that had the highest graduation percentage in the Big Ten.

4 Click cell **A1** to remove the highlighting.

Labeling Data Points

You can also use the StatPlus add-in to attach labels to all of the points in the data series. These labels can be linked to text in the worksheet so that if the text changes, the labels are automatically updated. Use StatPlus now to add the university name to each point in the chart.

To add labels to the chart:

1 Right-click the data series in the BigTen scatterplot and choose **Label points** from the pop-up menu.

2 Click the **Labels** button.

Most StatPlus commands give you the choice of entering range names or range references. Because range names have already been created for this workbook, you can select the appropriate range name from a list box. In this case, you'll use the text entered into the University column from the worksheet.

3 Click the **Use Range Names** option button.

4 Scroll down the list box and click **University**.

5 Click the **OK** button.

6 Click the **Link to label cells** checkbox.

By linking to the label cells, you ensure that any changes you make to text in the University column will be automatically reflected in the labels in the scatterplot.

7 Click the **OK** button.

8 Click outside the chart to deselect. Your chart should resemble the one shown in Figure 3-18.

**Figure 3-18
Attaching
labels to the
data points**

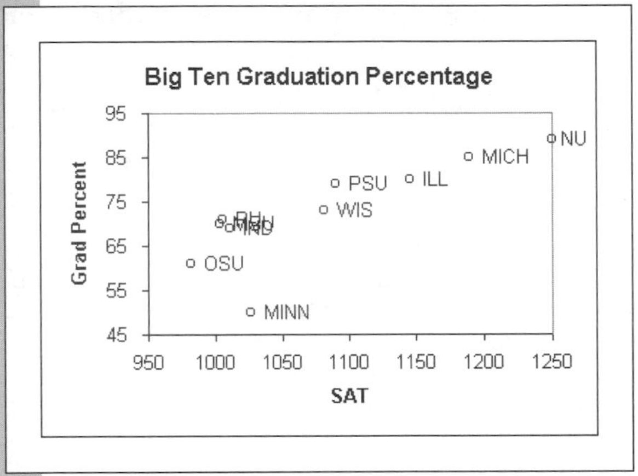

Big Ten Graduation Percentage

STATPLUS TIPS

- If you want to use the same text font and format in the worksheet and in the labels, click the "Copy Label Cell Format" checkbox in the Label Point(s) dialog box.

- If you want to replace the plot symbols with labels, click the "Replace points with labels" checkbox in the Label Point(s) dialog box. Be aware, however, that once you do this, you cannot go back to displaying the plot symbols.

- To label a single point rather than all of the points in the data series, click the data series containing the point, pause a second (do not double-click), and then click the point you're interested in. You can then right-click the point to display the pop-up menu containing the Label Points command.

When you label every data point, there is often a problem with overcrowding. Points that are close together tend to have their labels overlap, as is the case with the Indiana, Michigan State, and Purdue labels. This is not necessarily bad if you're interested mainly in points that lie outside the norm.

Formatting Labels

The Big Ten university that has a low graduation rate relative to the average SAT score of its freshman class is Minnesota. You might wonder why the graduation rate is so much lower than the rates for the other universities. On the basis of the values in the chart, you would expect a graduation rate between 60% and 75% for an average SAT score of around 1026, not one as low as 50%. Perhaps it is because Minneapolis–St. Paul is the largest city among Big Ten towns, and students might have more distractions there, or the composition of the student body might be different. Columbus is the next largest city, and Ohio State is next to last in graduation rate, which seems to verify this hypothesis. On the other hand, Northwestern is in Evanston, right next door to Chicago, the biggest Midwestern city, so you might expect it to have a low graduation rate too. However, Northwestern is also a private school with an elite student body and high admission standards. Minnesota's graduation rate still seems curious. You decide to "mark" this point for further study by changing the color of the label to boldface red.

To format a label:

1 Click any label in the chart to select all of the labels in the data series.

Note that selection handles appear around each label. If you wanted to format all of the labels, you could double-click these labels right now to bring up the Format Data Labels dialog box. However, you want to format only one of the labels.

2 Click the **MINN** label.

A gray selection box appears around the label to indicate that it is selected.

3 Click **Format > Selected Data Labels** from the Excel menu bar.

4 Click the **Font** dialog sheet tab, if necessary.

5 Click the **Color** drop-down list box and click the **red** checkbox from the color matrix (located in the first column and third row).

6 Click **Bold** from the Font Style list box.

7 Click the **OK** button to close the Format Data Labels dialog box.

8 Click outside the chart to deselect it. The color of the MINN data label should now be boldface red.

Creating Bubble Plots

Let's examine one possible explanation for Minnesota's low graduation rate: Does the size of the university have an impact on the relationship between graduation rate and SAT average? Perhaps Minnesota is greatly larger (or smaller) than other universities in the Big Ten, and perhaps this affects graduation rates. We want to create a chart that will display three variables: SAT score, graduation rate, and enrollment. One way of observing the relationship among the three variables is through a bubble plot. A **bubble plot** is similar to a scatterplot, except that the size of each point in the plot is proportional to the size of a third value. In this case, we'll create a bubble plot of graduation rate versus SAT average, the size of each plot symbol being determined by the size of the university. Note that we won't *prove* that the size of the university affects the graduation rate; we are merely exploring whether there is graphical evidence of such a relationship. We start creating the bubble plot with the Chart Wizard.

To start creating the bubble plot:

1 Click cell **A1**.

2 Click **Insert > Chart** from the Excel menu.

3 Select **Bubble** from the Chart type list box and verify that the first subtype is selected. See Figure 3-19.

Figure 3-19
Selecting a
bubble plot

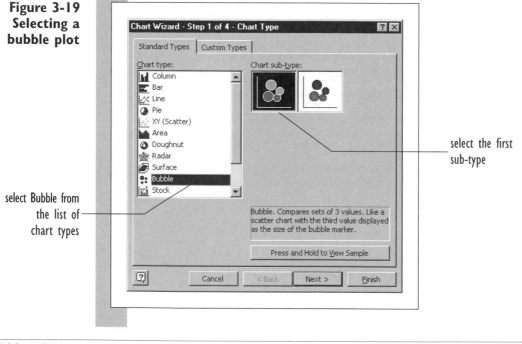

select Bubble from
the list of
chart types

select the first
sub-type

4 Click the **Next** button.

As you did earlier in creating the scatterplot, you'll now have to specify the location for the three different variables in the Bubble plot. Once again, Excel has inserted a default data series that you'll have to remove before adding your own.

To insert the data series:

1 Click the **Remove** button to remove the default data series and then click the **Add** button.

2 Click the **Name** box and select the cell range **D1**.

3 Click the **X Values** box and select the cell range **B2:B12**.

4 Click the **Y Values** box and select the cell range **D2:D12**.

5 Click the **Sizes** box and select the cell range **I2:I12**.

The completed dialog box should appear as shown in Figure 3-20.

**Figure 3-20
Specifying
the location
of the
bubble
plot data**

Preview of the
bubble plot

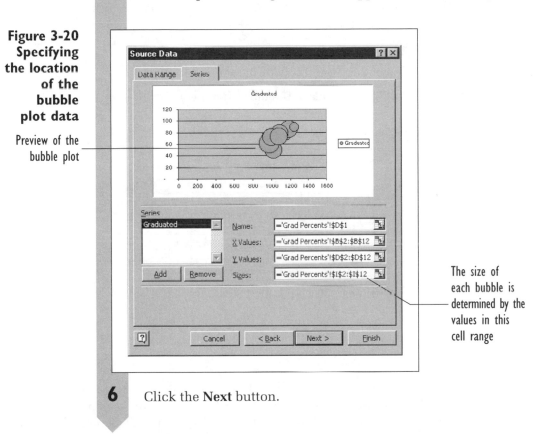

The size of
each bubble is
determined by the
values in this
cell range

6 Click the **Next** button.

Now add titles to your chart and remove the legends and gridlines.

To modify the chart's appearance:

1 Click the **Titles** dialog tab (if necessary) and enter **Graduation Rates for Different Enrollments** in the Chart Title box, **SAT** in the Value (X) Axis box, and **Grad Percent** in the Value (Y) Axis box.

2 Remove the gridlines and legend from the plot.

3 Click the **Next** button.

Now place the plot on an empty spot in the worksheet.

To finish the bubble plot:

1 Verify that the **As object in** "Grad Percents" option button is selected and click the Finish button.

2 Move the bubble plot to the right of the scatterplot and resize it so that it covers the cell range **G13:K28** (if you are using a monitor resolution other than 800 × 600, you may choose a different size to fit your monitor).

3 Change the x-axis scale to the range **950:1300**.

4 Change the y-axis scale to the range **45:100**.

Figure 3-21 displays the current appearance of your bubble plot.

**Figure 3-21
The bubble plot**

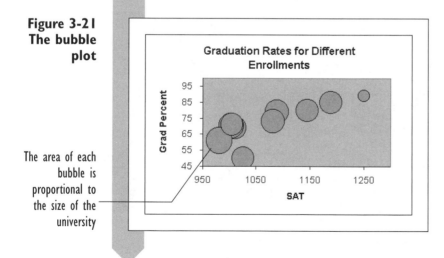

The area of each bubble is proportional to the size of the university

The area of each bubble in the plot is proportional to the school's enrollment. You can change this so that the *width* of each bubble is proportional to the enrollment. In some situations, this works better in displaying differences between points. You can also rescale the sizes of the bubbles to make them larger or smaller.

To edit the bubbles' appearance:

1 Right-click any bubble in the plot and click **Selected Object** from the pop-up menu.

2 Click the **Options** dialog sheet tab.

3 Click the **Width of bubbles** option button.

4 Click the **Scale bubble size to** spin box to change the scale from 100% of default to 50% of default. See Figure 3-22.

**Figure 3-22
Changing
the bubble
size and
style**

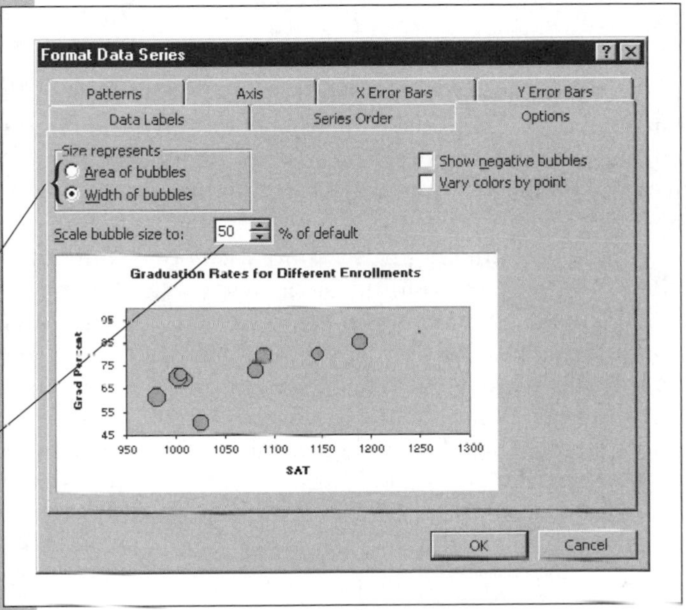

click to specify
whether the size
of the variable is
indicated by the
bubble's area or
the bubble's width

click here to
rescale the
bubble sizes

Because some are hidden, you can remove the interior color to allow you to see each bubble.

5 Click the **Patterns** dialog tab.

6 Click the **None** option button above the color matrix.

7 Click the **OK** button. Figure 3-23 shows the final form of your bubble plot.

Figure 3-23
The final
bubble plot

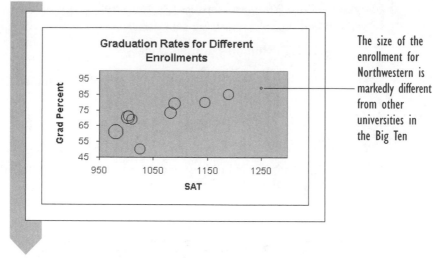

The size of the enrollment for Northwestern is markedly different from other universities in the Big Ten

Let's evaluate what we've created. In interpreting bubble plots, the statistician looks for a pattern in the distribution of the bubbles. Are bubbles of similar size all clustered in one area on the plot? Is there a progression in the size of the bubbles? For example, do the bubbles increase in size as we proceed from left to right across the plot? Is there a bubble that is markedly different from the others? In this plot, we notice immediately that all of the bubbles are pretty close to the same size except for the one on the far right, which is much smaller than the rest. This is, of course, the bubble for Northwestern, and the plot tells us that Northwestern has the highest graduation rate, the highest average SAT score for its freshmen, and also the smallest enrollment. What about Minnesota? The bubble for Minnesota is pretty close in size to the other bubbles, indicating that it is close in enrollment size to other Big Ten schools (aside from Northwestern). Thus, if we are looking for a reason for Minnesota's low graduation rate, we haven't seen it in the size of the enrollment.

Breaking a Scatterplot into Categories

Another approach we can take is to divide the universities into categories. For example, how do the larger universities (with total enrollments of 40,000 or greater) compare to the smaller universities with respect to the graduation rates and SAT averages of their freshman classes? One way of displaying this information is to use one plot symbol for the larger universities and a different symbol for the smaller universities. To do this with Excel, you have to copy the values for these universities into two separate columns and then recreate the scatterplot, plotting two data series instead of one. That can be a time-consuming process. To save time, you can use StatPlus to break the scatterplot into categories for you. You'll try this now, using the Size column to determine the category values (< 40,000 or >= 40,000). First, you'll make a copy of the scatterplot you created earlier.

To copy the scatterplot:

1 Click the scatterplot you created earlier in this chapter to select it. (Note: Make sure you've selected the scatterplot itself and not a specific element of the plot such as the chart title.)

2 Click the **Copy** button 🖹.

3 Click an empty cell directly below the lower-left corner of the scatterplot.

4 Click the **Paste** button 🖺.

Excel pastes a copy of the scatterplot directly below the first one.

5 Click any one of the plot labels in the scatterplot and press the **Delete** key.

The plot labels disappear (you won't be using them in the plot you'll create next).

Now, break the points in the scatterplot into two categories based on the values in the Size column.

To break the scatterplot into categories:

1 Right-click any point in the scatterplot and click **Display by Category**.

2 Click the **Categories** button.

3 Click the **Use Range Names** option button.

4 Scroll down the Range Names list box, click **Size**, and then click the **OK** button.

5 Click the **Bottom** option button to display the categories legend at the bottom of the scatterplot.

6 Click the **OK** button. Figure 3-24 displays the scatterplot broken down by categories.

Figure 3-24
Breaking
the
scatterplot
into
categories

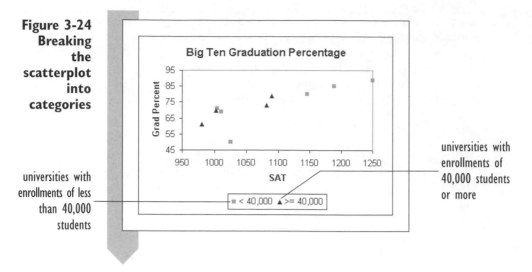

universities with
enrollments of less
than 40,000
students

universities with
enrollments of
40,000 students
or more

STATPLUS TIPS

- Once you break a scatterplot into categories, you cannot go back to the original scatterplot (without the categories). If this is a problem, create a copy of the scatterplot *before* you break it down.

With the data series now broken down into categories, we can compare the universities with enrollments of less than 40,000 students to the larger universities. Note that the top three universities in graduation rate and SAT average all had enrollments of less than 40,000 students. Does this indicate a difference between the larger and smaller universities? Note as well that other universities with less than 40,000 students, like Minnesota and Indiana, had *lower* graduation rates and SAT scores, so the data is not clear-cut. One must be careful not to overanalyze data—there is a great temptation to assign significant results to what are merely statistical aberrations.

So what have you learned? You've learned that Big Ten freshman classes with high SAT scores tend to have higher graduation rates (but not in every case). You've seen that there may be other factors at work, such as the size of the university and whether the university is a private school or a public school. But your observations are based on only ten universities. What if you increased the number of universities in the database? With more data, some of the issues could be resolved. Of course, with a larger study come additional questions: Does geographic location or conference affiliation matter? Is there a difference between Big Ten schools and schools in the Pac Ten? In examining those questions, you'll have to make use of many of the types of plots and charts you created in this chapter.

Plotting Several Variables

Before finishing, let's explore one more question. The data includes graduation rates for the athletes in the freshman class broken down by gender and race. How do these graduation rates compare? A scatterplot displaying these results needs to have several data series. You can create such a plot with the Chart Wizard. It's simpler, in this case, to select the data before we start the wizard. The wizard will then use our selection to create each data series. One important consideration: The Chart Wizard will assume that the *left-most* column of your selection contains the values for the x-axis, so you must have your data columns ordered correctly. In this example, you'll plot the graduation rates for white male and female athletes against the class's average SAT score.

To create the scatterplot:

1   Select the range **B1:B12**, hold down the **CTRL** key, select the range **E1:E12**, and then select the range **G1:G12**. Release the **CTRL** key.

2   Click **Insert > Chart**.

3   Click **XY(Scatter)** from the list of chart types and verify that the first chart sub-type is selected. Click the **Next** button.

4   Click the **Series** tab and note that Excel has automatically created two data series for you named "White Males" and "White Females". Click the **Next** button.

5   Click the **Titles** dialog sheet tab and enter **Athlete Graduation Rates** for the Chart title, **SAT** for the Value (X) Axis title, and **Grad Percent** for the Value (Y) axis title.

6   Click the **Legend** dialog tab and click the **Bottom** option button to place the chart legend at the bottom of the scatterplot.

7   Click the **Gridlines** dialog sheet tab and remove the gridlines from the plot. Click the **Next** button.

8   Click the **Finish** button to complete the Chart Wizard.

Now edit the chart so that the axis ranges match the range of the observed values.

To finish editing the scatterplot:

1   Change the scale of the x-axis to **950:1250** to match the other plots on the worksheet.

2   Change the scale of the y-axis to **45:100**.

3 Move the scatterplot to an empty location below the bubble chart you created earlier.

4 Increase the size of the chart so that it covers the range **G29:K44** (or to a size that seems appropriate to you).

Figure 3-25 shows the completed plot.

Figure 3-25 Plotting two sets of data series

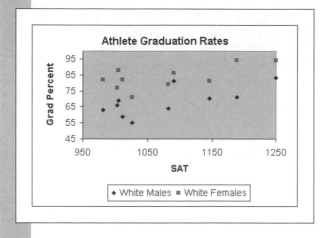

The scatterplot shows that the white female athletes generally have higher graduation rates than the white male athletes. Does this tell us something about white female and male athletes? Perhaps, but we should bear in mind that this chart plots the average graduation rates for these two groups against the average SAT score for the *entire* class of incoming freshmen. We don't have data on the average SAT score for incoming freshman male athletes or incoming freshman female athletes. It's possible that the female athletes also had higher SAT scores than their male counterparts, and thus we would expect them to have higher graduation rates. On the other hand, if their SAT scores are comparable, we might look at the college experiences of male and female athletes at these universities to see whether this would have an effect on graduation rates. Are the demands on male athletes different from those on female athletes, and does this affect the graduation rates?

You've completed your work on the Big Ten scatterplots. Save your changes to the workbook.

To complete your work:

1 Click **File > Save** to save your work.

2 Click **File > Exit** to exit Excel.

Exercises

1. Reopen the BigTen workbook and use a scatterplot to investigate further the relationship between graduation rate and total enrollment.

   a. Create a scatterplot with the graduation rate on the y-axis and the total enrollment on the x-axis. Title the chart "Graduation Rates." Title the y-axis "Grad Percent" and the x-axis, "Enrollment." Remove all legends and gridlines from the plot. Save your scatterplot in a chart sheet named "Grad Rate Chart."

   b. Edit the symbols in your scatterplot, changing the symbol to a red triangle 8 points in size.

   c. Add labels to the scatterplot, identifying each university.

   d. Print your chart. On the printout, draw a straight line that you think best describes the general relationship between the graduation rate and enrollment.

   e. Edit the scale of the scatterplot, changing the lower range of the Enrollment axis scale to 25,000 (this will exclude the Northwestern point from the chart). Print this new chart. Draw on this printout a straight line that you think best describes the relationship between the graduation rate and enrollment.

   f. Compare the two printouts. How do your two straight lines differ? What accounts for this? Discuss your findings in terms of the importance of a single observation affecting the conclusions of an entire group of observations.

   g. Save your workbook as E3BIGTEN.XLS.

2. The JRCOL.XLS workbook contains salary data for 81 faculty members at a junior college. It also includes information on the number of years employed and gender. The data were used in a legal action that was eventually settled out of court. Female professors sought help from statisticians to show that they were underpaid relative to their male counterparts.

   a. Open the workbook and create a scatterplot with Salary on the y-axis and Years on the x-axis. Title the scatterplot "Employee Salaries." Title the y-axis "Salary" and the x-axis "Years Employed." Remove the legend and gridlines from the plot. Save the plot as a chart sheet named "Salary Chart."

   b. Break the scatterplot points into two categories based on gender.

   c. Print the scatterplot and interpret the results. Is there evidence to suggest that the males' salaries are higher than females' salaries for comparable years of employment?

   d. Examine the list of other variables in the workbook. Are there other variables in that list which should be taken into account before coming to a conclusion about the relationship between gender and salary?

   e. Save your workbook as E3JRCOL.XLS.

3. The CALC.XLS data set includes grades (the range name is Calc) for the first semester of calculus, along with various scores. [See Edge and Friedberg (1984).] The scores include high school rank (HS Rank), American College Testing Mathematics test score (ACT Math), and an algebra placement test score (Alg Place) from the first week of class. Because admission decisions are often based on ACT math scores and high school rank, you might expect that they would be related to success in college.

   a. Open the workbook and create a scatterplot, on a separate chart sheet, plotting Calc on the *y*-axis and ACT Math on the *x*-axis. Label the axis appropriately.
   b. Break down the scatterplot by gender. Is there evidence of a difference in calculus scores based on gender?
   c. Print your scatterplot.
   d. Repeat steps a through c for a scatterplot relating calculus grades to the algebra placement test.
   e. Save your workbook as E3CALC.XLS.

4. The ALUM.XLS data record the mass and volume of eight chunks of aluminum, as measured in a high school chemistry lab.

   a. Open the workbook and create a scatterplot with mass values on the *y*-axis and volume values on the *x*-axis. Add major gridlines for both the *x*-axis and the *y*-axis.
   b. Is there a point that seems out of place? Mark this point by changing its color to red.
   c. Do the other points seem to form a nearly straight line? The ratio of mass to volume is supposed to be a constant (the density of aluminum), so the points should fall on a line through the origin. Draw the line, and estimate the slope (the ratio of the vertical change to the horizontal change) along the line. What is your estimate for the density of aluminum?
   d. Save your workbook as E3ALUM.XLS.

5. The WHEAT.XLS workbook contains information on wheat products.

   a. Open the workbook and create a scatterplot with Protein on the *y*-axis and Carbo on the *x*-axis. Label the axes appropriately. Save the chart into a chart sheet named "Protein Chart."
   b. Label each point in the scatterplot based on brand name. Print the chart.
   c. On the plot, the data seem to fall in two groups, with low Carbo on the left and high Carbo on the right. Which brand has the highest protein in the low-Carbo group? Which brand has the lowest protein in the high-Carbo group?
   d. Save your workbook as E3WHEAT.XLS.

6. The BASE.XLS workbook includes salaries and statistics for major league baseball players at the start of the 1988 season.

   a. Open the workbook and create a scatterplot with Salary on the *y*-axis and Aver Career (career batting average) on the *x*-axis. Label the axis appropriately and save the plot in a chart sheet named "Salary Plot."

b. Change the lower range of the *x*-axis scale to 0.15. Print the chart.

c. Edit the properties of the *y*-axis to use a logarithmic scale. Print the chart. How does the logarithmic chart differ from the first chart? *(Hint: Discuss the difference in terms of the amount of "scatter" between the points on the plot.)*

d. Identify on the plot the last name of the player with the highest batting average (label only this single point, not *all* of the points).

e. Create a scatterplot of Salary versus Years. Name the chart sheet "Salary Plot 2." Change the *y*-axis to a logarithmic scale. Print the chart sheet.

f. Do the highest salaries exist for players at the beginning of their careers, at the end, or somewhere in the middle?

g. Identify on the plot the last name of the player who has been in the major leagues the longest (label only this single point).

h. Create a bubble plot of salary versus batting average, the size of each bubble being determined by the number of years played. Save the bubble plot as a chart sheet named "Salary Bubble Plot." Change the scale of the *y*-axis to a logarithmic scale.

i. Edit the bubble plot so that the number of years played is represented by the width of each bubble. Rescale the bubble size to 50% of the default and remove the interior color of the bubble.

j. Some players with high batting averages (0.300 and above) are low on the salary scale. How does the bubble plot help to explain why this is so? Base your answer on the values you observe in the plot.

k. Save your workbook as E3BASE.XLS.

7. The LONGLEY.XLS workbook has seven columns related to the economy of the United States from 1947 to 1962. The last column is Total, the U.S. total employment in thousands. Another column, Armforce, is the total number of people in the armed forces, again in thousands.

a. Open the workbook and create a scatterplot of Total (on the *y*-axis) against Armforce (on the *x*-axis). Label the axes appropriately and save the chart as a chart sheet named "Employment Chart."

b. Label each point on the plot, identifying the year.

c. Edit the labels, reducing the font size to 8-point.

d. Use your knowledge of history to explain why four points on the lower loft of the scatterplot stand out.

e. Aside from the four points in the lower left corner of the plot, describe the general relationship between total employment and the number of people in the armed forces.

f. Save your workbook as E3LONGLEY.XLS.

8. The RACE.XLS workbook contains race results and reaction times (the time it takes for the runner to leave the starting block after hearing the sound of the starting gun) from the first round of the 100-meter heats at the 1996 Summer Olympic games in Atlanta.

a. Open the workbook and create a scatterplot on a separate chart sheet of Time versus Reaction. Use the default axis scales as selected by Excel.

Do you see a trend that would indicate that runners with faster reaction times have faster race times?

b. There is a point that lies away from the others. Label this point with the name of the runner. Print your chart.

c. Edit the *x*-axis and *y*-axis scales of the scatterplot. Set the *x*-axis range to 0.12 to 0.24 second. Set the *y*-axis scale to 9.5 to 12.5 seconds. Is there any more indication that a relationship exists between reaction time and race time? Print the chart.

d. Comment on how the scale used in plotting data can affect your perception of the results.

e. Save your workbook as E3RACE.XLS.

9. The PCSURV.XLS workbook contains the results of questionnaires given to 17,000 people, asking them to rate their PC manufacturer. Respondents rated their company on reliability, repairs, support, and whether they would buy again from the company.

a. Open the workbook and create a scatterplot on a separate chart sheet of Support versus Repairs. Rescale the scatterplot's axes so that they more closely follow the range of the observed data values.

b. Does the plot suggest that manufacturers with high support ratings also have high ratings for their repair service?

c. Label each point with the company name. One manufacturer has a lower-than-expected support rating, given the company's very high repair rating. Which is it? Print your chart.

d. Create a bubble plot on a separate chart sheet with Support displayed on the *y*-axis, Repair displayed on the *x*-axis, and the size of each bubble determined by the Buy Again column.

e. Edit the bubble plot so that the size of each bubble is rescaled to 40% of the default and the size is represented by the bubble's width. Print the chart.

f. Is there anything in the bubble plot to suggest that high Buy Again values are related to either Support values or Repair values?

g. Save your workbook as E3PCSURV.XLS.

10. The CARS.XLS workbook contains information on 392 car models. Data in the workbook include the miles per gallon (MPG) of each car as well as the acceleration, weight, and horsepower.

a. Open the workbook and create a scatterplot on a separate chart sheet of MPG (on the *y*-axis) versus horsepower (on the *x*-axis).

b. One model has a higher MPG than expected, given its horsepower. Label this *single* point with the name of the model (make sure that you select only that point rather than all of the points—or else you'll wait a long time for all 392 of the labels to be added to the scatterplot). Print your chart.

c. Copy your scatterplot to a second chart sheet. Break down the plot by region of origin. Is there any relationship between region of origin and MPG?

d. Add to the Car Data worksheet a new column of values containing the number of years beyond 1970 that the car was manufactured.

e. Create a bubble plot on a separate chart sheet with MPG on the *y*-axis, horsepower on the *x*-axis and the size of each bubble determined by the number of years past 1970 that the car was manufactured.

f. Rescale the bubbles to 50% of the default and relate the size of the Years Past 70 column to the bubbles' width. Print the chart.

g. Examine those cars with horsepowers of 150 or greater by rescaling the *x*-axis. Given that older model cars are represented by smaller bubbles on the plot, is there evidence that the year the car was created can affect the relationship between MPG and horsepower? Print the rescaled chart.

h. Save your workbook as E3CARS.XLS.

11. Voting results for two presidential elections have been stored in the VOTING.XLS workbook. The workbook contains the percentage of the vote won by the Democratic candidate in 1980 and 1984, broken down by state.

a. Open the workbook, save it as E4VOTING.XLS, and create a scatterplot of Dem1984 versus Dem1980 on a separate chart sheet.

b. Rescale the axes so that the minimum value for the *x*- and the *y*-axis is 20.

c. Examine the scatterplot. Does the voting percentage in 1984 generally follow the voting percentage from 1980? In other words, if the Democratic candidate received a large percentage of the vote from a particular state in 1980, did he do as well in 1984?

d. In one state, the candidate had a large percentage of the vote in 1980 (above 55%) but a small percentage of the vote in 1984 (about 40%). Label this point on your plot with the state's abbreviation. Print the chart.

e. Create a copy of the plot on a separate chart sheet. Break down this new scatterplot by region. Print the chart.

f. Examine the location of the Southern states in the scatterplot. Do they follow the general pattern shown by the other points in the plot? Interpret your answer in light of what you know of the 1980 and 1984 elections. (*Hint:* Consider whether the fact that the 1980 election involved a Southern Democratic candidate and the 1984 election involved a Midwestern Democratic candidate caused a change in the voting percentages of the Southern states.)

PART 2

Fundamentals of Statistics

Describing Your Data

Objectives

In this chapter you will learn:

- About different types of variables

- How to create tables of frequency, cumulative frequency, percentages, and cumulative percentages

- How to create histograms and break histograms down by groups

- About creating and interpreting stem and leaf plots

- How to calculate descriptive statistics for your data

- How to create and interpret boxplots

Chapter 4 introduces the different tools that statisticians use to describe and summarize the values in a data set. You'll work with frequency tables in order to see the range of values in your data. You'll use graphical tools like histograms, stem and leaf plots, and boxplots to get a visual picture of how the data values are distributed. You'll learn about descriptive statistics, such as the mean and standard deviation, that reduce the contents of your data to a few values. Applying these tools is the first step in the process of evaluating and interpreting the contents of your data set.

Variables and Descriptive Statistics

In this chapter you'll learn about a branch of statistics called descriptive statistics. In **descriptive statistics** we use various mathematical tools to summarize the values of a data set. Our goal is to take data that may contain thousands of observations and reduce it to a few calculated values. For example, we might calculate the average salaries of employees at several companies in order to get a general impression about which companies pay the most, or we might calculate the range of salaries at those companies to convey the same idea.

Note that we should be very careful in drawing any general conclusions or making any predictions based on our descriptive statistics. Those tasks belong to a different branch of statistics called **inferential statistics**, a topic we'll discuss in later chapters. The goal of descriptive statistics is to describe the contents of a specific data set, and we don't have the tools yet to evaluate any conclusions that might arise from examining those statistics. When the statistics involve only a single variable, we are employing a branch of statistics called **univariate statistics**. The statistics you'll learn about in this chapter are all univariate statistics.

We've used the term *variable* several times in this book. What is a variable? A **variable** is a single characteristic of any object or event. In the last chapter, you looked at data sets that contained several variables describing graduation rates of the Big Ten universities. Each column in that worksheet contained information on one characteristic, such as the university's name or total enrollment, and thus was a single variable.

Variables come in two main categories: quantitative and qualitative. **Quantitative variables** involve values that come in meaningful (not arbitrary) numbers. Examples of quantitative variables include age, weight, and annual income—anything that can be measured in terms of a number. Quantitative variables themselves come in two classes: discrete and continuous. **Discrete variables** are quantitative variables that assume values from a defined list of numbers. The numbers on a die come in discrete values (1, 2, 3, 4, 5, or 6). The number of children in a household is discrete, consisting of only positive integers. **Continuous variables**, on the other hand, have values from a wide range of possible values. An individual's weight could be 185, 185.5, or 185.5627 pounds. To be sure, there is some blurring in the distinction between discrete and continuous variables. Is salary a discrete variable or a continuous variable? From one point of view, it's discrete: The values are limited to dollars and cents, and there is a practical upper limit to how high a specific salary could go. However, it's more natural to think of salary as continuous.

The second type of variable is the qualitative variable. **Qualitative** or **categorical** variables are variables whose values fall into some category, indicating a quality or property of an object. Gender, ethnicity, and product name are all examples of qualitative variables. Qualitative variables are generally expressed in text strings, but not always. Sometimes a qualitative variable will be coded using numerical values. A common "gotcha" for people new to statistics is to analyze these coded values as quantitative variables. Consider the qualitative data values from Table 4-1.

Table 4-1 Qualitative Variables

Patient ID Number	Gender (0=Male; 1=Female)	Race (0=Caucasian, 1=African American; 2=Asian; 3=Other)
3458924065	1	0
3489109294	0	3
4891029494	0	1

Now all of these values were entered as numbers, but does it make sense to say that the average gender is $\frac{1}{3}$? Or that the sum of the races is 4? Of course not, but if you're not careful, you may find yourself doing things like that in other, more subtle cases. The point is that you should always understand what type of variables your data set contains before applying any descriptive statistic.

Qualitative variables can also be classified into two types: ordinal and nominal. An **ordinal variable** is a quantitative variable whose categories can be put into some natural order. For example, users are asked to fill out a survey ranking their product satisfaction from "Not satisfied" all the way up to "Extremely satisfied." These are categorical values, but they have a clear order of ascendancy. **Nominal variables** are quantitative variables without any such natural order. Race, state of residence, and gender are all examples of nominal variables. Table 4-2 summarizes properties of the different types of variables we've been discussing.

Table 4-2 Summary of variable properties

Quantitative: Variables whose values come in meaningful (not arbitrary) numbers.	**Discrete**: Variables that assume values only from a list of specific numbers
	Continuous: Variables that can assume values from a within a continuous range of possible values
Qualitative: Variables whose values fall into categories	**Ordinal**: Variables whose categories can be assigned some natural order
	Nominal: Variables whose categories cannot be put into any natural order

In this chapter, we'll work primarily with continuous quantitative variables. You'll learn how to work with qualitative variables later, in Chapter 7, when you work with tables and categorical data.

Looking at Distributions with Frequency Tables

To learn about distributions and descriptive statistics, you'll work with the HOMEDATA.XLS workbook. This workbook contains a sample of 117 housing prices in Albuquerque, New Mexico, with the variables shown in Table 4-3.

Table 4-3 Housing prices in Albuquerque

Range Name	Range	Description
Price	A2:A118	The selling price of each home.
Square_Feet	B2:B118	The square footage of the home.
Age	C2:C118	The age of the home in years.
Features	D2:D118	Number out of 11 features available in the home (dishwasher, refrigerator, microwave, disposal, washer, intercom, skylight(s), compactor, dryer, handicapped-accessible, cable TV access.)
NE_Sector	E2:E118	Located in the northeast sector of the city.
Corner_Lot	F2:F118	Corner location or not.
Offer_Pending	G2:G118	Offer pending on the home or not.
Annual_Tax	H2:H118	Estimated annual tax paid on the home.

To open the HomeData.xls workbook:

1 Start Excel.

2 Open the **HomeData** workbook from the folder containing your student files.

3 Click **File > Save As** and save the workbook as **HomeData2**. The workbook appears as shown in Figure 4-1.

Figure 4-1 The HomeData workbook

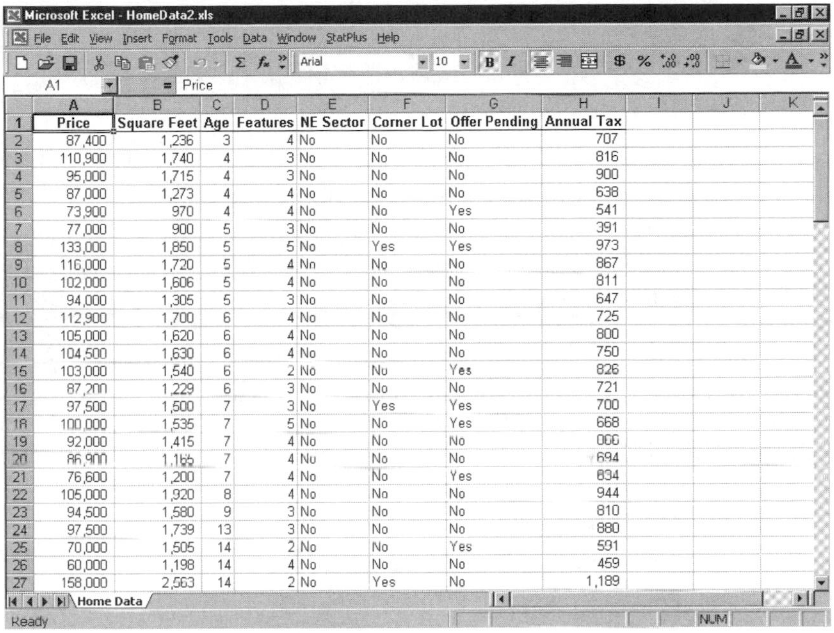

	Price	Square Feet	Age	Features	NE Sector	Corner Lot	Offer Pending	Annual Tax
2	87,400	1,236	3	4	No	No	No	707
3	110,900	1,740	4	3	No	No	No	816
4	95,000	1,715	4	3	No	No	No	900
5	87,000	1,273	4	4	No	No	No	638
6	73,900	970	4	4	No	No	Yes	541
7	77,000	900	5	3	No	No	No	391
8	133,000	1,850	5	5	No	Yes	Yes	973
9	116,000	1,720	5	4	No	No	No	867
10	102,000	1,606	5	4	No	No	No	811
11	94,000	1,305	5	3	No	No	No	647
12	112,900	1,700	6	4	No	No	No	725
13	105,000	1,620	6	4	No	No	No	800
14	104,500	1,630	6	4	No	No	No	750
15	103,000	1,540	6	2	No	No	Yes	826
16	87,200	1,229	6	3	No	No	No	721
17	97,500	1,500	7	3	No	Yes	Yes	700
18	100,000	1,535	7	5	No	No	Yes	668
19	92,000	1,415	7	4	No	No	No	666
20	86,900	1,165	7	4	No	No	No	694
21	76,600	1,200	7	4	No	No	Yes	834
22	105,000	1,920	8	4	No	No	No	944
23	94,500	1,580	9	3	No	No	No	810
24	97,500	1,739	13	3	No	No	No	880
25	70,000	1,505	14	2	No	No	Yes	591
26	60,000	1,198	14	4	No	No	No	459
27	158,000	2,563	14	2	No	Yes	No	1,189

Creating a Frequency Table

One of the first things we'll examine when studying this data set is the distribution of its values. The **distribution** is the way the observations are spread out across a range of values. If you were thinking about moving to Albuquerque, you might be interested in the distribution of home prices in the area. What is the range of housing prices in the area? What percentage of houses list for under $125,000?

As a first step in answering these types of questions, we'll create a frequency table of the home prices. A **frequency table** is a table that tabulates the number of occurrences or counts of a particular variable. Excel does not have a built-in command to create such a frequency table, but you can use the one supplied with the StatPlus add-in. Try creating a frequency table now.

To create a frequency table of home prices:

1. Click **StatPlus > Frequency Tables** from the Excel menu.
2. Click the **Data Values** button, click the **Use Range Names** option button, and click **Price**. Click the **OK** button.

   The Frequency Table command gives you three options in organizing your table. You can use discrete values so that the table is tabulated over individual price values, or you can organize the values into "bins" (you'll learn about bins shortly). For now, leave "Discrete" as the selected option.
3. Click the **Output** button, click the **New Worksheet** option button, and type **Discrete Table** in the New Worksheet name box. Click the **OK** button.
4. Click **OK** to start generating the frequency table.

   Figure 4-2 displays the completed table.

Figure 4-2 Calculating a frequency table based on discrete values

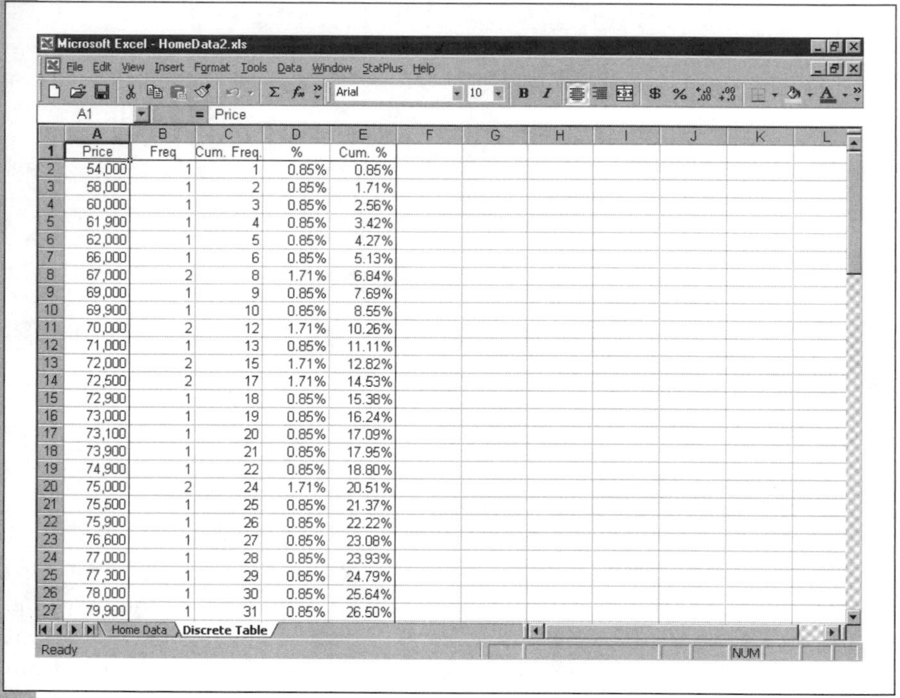

Price	Freq	Cum. Freq.	%	Cum. %
54,000	1	1	0.85%	0.85%
58,000	1	2	0.85%	1.71%
60,000	1	3	0.85%	2.56%
61,900	1	4	0.85%	3.42%
62,000	1	5	0.85%	4.27%
66,000	1	6	0.85%	5.13%
67,000	2	8	1.71%	6.84%
69,000	1	9	0.85%	7.69%
69,900	1	10	0.85%	8.55%
70,000	2	12	1.71%	10.26%
71,000	1	13	0.85%	11.11%
72,000	2	15	1.71%	12.82%
72,500	2	17	1.71%	14.53%
72,900	1	18	0.85%	15.38%
73,000	1	19	0.85%	16.24%
73,100	1	20	0.85%	17.09%
73,900	1	21	0.85%	17.95%
74,900	1	22	0.85%	18.80%
75,000	2	24	1.71%	20.51%
75,500	1	25	0.85%	21.37%
75,900	1	26	0.85%	22.22%
76,600	1	27	0.85%	23.08%
77,000	1	28	0.85%	23.93%
77,300	1	29	0.85%	24.79%
78,000	1	30	0.85%	25.64%
79,900	1	31	0.85%	26.50%

The table contains five columns. The first column, Price, lists in ascending order all of the home prices in the sample of 117 homes. Prices in this sample range from a minimum of $54,000 to a maximum of $215,000. The second column, Freq, counts the frequency, or number of occurrences, for each value in the price column. Many prices are unique and have frequencies of 1, but other prices (such as $75,000) occur for multiple homes. The third column contains the cumulative frequency, counting the total number of homes at or less than a given price. By examining the table you can quickly see that 24 of the homes in the sample have a price of $75,000 or less. The fourth column lists the percent occurrence of each home price out of the total sample. For example, 1.71% of the homes are listed for exactly $75,000. Finally, the fifth and last column of the table calculates the cumulative percentage for the home prices. In this case, 24.79%—almost one-quarter of the homes—list for $77,300 or less. A table of this kind can help you in evaluating the market. For example, if you were interested in homes that list for $125,000 or less, you could quickly determine that almost 80% of the homes in this database, or 93 different listings, met that criterion.

EXCEL TIPS

- If you don't have StatPlus handy, Excel comes with an add-in called the "Data Analysis ToolPak" that you can use to create a frequency table. The ToolPak does not have all the frequency table options that StatPlus contains.

- If you want to count how many values in a column are equal to a specific value, you can use Excel's COUNTIF function.

- You can also create a frequency table using Excel's FREQUENCY function. This function uses Excel's array feature, which you can learn about by using the online Help.

Using Bins in a Frequency Table

By creating a frequency table, you got a clear picture of the distribution of prices in the Albuquerque area. However, displaying individual values would be cumbersome if the sample contained 1000 or 10,000 observations. To get around this problem, you can have the frequency table group the values by placing them in **bins**, where each bin covers a particular range of values. Thus the frequency table would count the number of values that fall in each bin. There are three ways of counting values in bins: 1) count values that are ≥ bin value and < the next bin value, 2) count those values centered around the bin value, or 3) count those values that ≤ bin value but > the previous bin value. See Figure 4-3. To interpret your frequency table correctly, you need to know which of these methods is used in calculating the counts.

**Figure 4-3
Counting
within a bin**

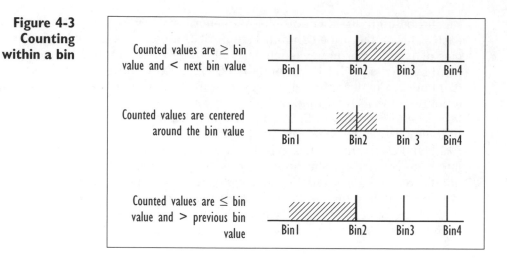

Counted values are ≥ bin
value and < next bin value

Bin1   Bin2   Bin3   Bin4

Counted values are centered
around the bin value

Bin1   Bin2   Bin 3   Bin4

Counted values are ≤ bin
value and > previous bin
value

Bin1   Bin2   Bin3   Bin4

To see how this works, we'll create another frequency table, this time breaking the data down into 15 equally spaced bins.

To create a frequency table with bins:

1 Click **StatPlus > Frequency Tables.**

2 Click the **Data Values** button and select **Price** as the data variable.

3 Click the **Create 15 equally spaced bins** option button.

Note that the first option button has been selected, so that counts can be calculated for values that are ≥ the bin value and < the succeeding bin value.

4 Click the **Output** button and click the **New Worksheet** option button. Type **Table with Bins** as the worksheet name and click the **OK** button.

5 Click the **OK** button to start generating the frequency table with bins.

See Figure 4-4.

Figure 4-4
Frequency
table with
equally
spaced bins

equally spaced
bin values

	A	B	C	D	E	F
1	Price	Freq	Cum. Freq.	%	Cum. %	
2	54,000	5	5	4.27%	4.27%	
3	64,733	19	24	16.24%	20.51%	
4	75,467	14	38	11.97%	32.48%	
5	86,200	21	59	17.95%	50.43%	
6	96,933	17	76	14.53%	64.96%	
7	107,667	11	87	9.40%	74.36%	
8	118,400	7	94	5.98%	80.34%	
9	129,133	6	100	5.13%	85.47%	
10	139,867	2	102	1.71%	87.18%	
11	150,600	4	106	3.42%	90.60%	
12	161,333	1	107	0.85%	91.45%	
13	172,067	1	108	0.85%	92.31%	
14	182,800	2	110	1.71%	94.02%	
15	193,533	1	111	0.85%	94.87%	
16	204,267	6	117	5.13%	100.00%	
17						

This frequency table gives us a little clearer picture of the distribution of housing prices. Note that almost 80% of the prices are clustered within the first seven bins of the table (representing homes costing about $129,000 or less). Moreover, there are relatively few homes in the $160,000 – $200,000 price range (only about 4% of the sample). There is, however, a small group of homes priced above $205,000.

Defining Your Own Bin Values

One problem with the frequency table from Figure 4-4 is that the bin values are awkward to work with. It would be much easier to use even numbers: $50,000, $60,000, and so forth. In situations like this, you can specify your own bin values, which may make more economic sense. Try this now, by creating a frequency table of home prices in $10,000 increments—starting at $50,000. You will first have to enter the bin values into cells in the workbook.

To create your own bin values:

1 Click cell **G2**, type **50,000**, and press **Enter**.

2 Type **60,000** in cell G3 and press **Enter**.

3 Select the range **G2:G3**, drag the fill handle down to cell **G20**, and release the mouse button.

The values 50,000 – 230,000, should be now entered into the cell range G2:G20.

4 Click **StatPlus > Frequency Tables** from the menu.

5 Click the **Data Values** button and select **Price** as the data variable.

6 Click the **Bin Values** button.

7 Click the **Use Range References** option button and then select the range **G2:G20**.

8 Because there is no column label for this data, deselect the **Range includes a row of column labels** checkbox and click the **OK** button.

9 Click the **<= bin and > previous bin** option button to control how the bin counts are determined.

10 Click the **Output** button, click the **Cells** option button, and select cell **G1**. Click the **OK** button.

11 Click the **OK** button to start generating the frequency table with your customized bin values.

Figure 4-5 displays the new frequency table.

**Figure 4-5
Frequency table with user-defined bins**

Price	Freq	Cum. Freq.	%	Cum. %
50,000	0	0	0.00%	0.00%
60,000	3	3	2.56%	2.56%
70,000	9	12	7.69%	10.26%
80,000	19	31	16.24%	26.50%
90,000	18	49	15.38%	41.88%
100,000	17	66	14.53%	56.41%
110,000	12	78	10.26%	66.67%
120,000	10	88	8.55%	75.21%
130,000	9	97	7.69%	82.91%
140,000	3	100	2.56%	85.47%
150,000	2	102	1.71%	87.18%
160,000	4	106	3.42%	90.60%
170,000	1	107	0.85%	91.45%
180,000	1	108	0.85%	92.31%
190,000	2	110	1.71%	94.02%
200,000	1	111	0.85%	94.87%
210,000	3	114	2.56%	97.44%
220,000	3	117	2.56%	100.00%
230,000	0	117	0.00%	100.00%

user-defined bin values

This table is a lot easier to interpret. Looking at the table, it's easy to discover that there are only 2 houses in the sample in the $140,000-to-$150,000 price range. If you've set yourself a certain spending goal, it's useful to know how many homes in the listing match it.

STATPLUS TIPS

- You can use the "Frequency Table" command to create tables that are broken down into categories based on a qualitative variable. To do so, click the "By" button in the Frequency Table dialog box and choose the range name or range reference containing the values of the qualitative variable.

- If you forget how the bin counts are determined, place your cursor over the column title for the bin value. A pop-up comment box will appear indicating the method used.

Working with Histograms

Frequency tables are good at conveying specific information about a distribution, but they often lack visual impact. It's hard to get a good impression about how the values are clustered from the counts in the frequency table. Many statisticians prefer a visual picture of the distribution in the form of a histogram. A **histogram** is a bar chart in which each bar represents a particular bin and the height of the bar is proportional to the number of counts in that bin. Histograms can be used to display frequencies, cumulative frequencies, percentages, and cumulative percentages. Most histograms display the frequency or counts of the observations.

Creating a Histogram

Excel does not have the capability to create a histogram, but you can create one using either the Data Analysis ToolPak supplied with Excel or StatPlus. Create one now for the home price data.

To create a histogram of the home prices:

1   Click **StatPlus > Single Variable Charts > Histograms**.

2   Click the **Data Values** button and select **Price** from the list of range names.

As with the Frequency Table command, you can specify options for the bins.

3   Click the **Chart Options** dialog tab.

4   Click the **Values** button and then click the **Use Range References** button. Select the range **G1:G20** in the Table with Bins worksheet. Click the **OK** button.

5   Click the **Right** option button to control how bin counts are determined. See Figure 4-6.

**Figure 4-6
Specifying
bin options
for the
histogram**

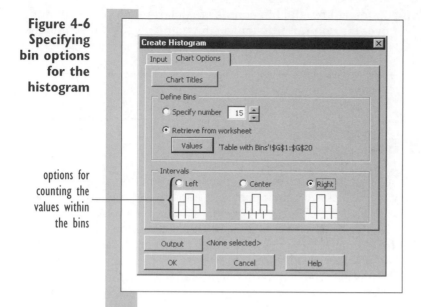

options for
counting the
values within
the bins

6 Click the **Input** dialog tab to view other options for the histogram.

7 Click the **Output** button.

8 Verify that the **As a new chart sheet** option button is selected and then type **Price Histogram** in the accompanying text box.

This will send the histogram to a chart sheet named "Price Histogram."

9 Click the **OK** button.

Figure 4-7 shows the completed Histogram dialog box. Note that this command allows you to create histograms of the frequency, cumulative frequency, percentage, or cumulative percentage. In most cases, histograms will display the frequency of a particular variable.

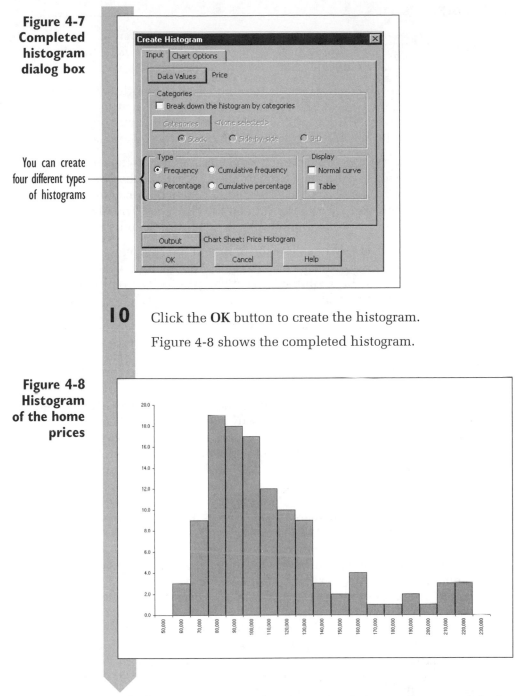

Figure 4-7 Completed histogram dialog box

You can create four different types of histograms

Figure 4-8 Histogram of the home prices

10 Click the **OK** button to create the histogram.

Figure 4-8 shows the completed histogram.

The histogram gives us the strong visual picture that most of the home prices in the sample are ≤ $130,000 and that most are in the $70,000-to-$100,000 range. There does not seem to be any clustering of values beyond $130,000; rather, the data values are clustered toward the lower end of the price scale.

- You also create separate histograms for the different levels of a categorical variable (or for different variables) by using the "StatPlus > Multi-variable Charts > Multiple Histograms" command.

- The Histogram command includes a Chart Titles button located on the Chart Options dialog sheet. By clicking this button, you can enter titles for the chart, *x*-axis, and *y*-axis. You can also control some of the appearance of the *x*-axis and *y*-axis.

- The Left option button for the bin intervals in the Histogram command is equivalent to counting observations that are ≥ bin value and < next bin value. The Center option button counts observations that are centered around the bin value. The Right option button counts observations that are > bin value and ≤ next bin value.

- You can add a table to the output of the Histogram command by clicking the Table checkbox in the dialog box. This table contains count values, similar to what you would see in the corresponding frequency table.

Shapes of Distributions

The visual picture presented by the histogram is often referred to as the distribution's "shape." Statisticians classify various distributions on the basis of their shape. These classifications will become important later on as we look for an appropriate statistic to summarize the distribution and its values. Some statistics are appropriate for one distribution shape but not for another.

A distribution is **skewed** if most of the values are clustered toward either the left or the right edge of the histogram. If the values are clustered toward the left edge of the histogram, this shows **positive skewness**; clustering toward the right edge of the histogram implies **negative skewness**. Skewed distributions often occur where the variable is constrained to have positive values. In those cases, values may cluster near zero, but because the variable cannot have a negative value, the distribution is positively skewed. A distribution is **symmetric** if the values are clustered in the middle with no skewness toward either the positive or the negative side. See Figure 4-9 for examples of these three types of shapes.

Another important component of a distribution's shape is the distribution's **tails**—the values located to the extreme left or right edge. A distribution where large numbers of observations are located in one or both of the tails is said to be a **heavy-tailed distribution**.

The sample of home prices we've examined appears to be positively skewed with a heavy tail (because there are a number of houses located at the high end of the price scale). This is not surprising, because there is a practical lower limit for housing prices (around $50,000) but an exceedingly large upper limit.

**Figure 4-9
Distribution
shapes**

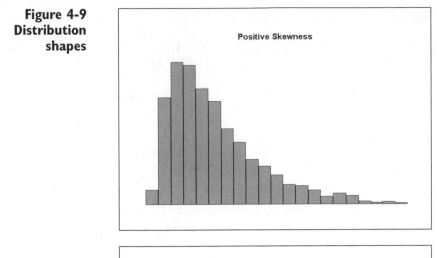

Positive Skewness

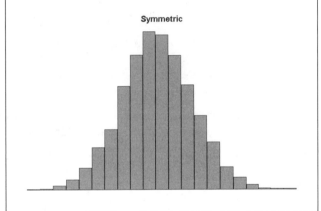

Symmetric

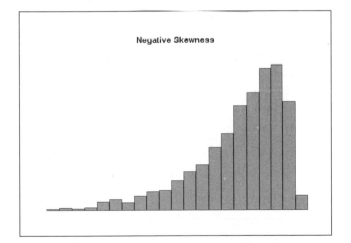

Negative Skewness

Breaking a Histogram Into Categories

You can gain a great deal of insight by breaking your histogram into categories. In the current example, we may be interested in knowing how the Albuquerque prices compare when broken down by location: Are certain locations more expensive than others? You've learned from a friend that one of the more desirable locations in Albuquerque is the northeast sector. Is this reflected in a histogram of the sample home prices? Let's find out.

To create a histogram broken down by categories:

1   Click the **Home Data** worksheet tab to return to the price data.

2   Click **StatPlus > Single Variable Charts > Histograms**.

3   Click the **Data Values** button and select **Price** from the list of range names as the source for the histogram.

4   Click the **Break down the histogram by categories** checkbox.

The various categories can be displayed in a histogram as stacked on top of each other, side by side, or in three dimensions. You'll see the effect of these choices on the histogram's appearance in a moment. For now, accept the default, "Stack."

5   Click the **Categories** button, click the **Use Range Names** option button and select **NE Sector**, and then click the **OK** button.

The NE Sector variable is a qualitative variable that is equal to "Yes" if the home is located in the northeast sector and is equal to "No" otherwise.

6   Now, define the options for the histogram's bins.

7   Click the **Chart Options** dialog tab.

8   Click the **Values** button, click the **Use Range References** option button, and then select the range **G1:G20** on the Table with Bins worksheet. Click the **OK** button.

9   Click the **Right** option button to set how bin values will be counted in the histogram.

10 Click the **Output** button and type **Price Histogram by NE Sector** in the Chart Sheet name box. Click the **OK** button.

11 Click the **OK** button to start creating the histogram. The completed chart appears in Figure 4-10.

Figure 4-10
The price
histogram
broken
down by the
NE sector
variable

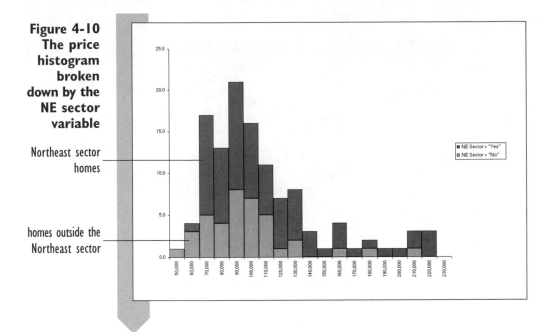

Northeast sector
homes

homes outside the
Northeast sector

In this histogram, the height of each bar is still equal to the total count of values within that bin, but each bar is further broken down by the counts for the various levels of the categorical variable. The counts are "stacked" on top of each other. The chart makes it clear that many higher-priced homes are located in the northeast sector, though there are still plenty of northeast sector homes in the $70,000 – $100,000 range.

How do the shapes of the distributions compare for the two types of homes? We can't tell from this chart, because the northeast sector homes are all stacked at uneven levels. To compare the distribution shapes, we can compare histograms "side by side."

To compare histograms side by side:

1 Click **Chart > Chart Type**.

2 Select the **Clustered Column** chart sub-type shown in Figure 4-11.

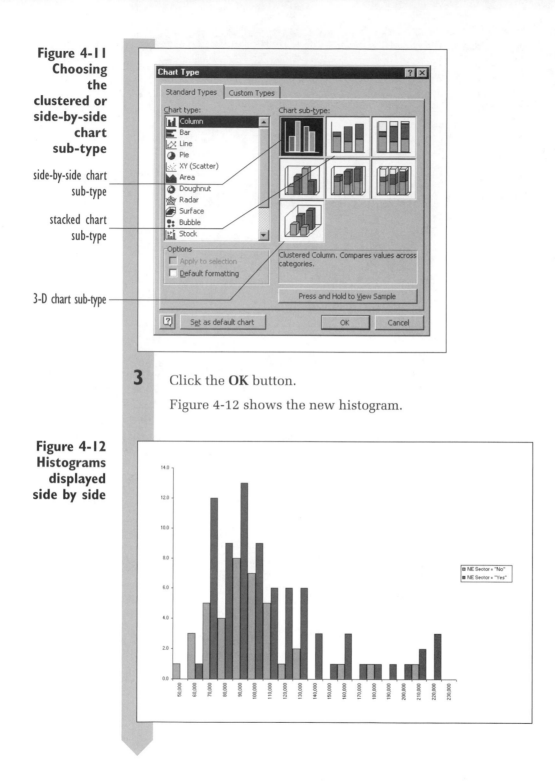

Figure 4-11
Choosing
the
clustered or
side-by-side
chart
sub-type

side-by-side chart
sub-type

stacked chart
sub-type

3-D chart sub-type

3   Click the **OK** button.

Figure 4-12 shows the new histogram.

Figure 4-12
**Histograms
displayed
side by side**

This chart shows us that the distribution of home prices is positively skewed (as we would expect) for both northeast sector and non-northeast sector homes. The primary difference is that in the northeast sector there are more homes at the high end. By selecting the appropriate chart sub-type, we can switch back and forth between side-by-side and stacked views of the histogram—we can even view the histogram in three dimensions!

Working with Stem and Leaf Plots

Stem and leaf plots are another way of displaying a distribution, while at the same time retaining some information about individual values. The stem and leaf plot is a bit of a holdover from earlier times before fast computers with nifty graphical output. It was used by statisticians as a quick way of generating a plot of the distribution. To create a stem and leaf plot, follow these steps:

1. Sort the data values in ascending order.
2. Truncate all but the first two digits from the values (i.e., change 64,828 to 64,000, change 14,048 to 14,000, and so forth). The first of the two digits is the **stem** and the second the **leaf**. In the case of a number like 64,000, the stem is 6 and the leaf is 4.
3. List the stems in ascending order vertically on a sheet and place a vertical dividing line to the right of the stems.
4. Match each leaf to its stem, placing the leaf values in ascending order horizontally to the right of the vertical dividing line.

For example, take the following numbers:

125, 189, 232, 241, 248, 275, 291, 311, 324, 351, 411, 412, 558, 713

Truncating all but the first two digits from the list, leaves us with

120, 180, 230, 240, 240, 270, 290, 310, 320, 350, 410, 410, 550, 710

The stem and leaf pairs are therefore (12), (18), (23), (24), (24), (27), (29), (31), (32), (41), (41), (55), and (71). Now, we list just the stems in ascending order vertically:

```
100× |
   1 |
   2 |
   3 |
   4 |
   5 |
   6 |
   7 |
```

At the top of the stem list, we've included a multiplier, so we know our data values go from 100 to 700. Note that we've added a stem for the value 6. We include this to preserve continuity in the stem list. Now we add a leaf to the right of each stem. The first stem and leaf pair is (12), so we add 2 to the right of the stem value 1, and so on. The final stem and leaf plot appears, as follows:

```
100× |
   1 | 28
   2 | 34479
   3 | 125
   4 | 11
   5 | 5
   6 |
   7 | 1
```

The stem and leaf plot resembles a histogram turned on its side. The plot has some advantages over the histogram. From the stem and leaf plot, you can generate the approximate values of all the observations in the data set by combining each stem with its leaves. Looking at the plot above, you can quickly see that the first two stem and leaf pairs are (12), (18). Multiplying these values by 100 yields approximate data values of 120 and 180. An added advantage is that the stem and leaf plot can be quickly generated by hand—useful if you don't have a computer handy.

The stem and leaf plot is at a disadvantage compared to the histogram in that you don't have control over the bin values. The size of each bin is directly determined by the data values themselves. Stem and leaf plots also don't work well for large data sets where each stem will need to display a large number of leaves. One way around this problem is to split the stems into two groups: those with leaves having values from 0 to 4 and those with leaves from 5 to 9. Doing this for the above chart yields the following stem and leaf plot:

```
100× |
   1 | 2
   1 | 8
   2 | 344
   2 | 79
   3 | 12
   3 | 5
   4 | 11
   4 |
   5 |
   5 | 5
   6 |
   6 |
   7 | 1
   7 |
```

Another modification to the stem and leaf plot is to truncate lower and upper values in order to reduce the range of stems in the plot. This is useful in situations where you have an extreme value whose presence would greatly elongate the plot's appearance. For example, if the data set above is changed to

125, 189, 232, 241, 248, 275, 291, 311, 324, 351, 411, 412, 558, 713, 2420

then the addition of the value 2420 will result in a large stem and leaf plot with a long list of empty stems. In this case, removing this value from the stem and leaf plot, but noting its value elsewhere, might be the best course of action. The plot might look as follows:

```
100× |
   1 | 28
   2 | 34479
   3 | 125
   4 | 11
   5 | 5
   6 |
   7 | 1
   _____
   2400
```

Let's create a stem and leaf plot for the home price data and compare it to the histogram we created earlier. As before, we'll break the stem and leaf plot down by home location.

To create a stem and leaf plot:

1 Return to the data set by clicking the **Home Data** worksheet tab.

2 Click **StatPlus > Single Variable Charts > Stem and Leaf**.

This command allows you to create plots of variables located in different columns or within a single column, broken down by category levels. You'll do the latter in this case.

3 Verify that the **Use column of category levels** option button is selected.

4 Click the **Data Values** button and select **Price** from the list of range names.

5 Click the **Categories** button and select **NE_Sector** from the list of range names.

6 Click the **Apply uniform stem values** checkbox.

This will apply the same stem values to home prices both in the northeast sector and elsewhere.

7 Click the **Add a summary plot** checkbox.

This will create a stem and leaf plot of prices for all of the homes, regardless of location.

8 Click the **Output** button, click the **New Worksheet** option button, and type **Price StemLeaf** in the New Worksheet name box. Click the **OK** button.

Figure 4-13 shows the completed Stem and Leaf dialog box.

Figure 4-13
The completed Stem and Leaf dialog box

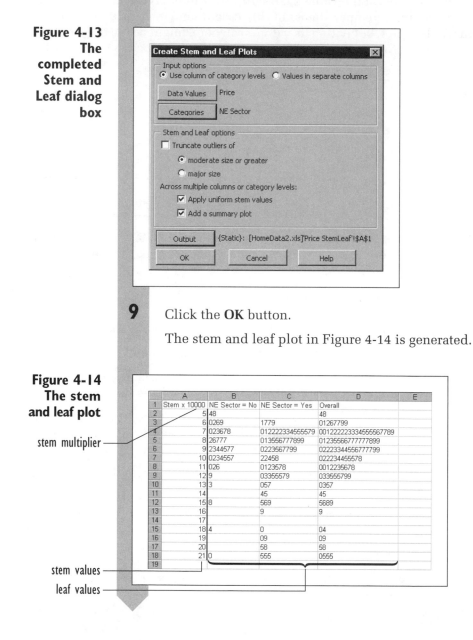

9 Click the **OK** button.

The stem and leaf plot in Figure 4-14 is generated.

Figure 4-14
The stem and leaf plot

stem multiplier

stem values

leaf values

In this plot, the stem values occupy the first column, and leaf values are placed in the following three columns for homes outside the northeast sector, in the northeast sector, and over all sectors. Note that cell A1 identifies the stem multiplier, indicating that each stem value must be multiplied by 10,000 in order to calculate the underlying data values. Let's see how this works. The first stem value is 5; this represents 50,000. The first leaf value is 4 (where the NE_Sector variable equals "No"), which would represent a value one decimal place lower, or 4,000. Thus the first data value in this plot equals the stem value plus the leaf value, or 54,000—which is equal to the value of the lowest-priced home in the sample. Using the same method, you can calculate the value of the highest-priced home to be $215,000. You can also see at a glance that there are no homes in the $170,000 – $179,000 price range, though there is one home priced at about $169,000 (actually $169,500). In addition to this information, you can also use the stem and leaf plot to make the same observations about the shape of the distribution that you did earlier with the histogram.

Distribution Statistics

You should always create a chart of the distribution when analyzing a data set, but once you've done that, you'll probably look for statistics that summarize key elements of the distribution. These values are sometimes called **landmark summaries** because they are used as landmarks, comparing individual values to whole populations, or whole populations to each other.

Percentiles and Quartiles

One of these landmark summaries is the **pth percentile**, which is the value in a given distribution such that p percent of the distribution is either less than or equal to that value. You've probably seen percentiles used in growth statistics, where the progress of a newborn child will place him or her in the 75th percentile or 90th percentile, meaning that the child's weight is equal to or above 75% or 90% of the population. In the Albuquerque data, percentiles could be used as a benchmark to compare one community to another. If you knew the 10th and 90th percentiles for home price, you would have a basis for comparison between the two communities.

Closely related to percentiles are **quartiles**, which are the values located at the 25th, 50th, and 75th percentiles (the quarters). These are commonly referred to as the first, second, and third quartiles. Statisticians are also interested in the **interquartile range**, which is the difference between the first and third quartiles. Because the central 50% of the data lie within the interquartile range, the size of this value gives statisticians an idea of the width of the distribution.

One way of calculating the percentiles and quartiles of a given distribution is to create a frequency table like the one shown earlier in Figure 4-2. From the column of cumulative percents, you can determine which values correspond to the 10th, 25th, 50th, 75th, and 90th percentiles and so on. However, if your data set is large, this can be a cumbersome and time-consuming process. To save time, Excel has several functions that will calculate these values for you. A list of these functions is shown in Table 4-4.

Table 4-4 Excel functions to calculate percentiles and quartiles

Function	Description
PERCENTILE(*array*, *k*)	Returns the *k*th percentile of an array of values or range reference, where *k* is a value between 0 and 1.
PERCENTRANK(*array*, *x*, *significance*)	Returns the percentile of a value taken from an array of values or range reference. The number of digits is determined by the *significance* parameter.
QUARTILE(*array*, *quart*)	Returns the quartile of an array of values or range reference, where *quart* is either 1, 2, or 3 for the first, second, or third quartile.
IQR(*array*)	Calculates the interquartile range of the values in an array or range reference. *StatPlus required*.

Excel allows you to work with percentiles in two different ways. You can use the PERCENTILE function to take a percentile and determine the corresponding data value, or, given the data value, you can use the PERCENTRANK function to determine its percentile.

You can create a table of percentile and quartile values by typing in the above Excel formulas, or you can have StatPlus do it for you with the Univariate Statistics command. The Univariate Statistics command also allows you to break down the variable into different levels of a categorical variable. The Univariate Statistics command allows you to create tables containing a wide variety of statistics. In this example you'll limit yourself to percentiles and quartiles. Create such a table now of the home prices broken down by location.

To create a table of percentile and quartile values:

1 Click **StatPlus > Univariate Statistics**.

2 Click the **Input** button and select **Price** from the list of range names.

3 Click the **Output** button, click the **New Worksheet** option button, and type **Percentiles** in the New Worksheet box. Click the **OK** button.

4 Click the **By** button and select **NE_Sector** from the list of range names.

5 Click the **Distribution** dialog tab.

6 Click each of the checkboxes for the different percentiles. See Figure 4-15.

Figure 4-15 The Univariate Statistics dialog box

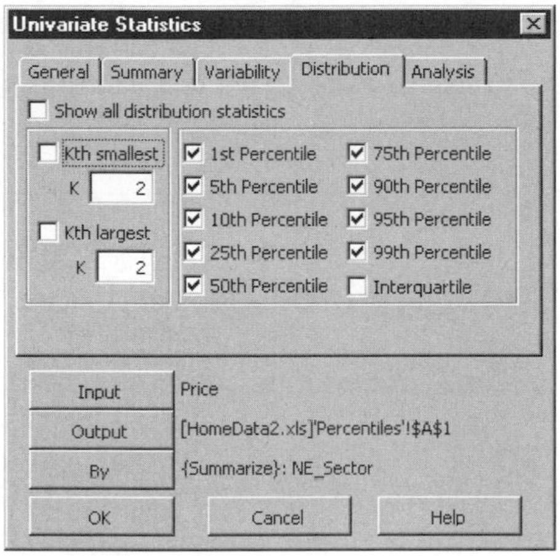

7 Click the **OK** button to create the table of percentiles.

Figure 4-16 shows the completed table.

Figure 4-16 Table of percentiles

	A	B	C	D	E
1		Univariate Statistics			
2		Price			
3		NE_Sector = "No"	NE_Sector = "Yes"	Overall	
4	1st Percentile	55,520	65,827	58,320	
5	5th Percentile	59,800	69,850	66,800	
6	10th Percentile	65,200	72,350	70,600	
7	25th Percentile	76,800	80,625	78,000	
8	50th Percentile	94,000	98,500	96,000	
9	75th Percentile	105,000	125,000	120,000	
10	90th Percentile	130,200	172,650	158,760	
11	95th Percentile	160,640	205,450	200,920	
12	99th Percentile	200,272	215,000	215,000	
13					

The table of percentiles gives us some additional information about the housing prices. The values are pretty close between the two locations up to the 50th percentile, after which large differences begin to appear. It's particularly striking to note that the 90th percentile for home prices outside the northeast sector is $130,200, whereas for northeast sector homes it's

$172,650—$40,000 more. We noted earlier that there are more high-priced homes in the northeast sector.

EXCEL TIPS _____

- You can also get a table of cumulative percents using the "Rank and Percentile" command in the Data Analysis ToolPak, an add-in packaged with Excel.

- You can also use the Data Analysis ToolPak to create a table of descriptive statistics.

_____

Measures of the Center: Means, Medians, and the Mode

The ultimate way to summarize a data set would be to calculate a statistic that summarized the contents into a single value. We could think of this value as the "typical" or "most representative" value. The table of percentiles suggests one such value: the 50th percentile, or **median**. Because the median is located at the 50th percentile, it represents the middle of the distribution: Half of the values are less than the median, and half are greater than the median. Based on the results from Figure 4-16, the median house price in the Albuquerque sample is $94,000 for non-northeast sector homes, $98,000 for northeast sector homes, and $96,000 overall.

The exact calculation of the median depends on the number of observations in the data set. If there is an odd number of values, the median is the middle value, but if there are an even number of values, the median is equal to the sum of the two central values divided by 2. If you were to calculate the median by hand, you would first have to sort the data values in order. This can be time-consuming.

Another commonly used summary measure is the average. The **average**, or **mean**, is equal to the sum of the values divided by the number of observations. This value is usually represented by the symbol $\bar{x}$ (pronounced "x-bar"), a convention we'll repeat throughout the course of this book. Expressed as a formula, this is

$$\bar{x} = \frac{\text{Sum of values}}{\text{Number of observations}}$$

$$= \frac{x_1 + x_2 + \cdots + x_n}{n}$$

$$= \frac{\sum\limits_{i=1}^{n} x_i}{n}$$

The total number of observations in the sample is represented by the symbol n, and each individual value is represented by x followed by a subscript. The first value is x_1, the second value is x_2, and so forth, up to the last value (the nth value), which is represented by x_n. The formula calls for us to sum all of these values, an operation represented by the Greek symbol Σ (pronounced "sigma"), a summation symbol. In this case, we're instructed to sum the values of x_i, where i changes in value from 1 up to n; in other words, the formula tells us to calculate the value of $x_1 + x_2 + \cdots + x_n$. The average, or mean, is equal to this expression divided by the total number of observations.

How do these two measures, the median and the mean, compare? One weakness of the mean is that it can be influenced by extreme values. Figure 4-17 shows a distribution of professional baseball salaries. Note that most of the salaries are less than $400,000 per year, but there are a couple of players who make more than a million dollars per year and two players who make almost 3 million dollars annually. What, then, is a "typical" salary? The median value for this distribution is $363,000, but the mean salary is $501,000. The median seems more representative of what the typical player makes, whereas the mean salary is almost $150,000 higher as a result of the influence of the two largest salaries. If you were a union representative negotiating a new contract, which figure would you quote?

**Figure 4-17
Distribution
of baseball
salaries**

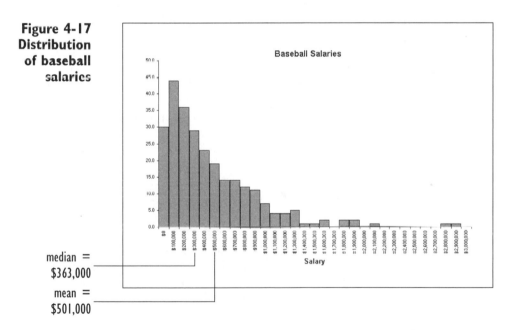

median =
$363,000

mean =
$501,000

The lesson from this example is that you should not blindly accept any single summary measure. The mean is sensitive to extreme values; the median overcomes this problem by ignoring the magnitude of the upper and lower values. Both approaches have their limitations, and the best approach is to examine the data, create a histogram or stem and leaf plot of the distribution, and thoroughly understand your data before attempting to summarize it. Even then, it may be best to include several summary measures to compare.

The mean and median are the most common summary statistics, but there are others. Let's examine those now.

One method of reducing the effect of extreme values on the mean is to calculate the trimmed mean. The **trimmed mean** is the mean of the data values calculated after excluding a percentage of the values from the lower and upper tails of the distribution. For example, the 5% trimmed mean would be equal to the average of the middle 90% of the data after exclusion of values from the lower and upper 5% of the range. The trimmed mean can be thought of as a compromise between the mean and the median.

Another commonly used measure of the center is the geometric mean. The **geometric mean** is the nth root of the product of the data values:

$$\text{Geometric mean} = \sqrt[n]{(x_1) \cdot (x_2) \cdot \cdots \cdot (x_n)}$$

Once again, the symbols x_1 to x_n represent the individual data values from a data set with n observations. The geometric mean is most often used when the data comes in the form of ratios or percentages (limited to values between 0 and 1). Certain drug experiments are recorded as percent changes in chemical levels relative to a baseline value, and those values are best summarized by the geometric mean. The geometric mean can also be used in situations where the distribution of the values is highly skewed in the positive or negative direction. The geometric mean *cannot* be used if any of the data values are negative or zero.

Another measure, not widely used today (though the ancient Greeks used it extensively) is the **harmonic mean**. The formula for the harmonic mean, H, is

$$\frac{1}{H} = \frac{1}{n} \sum_{i=1}^{n} \frac{1}{x_i}$$

The harmonic mean can be used to calculate the mean values of rates. For example, a car traveling at a rate of S miles per hour from point A to point B, and then at a rate of T miles per hour on the return trip travels at an average rate equal to the harmonic mean of S and T.

Our final measure of the center is the mode. The **mode** is the most frequently occurring value in a distribution. The mode is most often used when we are working with qualitative data or discrete quantitative data—basically any data in which there are a limited number of possible values. The mode is not as useful in continuous quantitative data, because if the data are truly continuous, we would expect few, if any, repeat values.

Table 4-5 displays the Excel functions used to calculate the various measures of the distribution's center.

Table 4-5 Excel functions to calculate the distribution's center

Function	Description
AVERAGE(*array*)	Returns the average, or mean, of the values in an array or data range.
GEOMEAN(*array*)	Returns the geometric mean of the values in an array or data range.
HARMEAN(*array*)	Returns the harmonic mean of the values in an array or data range.
MEDIAN(*array*)	Returns the median of the values in an array or data range.
MODE(*array*)	Returns the most frequently occurring value in an array or data range.
TRIMMEAN(*array*, *percent*)	Returns the trimmed mean of the values in an array or data range, trimming the lower and upper *percent* of the data values (*percent* must be between 0 and 1).

Now that you've learned a little about these functions, use the Univariate Statistics command from the StatPlus add-in to generate a table of their values.

To create a table of mean and median values:

1   Click **StatPlus > Univariate Statistics**.

2   Click the **Input** button and select **Price** from the list of range names.

3   Click the **Output** button, click the **New Worksheet** option button, and type **Means** in the New Worksheet box. Click the **OK** button.

4   Click the **By** button and select **NE_Sector** from the list of range names.

5   Click the **Summary** dialog tab.

6   Click the **Show all summary statistics** checkbox. See Figure 4-18.

Figure 4-18
Selecting
the
summary
statistics to
display

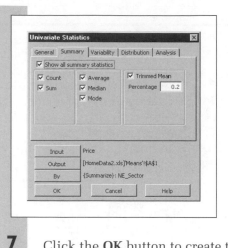

7 Click the **OK** button to create the table of values.

Figure 4-19 shows the completed table.

Figure 4-19
Summary
statistics for
the Price
variable

	A	B	C	D	E
1			Univariate Statistics		
2		Price			
3		NE_Sector = "No"	NE_Sector = "Yes"	Overall	
4	Count	39	78	117	
5	Sum	3,794,000	8,640,000	12,434,000	
6	Average	97,282.05	110,769.23	106,273.50	
7	Median	94,000	98,500	96,000	
8	Mode	(105000 ; 97500)	(215000 ; 125000)	97,500	
9	Trimmed Mean (0.2)	93,018.18	104,909.38	100,393.68	
10					

As we would expect, the average price of a home in Albuquerque is higher than the median value. This effect is more noticeable in the northeast sector homes because of the group of high-priced homes in that location. The mean home price is almost $12,000 greater than the median.

Measures of Variability

The mean and median do not tell the whole story about a distribution. It's also important to take into account the variability of the data. **Variability** is a measure of how much data values differ from one another or, equivalently, how widely the data values are spread out around the center. Consider the pair of histograms shown in Figure 4-20. The mean and median are the same for both distributions, but the variability of the data is much greater for the second figure.

Figure 4-20
Distribu-
tions with
low and
high
variability

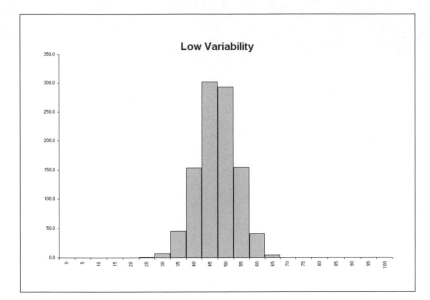

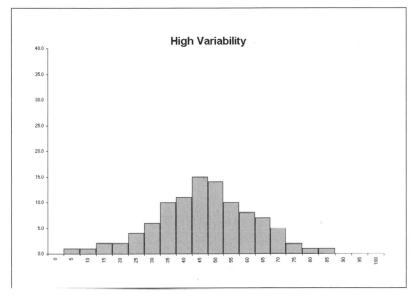

The simplest measure of variability is the **range**, which is the difference between the maximum value in the distribution and the minimum value. A large variability usually results in a large range of values. However, the range can be a poor and misleading measure of variability. As shown in Figure 4-21, two distributions can have the same range but be very different in the variability of their data.

**Figure 4-21
Distribu-
tions with
different
variability
but the
same range**

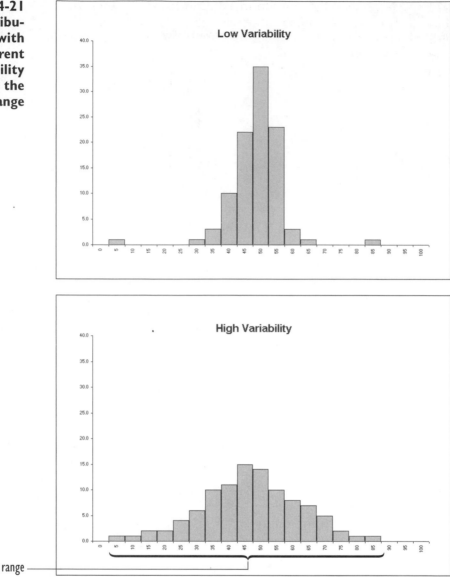

range

The most common measure of variability depends on the deviation of each data value from the average. For each data value x_i, calculate the deviation d_i, where

$$d_i = x_i - \overline{x}$$

Some of these deviations will be negative (where the data value is less than the mean), and some will be positive, so we cannot simply take the average of the deviations; the positive and negative values would cancel each other out. In fact, the sum of the deviations will always be zero, so the average deviation is also always zero. Instead of averaging the deviations, we'll square each deviation (to make it positive) and then sum those values and divide by the number of observations minus 1. This value is the **variance**, represented by the symbol s^2, and the formula for calculating s^2 is

$$s^2 = \frac{\text{Sum of squared deviations}}{\text{Number of observations} - 1}$$

$$= \frac{1}{n-1} \sum_{i=1}^{n} (x_i - \overline{x})^2$$

To measure variability, statisticians calculate the **standard deviation** (represented by the symbol s), which is the square root of the variance. The complete formula for s is

$$s = \sqrt{\frac{\sum_{i=1}^{n} (x_i - \overline{x})^2}{n-1}}$$

Why do we divide the total of the squared deviations by $n - 1$, rather than n? Recall that the sum of the deviations is known to be zero, so given the first $n - 1$ deviations, we can always calculate the remaining deviation. This means only $n - 1$ of the deviations can vary freely—the last value is constrained by the values of the preceding deviations. This figure, $n - 1$, is known as the **degrees of freedom** and is a value that will become more important in the chapters that follow.

The standard deviation represents the "typical" deviation of values from the average. A large value of s indicates a high degree of variability in the data. *High* is a relative term, and we usually speak about high degrees of variability only when comparing one distribution with another.

Table 4-6 lists the formulas provided by Excel to calculate the variability of values in a data set.

Table 4-6 Formulas to calculate variability of values in data sets

Function	Description
AVEDEV(*array*)	Returns the average of the absolute values of the deviations in an array or data range.
DEVSQ(*array*)	Returns the sum of the squared deviations in an array or data range.
MAX(*array*)	Returns the maximum value in an array or data range.
MIN(*array*)	Returns the minimum value in an array or data range.
STDEV(*array*)	Returns the standard deviation of the values in an array or data range.
VAR(*array*)	Returns the variance of the values in an array or data range.
RANGEVALUE(*array*)	Returns the range of the values in an array or range reference. *StatPlus required.*

Measures of Shape: Skewness and Kurtosis

Skewness is a measure of the lack of symmetry in the distribution of the data values.

A positive skewness value indicates a distribution with values clustered toward the lower range of values with a long tail extending toward the upper values range. A negative skewness indicates just the opposite, with the long tail extending toward the values lower in the data range. A skewness of zero indicates a symmetric distribution.

Kurtosis measures the heaviness of the tails in the distribution.

A positive kurtosis indicates more extreme values than expected in the distribution. A negative kurtosis indicates less extreme values than expected.

Table 4-7 shows the Excel functions used to calculate skewness and kurtosis.

Table 4-7 Excel functions to calculate skewness and kurtosis

Function	Description
KURT(*array*)	Returns the kurtosis of the values in an array or data range.
SKEW(*array*)	Returns the skewness of the values in an array or data range.

Use the Univariate Statistics command to calculate the variability and shape statistics for the prices of homes in the Albuquerque sample.

To create a table of variability and shape statistics:

1 Click **StatPlus > Univariate Statistics**.

2 Click the **Input** button and select **Price** from the list of range names.

3 Click the **Output** button, click the **New Worksheet** option button, and type **Variances** in the New Worksheet box. Click the **OK** button.

4 Click the **By** button and select **NE_Sector** from the list of range names.

5 Click the **Variability** dialog tab.

6 Click the **Show all variability statistics** checkbox. See Figure 4-22.

Figure 4-22
Selecting
the
variability
statistics to
display

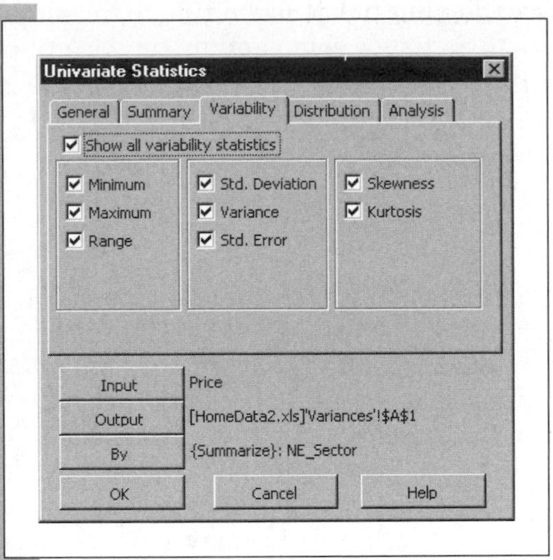

7 Click the **OK** button to create the table.

Figure 4-23 shows the output from the Univariate Statistics command.

Figure 4-23
Variability
statistics

	A	B	C	D	E
1			Univariate Statistics		
2		Price			
3		NE_Sector = "No"	NE_Sector = "Yes"	Overall	
4	Minimum	54,000	61,900	54,000	
5	Maximum	210,000	215,000	215,000	
6	Range	156,000	153,100	161,000	
7	Standard Deviation	32,039.522	40,154.213	38,043.699	
8	Variance	1,026,530,985.155	1,612,360,859.141	1,447,322,998.821	
9	Standard Error	5,130.430	4,546.569	3,517.141	
10	Skewness	1.722	1.233	1.375	
11	Kurtosis	4.009	0.809	1.447	
12					

Based on the output from Figure 4-23, we note that the variability of the home prices is higher in the northeast sector than outside of it (though it's interesting to note that the range of home prices is higher for non-northeast sector homes).

STATPLUS TIPS

- You can select all summary, variability, or distribution statistics by clicking the appropriate checkboxes in the General dialog sheet of the Univariate Statistics dialog box.

- The "Univariate Statistics" command can display the table with statistics displayed in rows or in columns.

- You can add your own custom title to the output from the Univariate Statistics command by typing a title in the Table Title box in the General dialog sheet.

Outliers

As the earlier discussion on means and medians showed, distribution statistics can be heavily affected by extreme values. It's difficult to analyze a data set in which a single observation dominates all of the others, skewing the results. These values, known as **outliers**, don't seem to belong with the others because they're too small, too large, or don't match the properties one would expect for them. As you've seen, a large salary can affect an analysis of salary values, pushing the average salary value upward. An outlier need not be an extreme value. If you were to analyze fitness data, the records of an extremely fit 70-year-old might not be remarkable compared to all of the values in the distribution, but it might be unusual compared to the values of others in his or her age group.

Outliers are caused by either of two things: (1) mistakes in data entry or (2) an unusual or unique situation. A mistake in data entry is easier to deal with: You discover and correct the mistake and then redo the analysis. If there is no mistake, you have a bigger problem. In that case you have to study the outlier and decide whether it really belongs with the other data values. For example, in a study of Big Ten universities, we might decide to remove the results from Northwestern because that school, unlike the other schools, is a small, private institution. In the Albuquerque data, we might remove a high-priced home from the sample if that house was a public landmark and thus uniquely expensive.

However, and this point cannot be emphasized too strongly, *its merely being an extreme value is not sufficient grounds to remove the observation.* Many advances in science have been made by scientists studying the observations that didn't seem to fit the "expected" distribution. Extreme values may be a natural part of the data (as with some salary structures). By removing those values you are removing an important aspect of the distribution.

One possible solution to the problem of outliers is to perform two analyses: one with the outliers and one without. If your conclusions are the same, you can be confident that the outlier had no effect. If the results are extremely different, you can report both answers with an explanation of the differences involved. In any case, you should not remove an observation without (1) good cause and (2) documentation of what you did and why.

What constitutes an outlier? How large (or small) must a value be before it can be considered an outlier? One accepted definition depends on the interquartile range, or IQR (recall that the interquartile range is equal to difference between the third and first quartiles).

1. If a value is greater than the third quartile plus 1.5 × IQR, or less than the first quartile minus 1.5 × IQR, it's a **moderate outlier**.
2. If a value is greater than the third quartile plus 3 × IQR, or less than the first quartile minus 3 × IQR, it's an **extreme outlier**.

A diagram displaying the boundaries for moderate and extreme outliers is shown in Figure 4-24.

**Figure 4-24
The range of
moderate
and extreme
outliers**

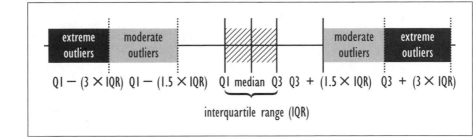

For example, if the first quartile equals 30 and the third quartile equals 80, the interquartile range is 50. Any value above 80 + (1.5 × 50), or 155, would be considered a moderate outlier. Any value above 80 + 150, or 230, would be considered an extreme outlier. The lower ranges for outliers would be calculated similarly.

This definition of the outlier plays an important role in constructing one of the most useful tools of descriptive statistics—the boxplot.

Working with Boxplots

In this section, we'll explore one of the more important tools of descriptive statistics, the boxplot. You'll learn about boxplots interactively with Excel, and then you'll apply what you've learned to the Albuquerque price data.

CONCEPT TUTORIALS:
Boxplots

Your Student disk contains several **instructional workbooks**. The instructional workbooks include interactive worksheets and macros to allow you to explore various statistical concepts on your own. The first of these workbooks that you will examine concerns the boxplot. Open this workbook now.

To start the Boxplots instructional workbook:

1 Open the file **Boxplots**, located in the Explore folder of your Student disk.

The workbook opens to the Contents page, describing the nature of the workbook. To the left side of the display area is a column of subject titles. You can move between subject titles either by clicking an entry in the column or by clicking the arrow icon located at the top of the page.

2 Click **What is a boxplot?** from the list of subject titles. The page shown in Figure 4-25 appears.

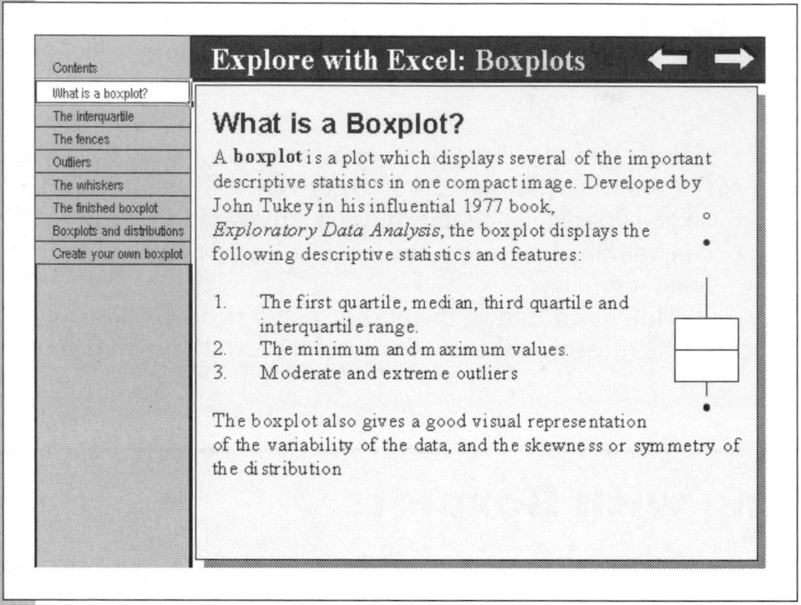

Figure 4-25
"What is a boxplot?"
from the
Boxplots
instructional
workbook

Boxplots are designed to display in a single chart several of the important descriptive statistics, including the quartiles of the distribution as well as the minimum and the maximum. Boxplots will also identify any moderate or extreme outliers (using the definition supplied above).

3 Click **The interquartile**.

**Figure 4-26
The "box"
from the
boxplot**

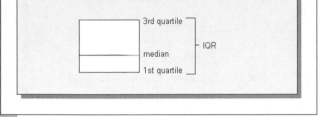

The interquartile

The first component of the boxplot is the **interquartile range**, the part of the distribution ranging from the first quartile to the third quartile. To create this section, draw a box extending from the first to the third quartile. Within this box, draw a horizontal line at the location of the median (the second quartile.)

The "box" part of the boxplot displays the interquartile range of the distribution, ranging from the first quartile to the third. The median is shown as a horizontal line within the box. Note that the median need not be in the center of the box. The box tells you where the central 50% of the data is located. By observing the placement of the median within the box, you can also get an indication of how those values are clustered within that central 50%. A median line close to the first quartile indicates that a lot of the values are clustered in the lower range of the distribution.

4 Click **The fences**.

Figure 4-27
The
"fences" of
the boxplot

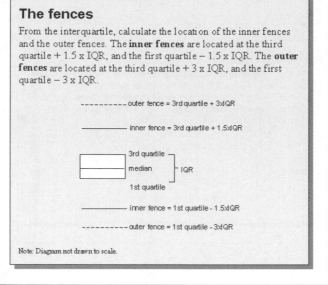

The inner and outer fences of the boxplot set the boundaries between "standard" observations, moderate outliers, and extreme outliers. Note that the formula for the fences matches the formula for moderate and extreme outliers discussed in the previous section.

5 Click **Outliers**.

Figure 4-28
Represent-
ing outliers
in a
distribution

Outliers

Any values that lie between the inner and outer fences are **moderate outliers** and are shown with the the symbol, •. Any values that lie beyond the outer fences are **extreme outliers** and are shown with the symbol, o.

```
                       o
            ----------  outer fence

            ----------  inner fence

              ┌─────┐ 3rd quartile
              │     │ median
              └─────┘ 1st quartile

            ----------  inner fence
                       •
            ----------  outer fence
```

Note: Diagram not drawn to scale.

If there are any moderate or extreme outliers in the distribution, they're displayed in the boxplot. Moderate outliers are displayed using a black circle, •. Extreme outliers are represented by an open circle, ∘. With a boxplot you can quickly see the outliers in your distribution and their severity.

6 Click **The whiskers**.

The whiskers

Vertical lines called **whiskers**, are drawn from the box to highest and lowest values in the distribution that lie within the inner fence. These are the highest and lowest values in the distribution, which are *not* considered outliers.

```
           o
     ----------- outer fence

           •
              ------ inner fence


                   3rd quartile
                   median
                   1st quartile


              ------ inner fence
           •
     ----------- outer fence
```

Note: Diagram not drawn to scale

The final component of the boxplot is the whiskers. These are lines that extend from the boxplot to the highest and lowest points that lie inside the moderate outliers. Thus the lines indicate the smallest and largest values in the distribution that are *not* considered outliers. The length of the whisker lines also gives you a further indication of the skewness of the distribution.

In the finished boxplot, the inner and outer fences are not shown. Figure 4-30 shows a typical boxplot.

Figure 4-30
The
completed
boxplot

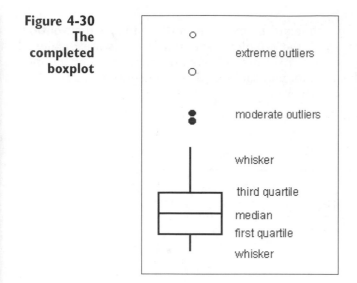

Figure 4-31 shows how a boxplot might look for distributions with positive or negative skewness and for symmetric distributions.

Figure 4-31
Boxplots for
various
distributions

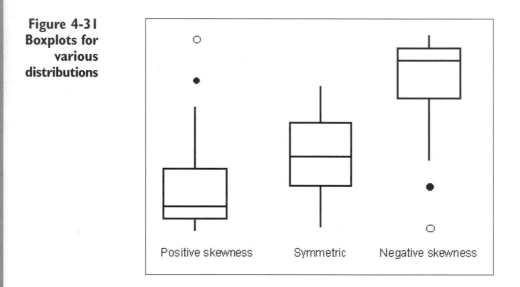

Now that you've learned about the structure of the boxplot, try creating a few boxplots on your own with some sample data.

To create your own boxplot:

1 Click **Create your own boxplot** from the list of topics.

2 Enter the following numbers into the green cells located to the left of the empty chart: **0, 1, 2, 2, 3, 3, 3, 3, 4, 4, 4, 5, 5, 6, 7**

As you enter the numbers, the chart is automatically updated to reflect the new distribution. The final form of the boxplot is shown in Figure 4-32.

**Figure 4-32
A boxplot
for some
sample data**

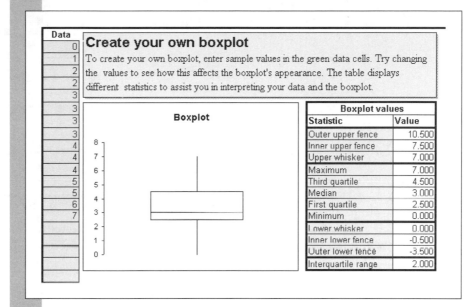

Create your own boxplot

To create your own boxplot, enter sample values in the green data cells. Try changing the values to see how this affects the boxplot's appearance. The table displays different statistics to assist you in interpreting your data and the boxplot.

Boxplot values	
Statistic	Value
Outer upper fence	10.500
Inner upper fence	7.500
Upper whisker	7.000
Maximum	7.000
Third quartile	4.500
Median	3.000
First quartile	2.500
Minimum	0.000
Lower whisker	0.000
Inner lower fence	-0.500
Outer lower fence	-3.500
Interquartile range	2.000

The central 50% of the data is found in the range from 2.5 to 4.5. The median value is 3, which is not in the middle of that central 50%. From the plot we can see that the values range from 0 to 7. There are no outliers in the distribution. Now let's see what happens if we change a few of those numbers.

3 Change the last two numbers in the sample data from 6 and 7 to **9** and **12**.

Figure 4-33 shows the updated chart.

Figure 4-33
The sample data with outliers added

outliers appear in the boxplot

new data values

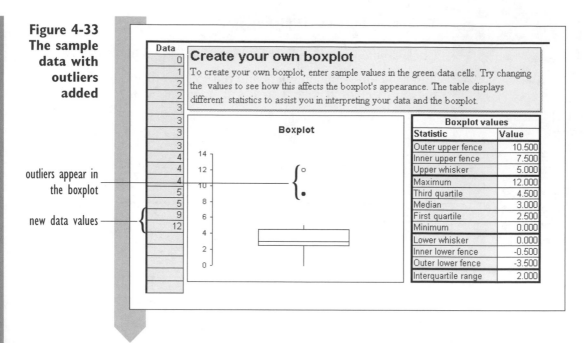

The two revised observations are considered moderate and extreme outliers. From the boxplot we can see that there is a large gap between the moderate outlier and the largest "standard" value.

Continue to explore boxplots with the instructional workbook. Try different combinations of values and different types of distributions.

When you're finished with the Boxplots workbook:

1 Click **File > Close**.

2 You do not have to save any of your changes.

3 Return to the HomeData2 workbook.

Creating a Boxplot for the Housing Data

Excel does not contain any commands to create boxplots, but you can create one using StatPlus. The StatPlus command includes the added feature of displaying a dotted line representing the average value for the distribution. This added information gives you the ability to compare the mean and median values graphically. Try creating a boxplot now for the northeast sector and other homes in the Albuquerque sample.

To create a boxplot of the price data:

1 Click **StatPlus > Single Variable Charts > Boxplots** from the Excel menu.

The Boxplots command allows you to create boxplots based on values in separate columns or within a single column broken down by the levels of a categorical variable. In this case you'll use a single column, Price, and a categorical variable, NE_Sector.

2 Verify that the **Use column of category levels** option button is selected.

3 Click the **Data Values** button and choose **Price** from the list of range names.

4 Click the **Categories** button and choose **NE_Sector** from the list of range names.

5 Click the **Output** button, click the **As a new chart sheet** option button, and type **Price Boxplot** in the adjacent name box. Click the **OK** button.

Figure 4-34 shows the completed dialog box. Click the **OK** button.

**Figure 4-34
The completed Boxplot dialog box**

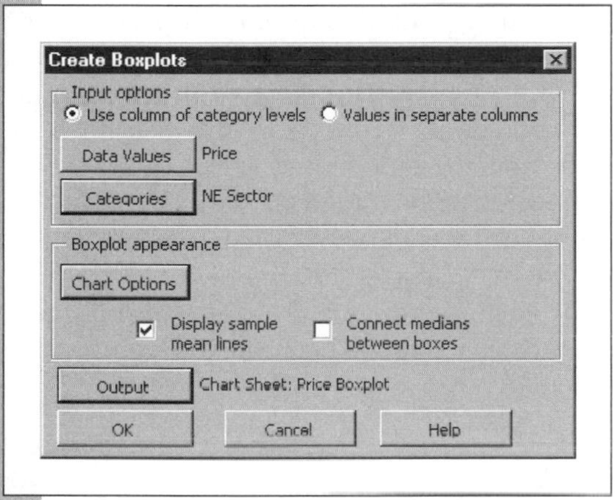

The output from the Boxplot command is shown in Figure 4-35.

Figure 4-35
Boxplot of
the
Albuquerque
home prices

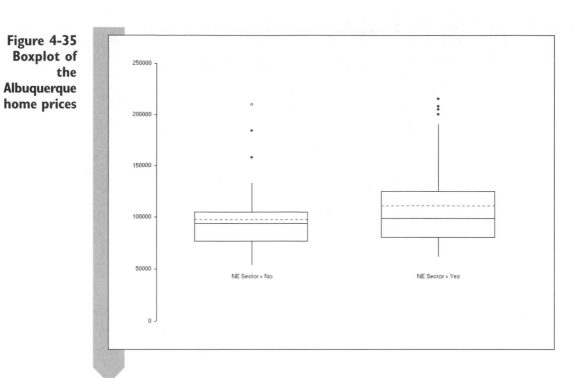

The boxplot gives us yet another visual picture of our data. Note that the three extreme prices in the non-northeast sector homes are all considered outliers, and that the range of the other values for that area extends from about $50,000 to around $140,000. Are those homes overpriced for their area? There may be something unusual about those three homes that would require further research. The range of values for the northeast homes is clearly much wider, and only homes above $200,000 are considered moderate outliers. The boxplot also gives us a visual picture of the difference between the mean and median for the northeast homes. This information would certainly caution us against blindly using only the mean to summarize our results.

STATPLUS TIP

- To specify the chart's title or a title for the *x*-axis or *y*-axis, click the "Chart Options" button in the Boxplot dialog box.

You've completed your work on the Albuquerque data. Save your workbook and close Excel.

Exercises

1. You see the following stem and leaf plot in a technical journal:

$$\underline{\text{Stem} \times 100 \mid \text{Leaf}}$$

```
0 | 336
1 | 01228
2 | 00111249
3 | 04
4 | 5
5 |
6 | 1
7 |
8 |
9 | 0
```

    a. What are the approximate values of the data set?
    b. Is the distribution positively skewed, negatively skewed, or symmetric?
    c. Give approximate values for the mean and the median.
    d. What values, if any, appear to be moderate and extreme outliers in the distribution?
    e. Save your answers in a workbook named E4STEMLEAF.XLS.

2. A data distribution has a median value of 22, a first-quartile value of 20, and a third-quartile value of 30. Five observations lie outside the interval from the first to the third quartile, with values of 17, 18, 40, 50, and 75.
    a. Draw the boxplot for this distribution.
    b. Is the skewness positive, negative, or zero?

3. Reopen the HOMEDATA.XLS workbook you examined in this chapter. You'll do further research on this workbook, examining the size of homes in the Albuquerque sample as well as the price per square foot of the homes.
    a. Create a table of univariate statistics for the size of the homes in square feet, including all distribution, variability, and summary statistics except the mode. Place the table on a worksheet named "SqF Stats 1."
    b. What are the smallest and largest houses in the sample?
    c. If you were interested only in houses that were 2,200 square feet or higher, what percentage of the houses in the sample meet the requirement? (*Hint*: Use the PERCENTRANK function).
    d. Create a boxplot of the size of the homes in square feet. Place the boxplot in a chart sheet named "SqF BP1."
    e. What value appears to be an extreme outlier in the boxplot?
    f. Create a second boxplot of house size, this time breaking the boxplot down by whether or not the home is a corner lot. Place the plot in a chart sheet named "SqF BP2."
    g. Interpret your boxplot in terms of the relationship between size of a house and whether or not it lies on a corner lot.

h. What happened to the extreme outlier you identified earlier? Discuss this in terms of the definition of *outlier* given in the text. Is this value an outlier or not?

i. Recreate the table of univariate statistics for house size, and this time break the table down by the Corner Lot variable. Place the table in a worksheet named "SqF Stats 2."

j. Create a new column containing the price per square foot of each house. Assign the values in the new column a range name. What type of variable is this?

k. Create a histogram with 20 evenly spaced bins of the price per square foot on a chart sheet named "PPSqF Hist" (you will have to reformat numbers on the horizontal axis, reducing the number of decimals displayed).

l. What is the shape of the distribution of price per square foot? Are there any outliers?

m. Print the chart sheets and tables in your workbook. Hand in the printouts along with your analysis.

n. Save your workbook as E4HOMEDATA.XLS.

4. The WBUS.XLS workbook contains data on 50 of the largest women-owned businesses in Wisconsin. Analyze and report the descriptive statistics on this data set.

a. Open the workbook and create a table of the distribution, variability, and summary statistics for the Employees variable. Store the table in a worksheet named "Employee Stats."

b. What is the average number of employees for the 50 businesses? What is the median amount? Which statistic do you think more adequately describes the size of these businesses? How does the average number of employees compare to the third quartile?

c. Create a boxplot of employees stored in a chart sheet named "Employee Bplot." How would you describe this distribution?

d. Create a new variable containing the Base 10 log of the Employees variable. Assign a range name to this new column and then create a boxplot of the log(Employees) in a chart sheet named "Log Employee Bplot." How does the shape of this distribution compare to the untransformed values?

e. Create a table of descriptive statistics for the log(Employees) and store the table in a worksheet named "Log Employee Stats." Compare the skewness and kurtosis values between the Employees and log(Employees) variables. Explain how the difference in the distribution shapes is reflected in these two statistics.

f. Calculate the mean log(Employees) value in terms of the number of people employed (in other words, transform this value back to the original scale). How does this value compare to the geometric mean of the number of employees in each company?

g. The geometric mean is used for values that either are ratios or are best compared as ratios. Which pair of companies is more similar in terms of size: a company totaling $50,000,000 in annual sales and a company with $10,000,000, or a company with $450,000,000 in sales and a company with $400,000,000? What are the differences in sales between the two sets of companies? What are the ratios? Does the difference or the ratio better express the similarity of the companies?

h. Hand in your analysis along with the tables and charts you created. Save the workbook as E4WBUS.XLS.

5. The PCINFO.XLS workbook contains price information on some PC models. Analyze this data to report on the distribution of the price values.

    a. Open the workbook and create a frequency table for prices in increments of 500 starting at $1,500 and going up to $7,500. Have the bin option set to display values ≥ the bin value and values < the previous bin value. Save the frequency table in a worksheet named "Price Table."

    b. Create a histogram of the prices using the same options you used in the frequency table. Save the histogram on a chart sheet named "Price Histogram."

    c. Create a second histogram showing cumulative percentages of the prices with the same bin options. Save the plot on a chart sheet named "Cumulative Prices."

    d. If you were limited to spending only $3,000 per computer, what percentage of computers in the sample are available to you?

    e. Print your tables and charts. Save the workbook as E4PCINFO.XLS.

6. The MORTGAGE.XLS workbook contains information on refusal rates from 20 lending institutions broken down by race and income status. This data was presented to a joint congressional hearing on discrimination in lending.

    a. Open the workbook and create a table of univariate statistics for the four data columns. Save the table in a worksheet named "Refusal Statistics."

    b. Create a boxplot of the refusal rates for the four data columns stored in a chart sheet named "Refusal Bplots." Label the chart appropriately.

    c. Including the descriptive statistics and boxplot you've created, write a report detailing your findings. What conclusion do you draw from the data? Is there any specific information that this data sample is lacking? Include a discussion of potential problems in this data set and how you would go about remedying them.

    d. Save your workbook as E4MORTGAGE.XLS.

7. Average teacher salary, public school spending per pupil, and the ratio of teacher salary to pupil spending for 1985 have been stored in the TEACHER.XLS workbook. The values are broken down by state and region.

    a. Open the workbook and create a table of univariate statistics for the teacher salaries broken down by region and overall. Save the table on the worksheet "Salary statistics."

    b. Create a boxplot, broken down by region, of the teacher salaries, on a chart sheet named "Salary Boxplots."

c. Discuss the distribution of the teacher salaries for each region. There is an extreme outlier in the West region. Which state is this? Discuss why salaries for teachers in this state might be so high.

d. Create a table of univariate statistics for the ratio of teacher salary to spending per pupil, broken down by region and overall. Save the table on a worksheet named "Salary Pupil Ratio Statistics."

e. Create, on a chart sheet named "Salary Pupil Ratio Boxplots," a boxplot of the ratio values broken down by region.

f. For the state that was an outlier in the West region in terms of teacher salary, check to see if it is also an outlier in terms of the ratio of teacher salary to public spending per pupil. Estimate the percentile of this state's salary/pupil ratio within the West region. How does that compare to its percentile for teacher's salary alone? If the cost of education per pupil is indicative of the cost of living in a state, are teachers in this particular state overpaid or underpaid relative to other states in the West region?

g. Print your tables and charts along with your analysis. Save the workbook as E4TEACHER.XLS.

8. The BASE.XLS workbook contains annual salary figures for major league baseball players (in terms of hundreds of thousands of dollars).

a. Open the workbook and create a histogram of the player salaries with bin intervals of 100 (for $100,000) ranging from $0 to $2,500 (or $2,500,000). Have the counts within each bin be ≥ the bin value and < the next bin value. Store the histogram on a chart sheet named "Salary Histogram."

b. Create a frequency table of the player's salaries, using the same bin intervals and options you used to create the histogram. Save the frequency table on a worksheet named "Salary Table."

c. Calculate the 10th and 90th percentiles of the salaries.

d. Using the value for the 90th percentile, filter the player data to show only those players who were paid in the upper 10% of the salary range. Print the list of players.

e. What is the average player's salary? What is the median player's salary? If a player made the average salary, at what percentile would he be ranked in the data?

f. Print the tables and charts you created and summarize your results. Save the workbook as E4BASE.XLS.

9. The workbook CANCERRATE.XLS contains data on all of the 50 states' cigarette use per household per capita and their rates of bladder cancer, kidney cancer, lung cancer, and leukemia per 100,000. An additional variable, Cig Use Category, breaks down the cigarette use into three categories: 0 for low rates of cigarette use, 1 for medium, and 2 for high.

a. Open the workbook and create boxplots of the rates of bladder cancer, kidney cancer, lung cancer, and leukemia broken down by cigarette use. Label the charts and chart sheets appropriately.

b. Create a table of univariate statistics for the rate of each illness broken down by cigarette use.

c. Does there appear to be any relationship between these illnesses and the level of cigarette use in the states? Defend your answer with your charts, statistics, and tables.

d. There is one state with a high level of cigarette use but a relatively low level of lung cancer. Identify this state.

e. Print the tables and charts you created, summarizing your results. Save your workbook as E4CANCERRATE.XLS.

10. Air quality data was collected by the Environmental Protection Agency in 1980 and from 1985 to 1989. The workbook POLU.XLS contains data on the number of unhealthy days for 14 major U.S. cities. There are two additional variables: Diff80 and Ratio80. Diff80 contains the difference between the average number of unhealthful days from 1985 to 1989 and the number of unhealthy days in 1980. Ratio80 contains the ratio of the average number of unhealthy days in 1985 to 1989 to those in 1980. A negative Diff80 value or a Ratio80 value that is less than 1 indicates an improvement in air quality. Is there evidence that air quality improved?

a. Create a table showing the mean and median values of Diff80 and Ratio80.

b. Create a boxplot of Diff80 and Ratio80. Identify the extreme outlier on the Diff80 boxplot. Which city does it come from?

c. Copy the air quality data to a new worksheet without the extreme outlier you noted in part b. Redo the table of statistics and boxplots with this new set of data.

d. What are your conclusions? Have the conclusions changed without the presence of the outlier? What effect did the outlier have on the Ratio80 statistics? On the Diff80 statistics?

e. Write a report summarizing your findings, including your charts and tables. Save the workbook as E4POLU.XLS.

11. The workbook REACT.XLS contains reaction times from the first-round heats of the 100-meter race at the 1996 Summer Olympic games. Reaction time is the time elapsed between the sound of the starter's gun and the moment the runner leaves the starting block. The workbook also contains the heat number, the order of finish, and the finish group (1st through 3rd, 4th through 6th, and so forth).

a. Open the workbook and calculate univariate descriptive statistics for the reaction times listed. What are the average, median, minimum, and maximum reaction times?

b. Create a boxplot of the reaction times. Are there any moderate or extreme outliers in the distribution? How would you characterize the shape of the distribution?

c. Create a stem and leaf plot of the reaction times.

d. Reaction times are important for determining whether a runner has false-started. If the runner's reaction time is less than 0.1 second, a false start is declared. Where would a reaction time of 0.1 second fall on your boxplot: as a "typical" value, a moderate outlier, or, an extreme outlier? Does this definition of a false start seem reasonable given your data?

e. Do reaction times affect the order of finish? Calculate descriptive statistics for the reaction times broken down by order of finish. Pay particular attention to the mean and the median.

f. Create a boxplot of the reaction times broken down by order of finish.

g. Is there anything in your descriptive statistics or boxplots to suggest that reaction time plays a part in how the runner finishes the race?

h. Repeat questions e through g breaking down the reaction times by finish group. Compare your results. Do you arrive at the same conclusion, or does the data suggest a different interpretation?

i. Print your tables and charts along with your analysis of the data. Save your workbook as E4REACT.XLS.

12. The workbook SALARY.XLS contains salary data from 50 universities for full, associate, and assistant professors. Salaries are shown in terms of thousands of dollars.

a. Open the workbook and create a table of descriptive statistics for the three types of professors. Which of the three types shows the greatest variability in salary?

b. Create a stem and leaf plot of the salaries for the three groups. Plot each group on the same scale, but do not include a summary plot. Describe the three distributions in terms of shape and skewness.

c. Create a boxplot of the full, associate, and assistant salaries. How much overlap is there in the salaries? What would account for this?

d. Determine for each level of professor, an appropriate range of salaries that would include the central 50% range of the salaries from the workbook.

e. If you are trying to recruit a new full professor who asks to be paid at least in the upper 5% range of her or his profession, what should your opening offer be and why?

f. Summarize your results and print your tables and charts. Save the workbook as E4SALARY.XLS.

13. The workbook WLABOR.XLS shows the change in the percentage of women in the labor force from 19 cities in the United States from 1968 to 1972. You can use this data to gauge the growing presence of women in the labor force during this time period.

a. Open the workbook and calculate descriptive statistics for both the 1968 and the 1972 values. What are the mean, median, and standard deviation for those years?

b. Calculate the geometric mean for the two years in question.

c. Create a boxplot of the 1968 and 1972 values. Are there any outliers present in the data? Identify which city the value comes from. What does the data tell you about the presence of women in the labor force?

d. Describe the shape of the distribution. Is the data positively or negatively skewed or symmetric? Can you use the mean to summarize the results from this study?

e. Print the charts and tables from the workbook along with your summary of the data. Save the workbook as E4WLABOR.XLS.

14. In 1970, draft numbers were determined by lottery. All 366 possible birth dates were placed in a rotating drum and selected one by one. The first birth date drawn received a draft number of 1, and men born on that date were drafted first; the second birth date received a draft number of 2; and so forth. Data from the draft number lottery can be found in the DRAFT.XLS workbook.

a. Open this workbook and create a box plot of the draft numbers broken down by month. Also create a table of counts, means, medians, and standard deviations. Is there any evidence of a trend in the draft numbers selected compared to the month?

b. Repeat question a, this time breaking the numbers down by quarters. Is there any evidence of a trend between draft numbers and the year's quarter?

c. Repeat question a, breaking the draft numbers by first half of the year versus second half. Is the typical draft number selected for the first half of the year close in value to the draft number for birthdays from the second half of the year?

d. Discuss your results. The draft numbers should have no relationship to the time of the year. Does this appear to be the case? What effect does breaking the numbers down into different units of time have on your conclusion?

e. Report your results, answering whether the date of one's birth affects the probability of one's being drafted in this system. Include your tables and charts in the report. Save the workbook as E4DRAFT.XLS.

Chapter 5

Probability Distributions

Objectives

In this chapter you will learn to:

- Work with random variables and probability distributions

- Generate random normal data

- Create a normal probability plot

- Explore the distribution of the sample average

- Apply the Central Limit Theorem

U p to now, you've used tools such as frequency tables, descriptive statistics, and scatterplots to describe and summarize the properties of your data. Now you'll learn about probability, which provides the foundation for understanding and interpreting these statistics. You'll also be introduced to statistical inference, which uses summary statistics to help you reach conclusions about your data.

Probability

Much of science and mathematics is concerned with prediction. Some of these predictions can be made with great precision. Drop an object, and the laws of physics can predict how long the object will take to fall. Mix two chemicals, and the laws of chemistry can predict the properties of the resulting mixture. Other predictions can be made only in a general way. Flip a coin, and you can predict that either a head or a tail will result, but you cannot predict which one. That doesn't mean that you can't say anything. If you flip the coin several times, you'll notice that roughly half the flips result in heads and half result in tails.

Flipping a coin is an example of a **random phenomenon**, in which individual outcomes are uncertain but follow a general pattern of occurrences. When we study random phenomena, our goal is to quantify that general pattern of occurrences in order to make general predictions. How do we do this? One way is through theory. We imagine an ideal coin with two sides: a head and a tail. Because this is an ideal coin, we assume that each side is equally likely to occur during our coin flip. There are two possible outcomes, so half of these will be heads and half will be tails. From this, we can define the **theoretical probability** of an event:

$$\text{Theoretical probability of an event} = \frac{\text{Number of possible ways of obtaining the event}}{\text{Total number of equally likely possible outcomes}}$$

In the coin-tossing example, there is one way to obtain a head and there are two possible outcomes, so the theoretical probability of obtaining a head is $\frac{1}{2}$, or 0.5.

Another way of quantifying random phenomena is through observation. To determine the probability of obtaining a head, we repeatedly toss the coin. From our observations, we calculate the **relative frequency** of tosses that result in heads, where

$$\text{Relative frequency} = \frac{\text{Number of times an event occurs}}{\text{Number of replications}}$$

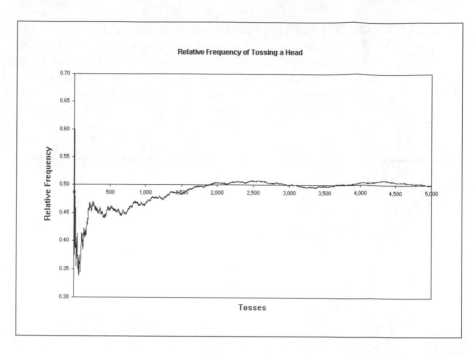

Figure 5-1
The relative frequency of tossing a head versus the number of tosses

Figure 5-1 shows a chart of the results of tossing such a coin 5,000 times. Early in the experiment, the relative frequency of heads jumps around quite a bit, hovering below 0.5. As the number of tosses increases, the relative frequency narrows around the 0.5 level. The **law of large numbers** states that as the number of replications increases, the relative frequency will approach the probability of the event, or, to put it another way, we can define the **probability** of an event as the value approached by the relative frequency after an indefinitely long series of trials.

Probability Distributions

The pattern of probabilities for a set of events is called a **probability distribution**. Probability distributions contain two important elements:

1. The probability of each event or combination of events must range from 0 to 1.
2. The sum of the probabilities of all possible events must be equal to 1.

In the coin-tossing example, there are two outcomes (head or tail), each with a probability of 0.5. The sum of both events is 1, so this is an example of a probability distribution. Two types of probability distributions are discrete and continuous.

Discrete Probability Distributions

In a **discrete probability distribution**, the probabilities are associated with a series of discrete outcomes. The probabilities associated with tossing a coin form a discrete distribution. If you toss a 6-sided die, the probabilities associated with that outcome also form a discrete distribution, where each side has a $\frac{1}{6}$ probability of turning up. We can write this as

$$p(y) = 1/6 \quad y = 1,2,3,4,5,6$$

where $p(y)$ means the "probability of y," which equals $\frac{1}{6}$ for values of y from 1 to 6.

Note that discrete does not mean "finite." There are discrete probability distributions that cover an infinite number of possible outcomes. One of these is the **Poisson** distribution, used when the outcome event involves counts within a specified period of time. The equation for the Poisson distribution is

Poisson Distribution

λ~ # events

$$p(y) = \frac{\lambda^y}{y!}e^{-\lambda} \quad y = 0,1,2,\ldots,$$

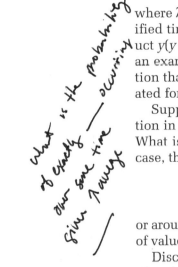

what is the probability of exactly # occurring over some time given # average

where λ (pronounced "lambda") is the average number of events in the specified time period, and $y!$ stands for "y factorial," which is equal to the product $y(y-1)(y-2)\ldots(3)(2)(1)$. For example, $4! = 4 \times 3 \times 2 \times 1 = 24$. Lambda is an example of a **parameter**, a term in the formula for a probability distribution that defines its shape and values. Let's see what probabilities are generated for a specific value of λ.

Suppose we want to determine the number of car accidents at an intersection in a given year, and we know that the average number of accidents is 3. What is the probability of exactly two accidents occurring that year? In this case, the value of λ is 3, $y = 2$, and the probability is

$$\frac{3^2}{2!}e^{-3} = \frac{9 \cdot 0.0498}{2 \cdot 1} = 0.224$$

or around 22%. Note that we can calculate probabilities for an infinite number of values, but we are limited to integers.

Discrete distributions are displayed with a bar chart in which the height of each bar is proportional to the probability of the event. To find the probability of a group of events, we simply add up the individual probabilities of the event in the group. Figure 5-2 displays the probability distribution from $y = 0$ to $y = 10$ accidents per year.

Figure 5-2
Poisson
probability
distribution
for car
accident
data

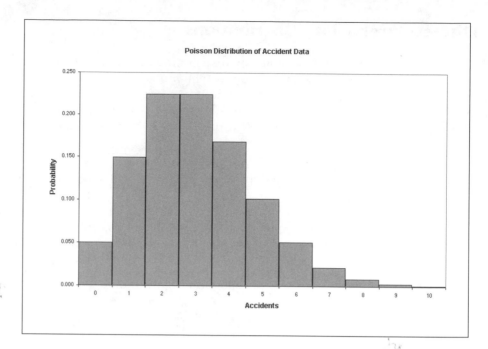

Continuous Probability Distributions

In **continuous probability distributions**, probabilities are assigned to a range of continuous values rather than to distinct individual values. For example, consider a marksman shooting at a target. The distribution of shots around the bull's eye follows a continuous distribution. If the marksman is good, the probability that the shots will cluster closely around the bull's eye is very high, and it is unlikely that a shot will miss the target entirely.

Note that unlike discrete distributions, we can't assign a positive probability to a specific value in a continuous distribution. The probability of any specific value is zero.

Continuous probability distributions are calculated using a **probability density function**, or **PDF**. When we plot a PDF against the range of possible values, we get a curve in which the curve's height indicates the position of the most likely values. Figure 5-3 shows a sample PDF curve.

**Figure 5-3
A sample
probability
density
function**

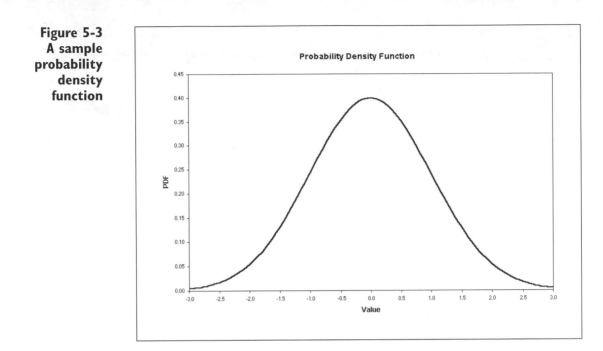

The probability associated with a range of values is equal to the area under the PDF curve. A single value has zero probability because its area under the curve is zero (it has a positive height, but zero width). The total area under any PDF curve must be equal to 1.

CONCEPT TUTORIALS:
PDFs

To see the relationship between probability and the area under the PDF curve, open the instructional workbook named Probability.

To open the Probability workbook:

1 Open the file **Probability**, located in the Explore folder of your Student disk.

The workbook opens to the Contents page, describing the nature of the workbook.

2 You can move through the sheets in the workbook, reviewing the material on probability discussed so far in this chapter.

3 Click **Explore a PDF** from the Table of Contents column.

The sheet displays the curve shown in Figure 5-4.

Figure 5-4
The
Probability
instructional
workbook

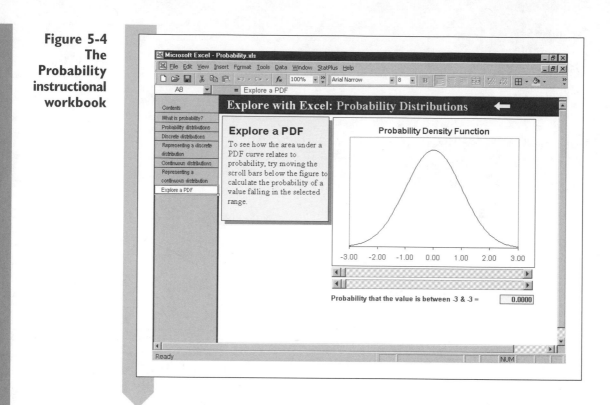

Notice the two horizontal scroll bars below the chart. You'll use these to set the range boundaries on the PDF curve. Use them now to select the range from −1 to 1 on the curve.

To set the range boundaries on the PDF curve:

1 Click or drag the **scroll button** of the bottom scrollbar until the right boundary equals 1.

2 Click or drag the **scroll button** arrow of the top scrollbar until the left boundary equals −1.

Figure 5-5 shows the PDF with the range from −1 to 1 selected. The area of this range is equal to 0.6827.

**Figure 5-5
Selecting
the range
from -I to I**

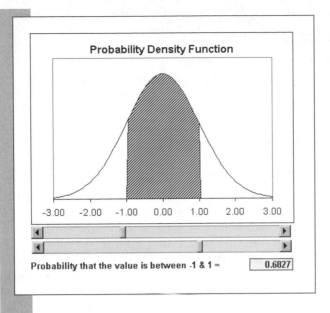

Because the area under the curve is equal to 0.6827, the probability of a value falling between −1 and 1 is equal to 68.27%. Try experimenting with other range boundaries until you get a good feel for the relationship between probabilities and areas under the curve.

3 Click the **left scroll arrow** until the left boundary equals −3.

4 Click the **right scroll arrow** until the right boundary equals 3.

Note that when you move the lower boundary to −3 and the upper boundary to 3, the probability is 99.73%, not 100%. That is because this particular probability distribution extends from minus infinity to plus infinity; the area under the curve (and hence the probability) is never 1 for any finite range.

5 Close the Probability workbook, but you do not have to save any changes.

Random Variables and Random Samples

One of the important concepts in statistics is the idea of the random variable. A **random variable** is a variable whose values occur at random, following a probability distribution. A **discrete random variable** comes from a discrete probability distribution, and a **continuous random variable** comes from a continuous probability distribution. Random variables are usually written with a capital letter, whereas we use a lower-case letter to denote a particular value that the random variable may attain. For example, if Y follows a Poisson distribution, the probability that Y is equal to y is written as follows:

$$P(Y = y) = \frac{\lambda^y}{y!}e^{-\lambda}$$

When the random variable actually attains a value (such as when we actually flip a coin or observe the number of traffic accidents in a year), that value is called an **observation**. A collection of several such observations is called a **sample**. If the observations are generated in a random fashion with no bias, the sample is known as a **random sample**.

In most cases, we want our samples to be random samples to give a true picture of the underlying probability distribution. For example, say we create a study of the weight of American males. We want to know what type of values we would be likely to get if we picked a man at random and weighed him (here weight is our random variable). However, if our sample is biased by selecting only men in their twenties, it would not be a true random sample. Part of the challenge of statistics is to remove all bias from sampling. This is difficult to do, and subtle biases can creep into even the most carefully designed studies.

By observing the distribution of values in a random sample, we can draw some conclusions about the underlying probability distribution. As the sample size increases, the distribution of the values should more closely approximate the probability distribution. To return to our example of the marksman shooting at the target, by observing the spread of shots around the bull's eye, we can estimate the probability distribution of the marksman and his ability.

CONCEPT TUTORIALS:
Random Samples

You can use the instructional workbook Random Samples to explore the relationship between a probability distribution and a random sample.

To use the Random Samples workbook:

1 Open the file **Random Samples**, located in the Explore folder of your Student disk. Enable any macros in the workbook.

2 Move through the sheets in this workbook, viewing the material on random variables and random samples.

3 Click **Explore a Random Sample** from the Table of Contents column.

In this worksheet, you can click the "Shoot" button to generate a random sample of shots at the target. You can select the underlying probability distribution and the number of shots the marksman takes. Try this now with the accuracy of the marksman set to "moderate" to create a sample of 50 random shots.

To generate a random sample of shots:

1   Click the **Shoot** button.

2   Click the **Moderate** button and click the spin arrow to reduce the number of shots to 50. See Figure 5-6.

3   Click the **OK** button.

Excel generates 50 random shots shown in Figure 5-7 (your random sample will be different).

**Figure 5-7
Randomly
generated
sample of
shots**

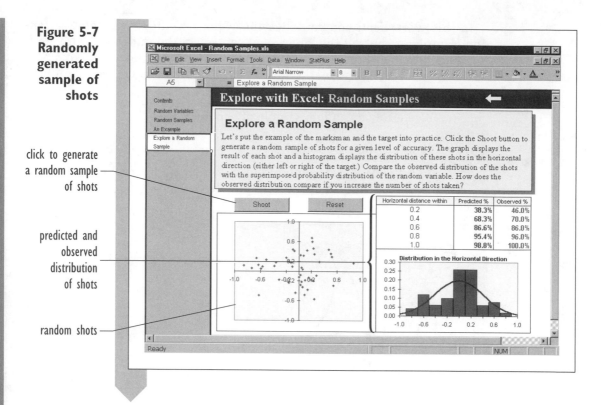

click to generate
a random sample
of shots

predicted and
observed
distribution
of shots

random shots

The *xy*-coordinate system on the target shows the bull's eye, located at the origin (0, 0). The distribution of the shots around the target is described by a **bivariate** density function because it involves two random variables (one for the vertical location and one for the horizontal location of each shot). We'll concentrate on the horizontal distribution of the shots.

Although many of the shots are near the bull's eye, about a third of them are farther than 0.4 horizontal unit away, either to the left or to the right of the target. Because this is random data, your values might be different. Based on the accuracy level you selected, a probability distribution showing the expected distribution of shots to the left or right of the target is also generated in the second column of the table. In this example, the predicted proportion of shots within 0.4 unit of the target is 68.3%, which is close to the observed value of 70%. In other words, the distribution predicts that a marksman of moderate ability is able to hit the bull's eye within 0.4 horizontal unit about 68% of the time. This marksman came pretty close.

You can also examine the distribution of these shots by looking at the histogram of the shots. For the purposes of this worksheet, a shot to the left of the target has a negative value, and a shot to the right of the target has a positive value. The solid curve is the probability density function of shots to the left or right of the target. After 50 shots, the histogram does not follow the probability density function particularly closely. As you increase the number of shots taken, the distribution of the observed shots should approach the predicted distribution.

To increase the number of shots taken:

1 Click the **Shoot** button again.

2 Click the **Moderate** button and click the spin arrow to increase the number of shots to 500.

Figure 5-8 compares the distribution of the random sample after 50 shots and 500 shots. Note that the larger sample size more closely follows the underlying probability distribution.

Figure 5-8 Distribution after 50 and after 500 shots

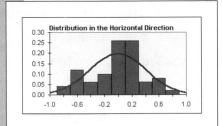

Distribution after 50 shots

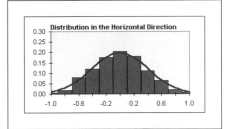

Distribution after 500 shots

3 Try generating some more random samples of various sample sizes. When you are finished, close the Random Samples workbook.

The Normal Distribution

In the Exploring Random Samples workbook, you worked with a distribution in the form of a bell-shaped curve, called the **normal distribution**. This common probability distribution is probably the most important distribution in **statistics**. There are many real world examples of normally distributed data, and normally distributed data is assumed in many statistical tests (for reasons you'll understand shortly). The probability density function for the normal distribution is

Normal Probability Density Function

$$f(y) = \frac{1}{\sigma\sqrt{2\pi}}e^{-(y-\mu)^2/2\sigma^2} \quad \sigma > 0, \; -\infty < \mu < \infty, \; -\infty < y < \infty$$

The expression $f(y)$ is a common way of displaying the probability density function. The value y represents a value on the horizontal axis of the PDF. The normal distribution has two parameters, μ (pronounced "mu") and σ (pronounced "sigma"). The μ parameter indicates the center, or mean, of the distribution. The σ parameter measures the standard deviation, or spread, of the distribution. To see how these parameters affect the distribution's location and shape, you can work with the instructional workbook named Distributions.

CONCEPT TUTORIALS:
The Normal Distribution

To explore the normal distribution:

1 Open the file **Distributions**, located in the Explore folder of your Student disk. Enable the macros that the workbook contains.

2 Click **Normal** from the Table of Contents column.

3 The Normal worksheet opens as shown in Figure 5-9.

**Figure 5-9
The normal
distribution
worksheet**

chart of the
normal probability
density function

click to change
the values of
μ and σ

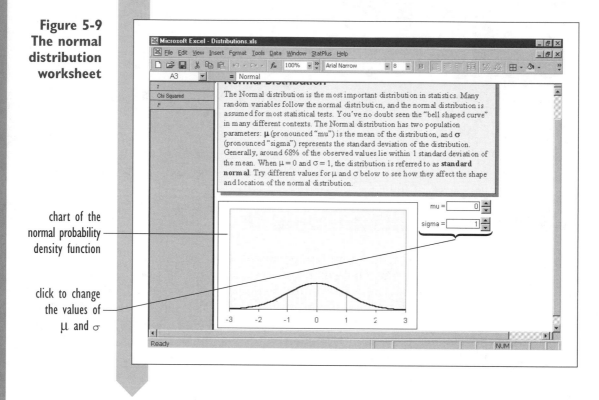

The Normal worksheet opens with μ set to a value of 0 and σ set to a value of 1. A normal distribution with these parameter values is referred to as **standard normal**. Now observe what happens as you alter the values of μ and σ.

To change the values of μ and σ:

1 Click the **up spin button** next to the mu box to change the value of μ to **2**. Note that the distribution shifts to the right as the center of the distribution now lies over the value 2.

2 Click the **down spin button** next to the mu box to change the value of μ back to **0**.

3 Click the **down spin button** next to the sigma box to reduce the value of σ to **0.3**. The distribution tightens around the center.

4 Click the **up spin button** next to the sigma box to increase the value of σ to **1.5**. The distribution spreads out, indicating a wider range of probable values.

Figure 5-10 shows the normal curve for a variety of μ and σ values.

Figure 5-10 The normal distribution for varying values of σ

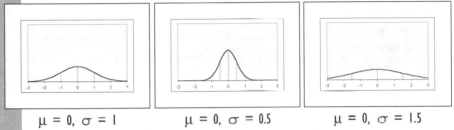

μ = 0, σ = 1   μ = 0, σ = 0.5   μ = 0, σ = 1.5

5 Change the values of μ and σ some more and then close the Distributions workbook without saving any changes.

In the normal distribution, about 68.3% of the values lie within 1 σ, or 1 standard deviation, of the mean, μ. About 95.4% of the values lie within 2 standard deviations of the mean, and more than 99% of the values lie within 3 standard deviations of the mean. See Figure 5-11. Because normally distributed data appears so often in statistical studies, these values become important rules of thumb. For example, if you are trying to calculate a range that will incorporate most of the data, taking the mean ± 2 standard deviations is a fast way of estimating that range.

Figure 5-11
Probabilities under the normal curve

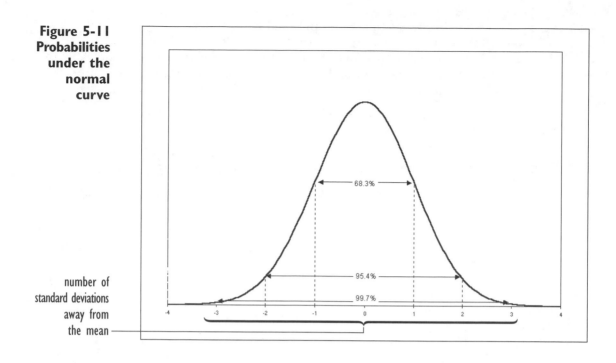

number of standard deviations away from the mean

Excel Worksheet Functions

Excel includes several functions to work with the normal distribution. Table 5-1 describes some of these functions.

Table 5-1 Functions with the Normal Distribution

Function	Description
NORMDIST(y, mean, sd, type)	Uses the normal distribution with $\mu = mean$ and $\sigma = sd$. Setting type = TRUE calculates the probability of $Y \le y$. Setting type = FALSE calculates the value of the probability density function at x.
NORMINV(p, mean, sd)	Returns the value y from the normal distribution for $\mu = mean$ and $\sigma = sd$, such that $P(Y \le y) = p$.
NORMSDIST(y)	Returns the probability of $Y \le y$ for the standard normal distribution.
NORMSINV(p)	Returns the value y from the standard normal distribution such that $P(Y \le y) = p$.
NORMBETW(lower, upper, mean, sd)	Calculates the probability from the normal distribution with $\mu = mean$ and $\sigma = sd$ for the range $lower \le y \le upper$. StatPlus required.

For example, if you want to calculate the probability of a random variable from a normal distribution with $\mu = 50$ and $\sigma = 4$ having a value ≤ 40, the Excel function is NORMDIST(40, 50, 4, TRUE). The value of the PDF at that point is NORMDIST(40, 50, 4, FALSE). On the other hand, if you want to calculate to the left of what point in a normal distribution with $\mu = 50$ and $\sigma = 4$ the total area under the curve (the probability) is equal to 0.90, the function is NORMINV(0.90, 50, 4).

Using Excel to Generate Random Normal Data

Now that you've learned a little about the normal distribution, you can use Excel to generate random samples of normal data. You'll start by creating a single sample of 100 observations coming from a normal distribution with $\mu = 100$ and $\sigma = 25$. To do this, you need to have the StatPlus add-in installed on Excel.

To create 100 random normal values:

1. Open a new blank workbook in Excel, click cell **A1**, and type **Normal Data**.

2. Click **StatPlus > Create Data > Random Numbers**.

   The Random Numbers command presents a dialog box from which you can create random samples from a large variety of distributions. In this case you'll choose the normal distribution.

3. Click **Normal** from the Type of Distribution list box.

4. Type **1** in the Number of Samples to Generate box.

5. Type **100** in the Size of Each Sample box.

6. Type **100** in the Mean box.

7. Type **25** in the Standard Deviation box.

8. Click the **Output** button, click the **Cell** option button, and select cell **A2** as your output destination. Click the **OK** button to close the Output Options dialog box.

   Figure 5-12 shows the completed Create Random Numbers dialog box.

Figure 5-12
The Create
Random
Numbers
dialog box

Figure 5-12 The Create Random Numbers dialog box

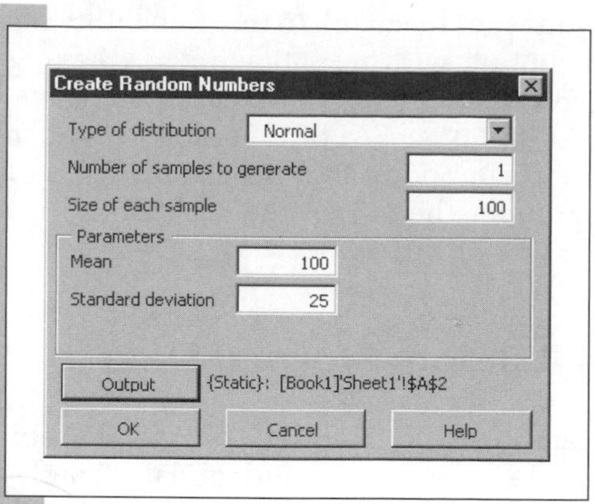

9 Click the **OK** button.

Excel generates a random sample of 100 observations following a normal distribution with mean 100 and standard deviation 25. See Figure 5-13. Because this is randomly generated data, your numbers will look different.

Figure 5-13
One
hundred
random
normal
observations

Figure 5-13 One hundred random normal observations

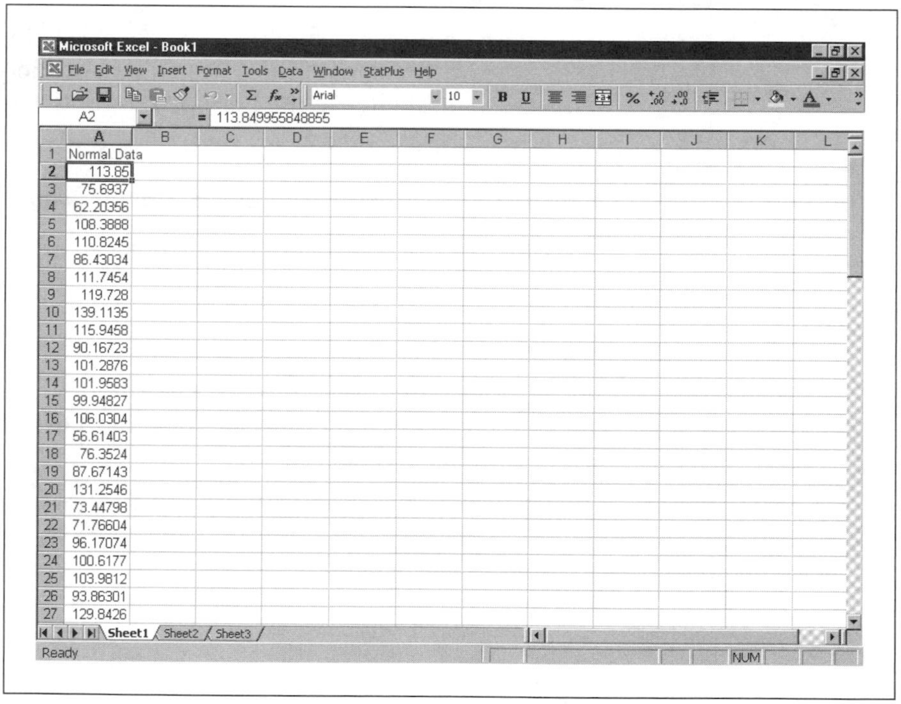

- You can also create random samples using the Analysis ToolPak add-in available with Excel. To create a random sample, load the Analysis ToolPak, click "Tools > Data Analysis" from the Excel menu, and double-click "Random Number Generation" from the Data Analysis dialog box.

- StatPlus adds several new functions to Excel to generate random numbers, including the RANDNORM command to create a random number from a normal distribution.

Charting Random Normal Data

Now that you've created a random sample of normal data, your next task is to create a histogram of the distribution. The StatPlus Histogram command also includes an option to overlay a normal curve on your histogram to compare the distribution of your data with the normal distribution.

To create a histogram of the random sample:

1 Click **StatPlus > Single Variable Charts > Histograms** from the Excel menu.

2 Click the **Data Values** button, click the **Use Range References** option button, and select the range **A1:A101**. Click the **OK** button.

3 Click the **Normal curve** checkbox.

4 Click the **Output** button and type **Normal Data** in the As a New Chart Sheet box to send the chart to a new chart sheet. Click the **OK** button.

A new chart sheet named Normal Data appears. The values on the x-axis show too many decimal places, so you should change this before going on.

5 Double-click one of the values on the x-axis. Click the **Number** tab and reduce the decimal places value from 13 to 3. Click the **OK** button. Figure 5-14 shows the final form of the histogram.

Figure 5-14
Histogram
of 100
random
normal
observa-
tions

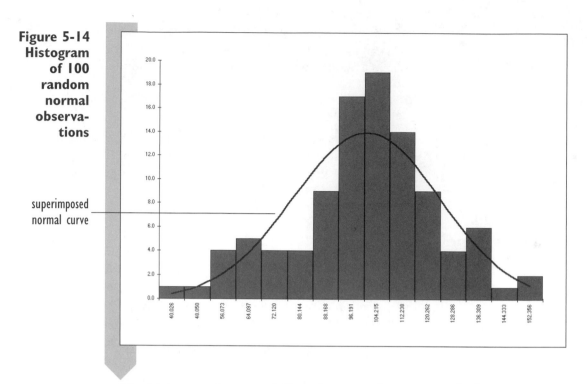

superimposed
normal curve

The histogram does not follow the normal curve exactly, but as you saw earlier, if you increase the size of the random sample, the distribution of the sample values approaches the underlying probability distribution. A sample size of 100 is perhaps still too small. Because you generated these values, you already know that the data is normally distributed, but suppose you observed these values in an experiment or study. Would the chart shown in Figure 5-14 convince you that you're working with normal data? It's not always easy to tell from a histogram whether your data is normal, so statisticians have developed some procedures to check for normality.

The Normal Probability Plot

To check for normality, statisticians compute normal scores for their data. A **normal score** is the value you would expect if your sample came from a standard normal distribution. As an example, for a sample size of 5, here are the five normal scores:

$$-1.163, -0.495, 0, 0.495, 1.163$$

To interpret these numbers, think of generating sample after sample of standard normal data, each sample consisting of five observations. Now, take the average of the smallest value in each sample, the second smallest value, and so forth up to the average of the largest value in each sample. Those averages are the normal scores. Here, we would expect the largest value from a random sample of five standard normal values to be 1.163 and the smallest to be −1.163.

Once you've generated the appropriate normal scores, plot the largest value in your data set against the largest normal score, the second largest value against the second largest normal score, and so forth. This is called a **normal probability plot**. If your data is normally distributed, the points should fall close to a straight line.

StatPlus includes a command to calculate normal scores and create a normal probability plot. Use it now to plot your random sample of normal data.

To create a normal probability plot:

1 Click **StatPlus > Single Variable Charts > Normal P-plots** from the Excel menu.

2 Click the **Data Values** button, click the **Use Range References** option button, and select the range **A1:A101** on your worksheet. Click the **OK** button.

3 Click the **Output** button and type **Normal P-plot** in the As a New Chart Sheet box to send the chart to a new chart sheet. Click the **OK** button.

4 Click the **OK** button to start creating the normal probability plot. Figure 5-15 shows the resulting plot (yours will look slightly different because you've generated a different set of random values).

**Figure 5-15
Normal
probability
plot of 100
random
normal
observations**

normal scores ———

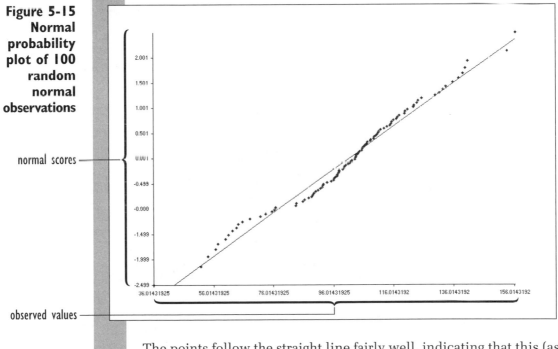

observed values ———

The points follow the straight line fairly well, indicating that this (as you already know) is normally distributed data.

5 Close your workbook. You do not have to save any of the random data or plots you created.

Let's apply this technique to some real data. The Base workbook on your Student disk contains information about baseball player salaries and batting averages. Do the batting averages follow a normal distribution? Let's find out. We'll start by creating a histogram of the batting average data.

To create a histogram of the batting average data:

1 Open the **Base** workbook located on your Student disk.

2 Save the workbook as **Base5** in your Exercises folder.

3 Click **StatPlus > Single Variable Charts > Histograms** from the Excel menu.

4 Click the **Normal Curve** checkbox.

5 Click the **Output** button and send the histogram to a chart sheet named **Batting Average**.

6 Click the **OK** button to start creating the histogram and normal curve. Figure 5-16 shows the resulting chart.

**Figure 5-16
Distribution
of the
batting
average
data**

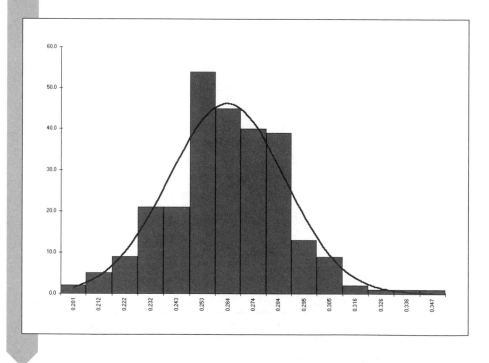

The distribution of the batting average values appears to follow the super-imposed normal curve pretty well (certainly no worse than the sample of random numbers generated earlier). There is no indication that the batting averages do not follow the normal distribution. Let's further check this assumption with a normal probability plot.

To create a normal probability plot of the batting average data:

1 Return to the Salaries worksheet.

2 Click **StatPlus > Single Variable Charts > Normal P-plots** from the Excel menu.

3 Click the **Data Values** button and select **Aver_Career** from the list of range names. Click the **OK** button.

4 Click the **Output** button and send the probability plot to a chart sheet named **Batting Average P-plot**.

5 Click the **OK** button to start creating the normal probability plot. See Figure 5-17.

Figure 5-17 Normal probability plot of the batting average data

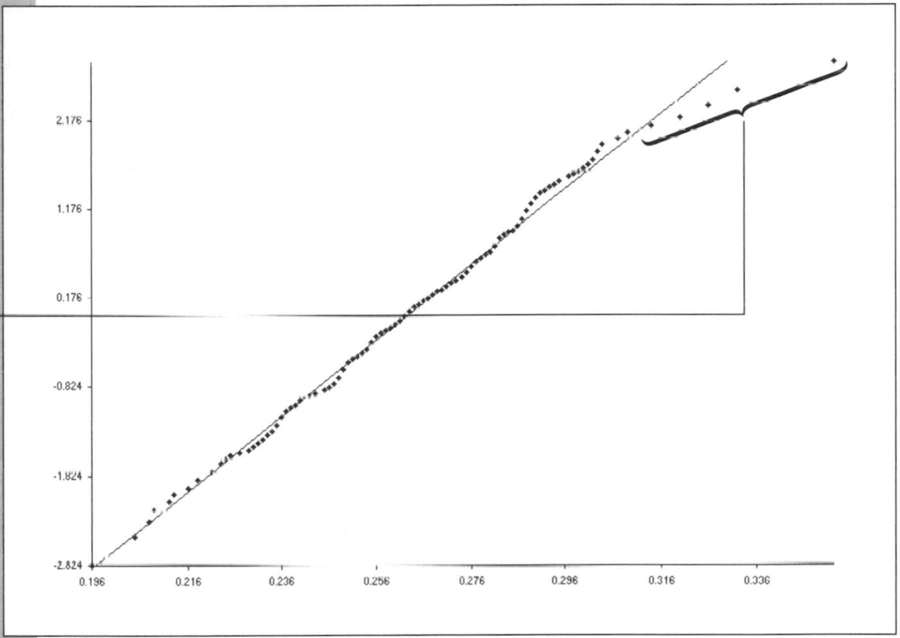

these values are higher than expected assuming normal data

The batting average data follows the straight line on the normal probability plot pretty well. The only serious departures from the line occur for the five largest batting averages. Because the values fall below the line, the expected batting averages are lower than the averages actually observed. To determine what the expected batting average values are, you must take the normal scores and convert them into the scale used by the batting average. This is done by multiplying the normal scores by the standard deviation of the observed values and then adding the sample average. In this case, the average batting average is 0.263, and the standard deviation is 0.0234. If the largest normal score is 2.824, this translates into an expected batting average of 2.824 * 0.0234 + 0.263 = 0.329—a value less than the observed maximum batting average of 0.352. Table 5-2 shows the five largest batting averages, the normal scores, and the expected batting average (assuming a normal distribution). You can determine the normal scores by holding your mouse pointer over individual points on the normal probability plot.

Table 5-2 Batting Average data

Observed Batting Average	Normal Score	Expected Batting Average (if Normal)
0.352	2.824	0.329
0.332	2.502	0.322
0.326	2.327	0.318
0.320	2.204	0.315
0.314	2.107	0.312

If your data is skewed in either the positive or the negative direction, this will be clearly displayed on a normal probability plot. Positively skewed data falls below the straight line on both ends of the plot, whereas negatively skewed data rises above the straight line at both ends of the plot. See Figure 5-18.

Figure 5-18 Normal probability plots for positively and negatively skewed data

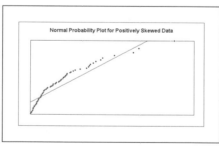
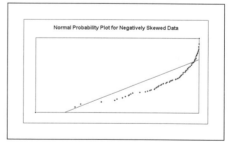

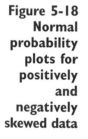

To finish your work with the batting average data:

1 Click **File > Close** to close the Base5 workbook.

2 Click **Yes** when asked whether you want to save the changes.

Parameters and Estimators

Up to now, we've assumed that the parameter values of our probability distributions are known. Most of the time we don't know the values of these parameters for our data, so we have to use the data to estimate them. Statistics called **estimators** provide values for these parameters.

In the normal distribution, we have two parameters: μ and σ. We can estimate the value of μ by calculating the sample average, $\bar{x}$, and the value of σ by calculating the standard deviation, s (see Chapter 4 for a description of these statistics).

The values $\bar{x}$ and s have a special and important property: They are not only estimators of μ and σ but are also **consistent estimators**, which means that as you increase the size of the random sample, the values of $\bar{x}$ and s come closer and closer to the true parameter values. With a large enough sample size, $\bar{x}$ and s will estimate the true values of μ and σ to whatever degree of precision you want. Figure 5-19 shows how the value of $\bar{x}$ approaches the value of μ as the sample size increases.

Figure 5-19 The sample average approaches the true parameter value as the sample size increases

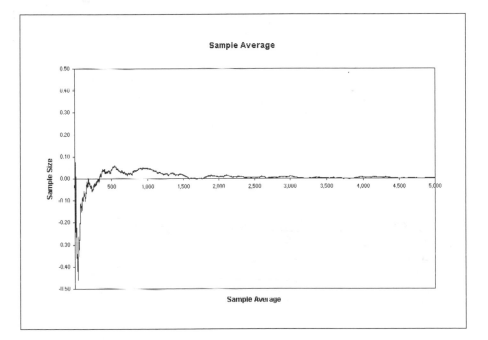

The key question for us is "How large must the sample be to estimate accurately the value of μ?" To answer this question, we'll have to examine the properties of the sample average.

The Sampling Distribution

Because the sample average is a function of random variables, it is itself a random variable with its own probability distribution called a **sampling distribution**. We'll start our investigation of the sample average by creating nine random samples of standard normal data, each sample containing 100 observations. From that random data, we'll create a new column of data containing the average of each of the samples.

To create 100 samples with nine observations each:

1 Open a new blank workbook in Excel.

2 Click **StatPlus > Create Data > Random Numbers** from the Excel menu.

3 Select **Normal** from the Type of Distribution list box.

Now we'll enter the number of samples and the size of each sample. This command places different samples in different columns. To make things easier, we have Excel place the samples in different rows by transposing the sample size and number of samples to generate. This will not affect our conclusions.

4 Enter **9** in the Number of Samples to Generate box.

5 Enter **100** in the Size of Each Sample box.

6 Enter **0** in the Mean box and **1** in the Standard Deviation box.

7 Click the **Output** button and select cell **A1** as the output cell on the current worksheet. Click the **OK** button.

The completed dialog box is shown in Figure 5-20.

Figure 5-20
The completed Create Random Numbers dialog box

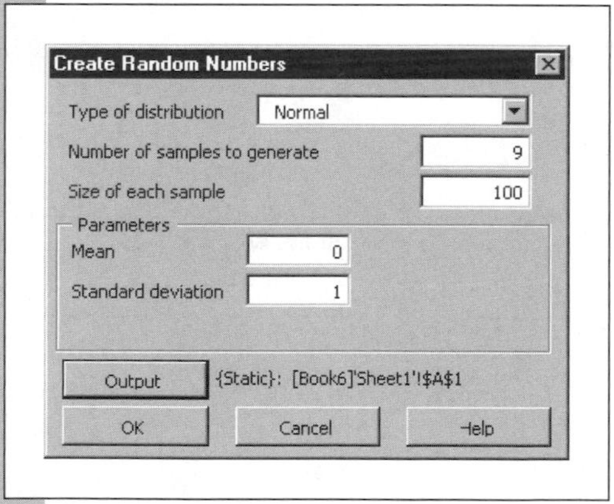

8 Click the **OK** button to start generating the random samples.

Depending on your computer's memory and processor speed, this might take several seconds. When Excel has generated the 900 standard normal values, create a 10th column in your worksheet that calculates the average of the first 9 columns.

To calculate the averages of the 1000 random samples:

1 Click cell **J1**, type the formula **=average(a1:i1)**, and press **Enter**.

2 Click cell **J1** again and drag the fill handle down to cover the range J1:J100.

3 Column J now contains the average of each of the 100 samples on the worksheet.

The column of averages you just created should be much less variable than each of the individual samples, because the average smooths out the highs and the lows. What kind of distribution does it have? Let's investigate by creating a histogram of the sample averages.

To create a histogram of the sample averages:

1 Click **StatPlus > Single Variable Charts > Histograms** from the Excel menu.

2 Click the **Data Values** button, click the **Use Range References** option button, and select the range **J1:J100**. *Deselect* the Range Includes Row of Column Labels checkbox. Click the **OK** button.

3 Click the **Normal Curve** checkbox.

4 Click the **Output** button and save the histogram to a chart sheet named "Sample Average Histogram."

5 Click the **OK** button to start creating the histogram.

6 Double-click one of the values on the *x*-axis. Click the **Number** tab and reduce the Decimal Places value to **3**. Click the **OK** button. Figure 5-21 shows the final form of the histogram (yours will appear different because it comes from a different random sample).

Figure 5-21
Histogram
of simulated
averages

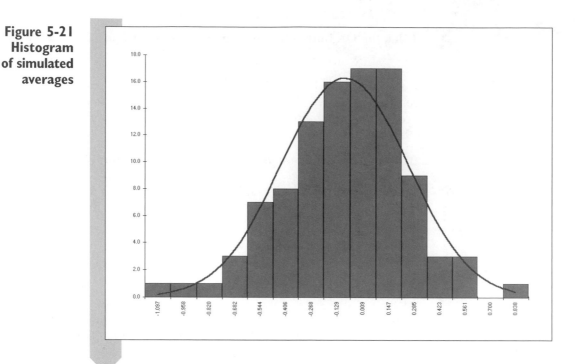

The distribution of the sample averages looks normal and, in fact, is normal. The distribution is centered at 0, as you would expect for averages based on samples taken from the standard normal distribution. The distribution differs from the standard normal in one respect: The sampling distribution is much narrower around the mean. We would expect that most of the standard normal values lie between −2 and 2, but here most of the values lie between −1 and 1. Clearly the standard deviation has decreased. To verify this, calculate descriptive statistics for the sample averages.

To calculate descriptive statistics for the sample averages:

1 Return to the worksheet containing the sample average values.

2 Click **StatPlus > Univariate Statistics** from the Excel menu.

3 Click the **Input** button, click the **Use Range References** option button, and select the range **J1:J100**. *Deselect* the Range Includes a Row of Column Labels checkbox. Click the **OK** button.

4 Click the **Summary** tab and click the **Count** and **Average** checkboxes.

5 Click the **Variability** tab and click the **Std. Deviation** checkbox.

6 Click the **Output** button, and then click the **New Worksheet** option button and type **Sample Average Statistics** for the new worksheet name. Click the **OK** button.

7 Click the **OK** button to generate the sample statistics.

8 Select the range **B4:B5** and reduce the number of decimal places to 3. Figure 5-22 shows the sample output (yours will be slightly different).

**Figure 5-22
Sample
average
statistics**

	A	B	C
1		Univariate Statistics	
2		Variable 1	
3	Count	100	
4	Average	-0.089	
5	Standard Deviation	0.336	
6			

9 Close the workbook with the random samples. You don't have to save your changes.

As you would expect, the average of the sample averages is near zero (in this example, the value is −0.089), close to the value of μ, which is zero. The standard deviation of the sample averages is about $\frac{1}{3}$ (0.336 in Figure 5-22). Thus, a sample size of 9 reduces the standard deviation from 1 to $\frac{1}{3}$. Is there a relationship between sample size and the value for the standard deviation?

CONCEPT TUTORIALS:
Sampling Distributions

You can use the instructional workbook Population Parameter to explore how sample size affects the distribution of the sample average.

To explore sampling distributions:

1 Open the file **Population Parameters**, located in the Explore folder on your Student disk, enabling the macros the workbook contains.

2 The workbook contains information on parameters and sampling distributions. Review the material.

3 Click **Exploring Sampling Distributions** from the Table of Contents.

4 The worksheet displays a histogram of 150 sample averages. A scroll bar allows you to change the size of each sample from 1 up to 16.

5 Move the scroll bar and observe how the shape of the distribution changes based on the different sample sizes. Also note how the value of the standard deviation changes. Figure 5-23 shows the sampling distribution for different sample sizes. Do you see a pattern in the values of the standard deviation compared to the sample size?

Figure 5-23 Sampling distribution of the average for different sample sizes

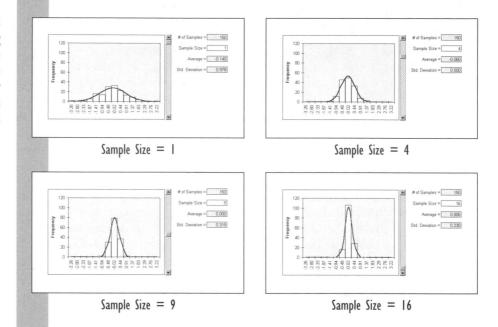

Sample Size = 1   Sample Size = 4

Sample Size = 9   Sample Size = 16

6 Continue working with the Population Parameters workbook and close it when you're finished. You do not have to save your changes.

These values illustrate an important statistical principle. If a sample is composed of n random variables coming from a normal distribution with mean μ and standard deviation σ, then the distribution of the sample average will be a normal distribution with mean μ but with standard deviation $\sigma/\sqrt{n}$. For example, the distribution of a sample average of 16 standard normals is normal with a mean of 0 and a standard deviation of $1/\sqrt{16} = \frac{1}{4}$.

The Standard Error

The standard deviation of $\bar{x}$ is also referred to as the **standard error** of $\bar{x}$. The value of the standard error gives us the information we need to determine the precision of $\bar{x}$ in estimating the value of μ. For example, suppose you have a sample of 100 observations that comes from a standard normal distribution, so that the value of μ is 0 and σ is 1. You've just learned that $\bar{x}$ is distributed normally with a mean of 0 and a standard deviation of 0.1 (because 0.1 = $1/\sqrt{100}$). Let's apply this to what you already know about the normal distribution, namely that about 95% of the values fall within 2 standard deviations of the mean. This means that we can be 95% confident that the value of $\bar{x}$ will be within 0.2 unit of the mean. For example, if $\bar{x}$ = 5.3, we can be 95% confident that the value of μ lies somewhere between 5.1 and 5.5. To be even more precise, we can increase the sample size. If we want $\bar{x}$ to fall within 0.02 of the value of μ 95% of the time, we need a sample of size 10,000 (because $1/\sqrt{10,000} = 0.01$). If $\bar{x}$ is 5.3 with a sample size of 10,000, we can be fairly confident that μ is between 5.28 and 5.32. Note that we can never discover *exactly* the value of μ, but we can with some high degree of confidence narrow the band of possible values to whatever degree of precision we wish.

The Central Limit Theorem

The preceding discussion applied only to the normal distribution. What happens if our data comes from some other probability distribution? Can we say anything about the sampling distribution of the average in that case? We can, by means of the Central Limit Theorem. The **Central Limit Theorem** states that if you have a sample taken from a probability distribution with mean μ and standard deviation σ, the sampling distribution of $\bar{x}$ is approximately normal with a mean of μ and a standard deviation of $\sigma/\sqrt{n}$. The remarkable thing about the Central Limit Theorem is that the sampling distribution of $\bar{x}$ is approximately normal, no matter what the original probability distribution is. As the sample size increases, the approximation to the normal distribution becomes closer and closer. Now you see why the normal distribution is so important in the field of statistics.

To see the effect of the Central Limit Theorem, you can use the instructional workbook named The Central Limit Theorem.

CONCEPT TUTORIALS:
The Central Limit Theorem

To use the Central Limit Theorem workbook:

1   Open the **Central Limit Theorem** file in the Explore folder of your Student disk. Enable the macros in the workbook.

2   Review, in the workbook, the concepts behind the Central Limit Theorem.

3   Click **Explore the Central Limit Theorem** from the Table of Contents.

The Central Limit Theorem worksheet opens. See Figure 5-24.

Figure 5-24
Central
Limit
Theorem
worksheet

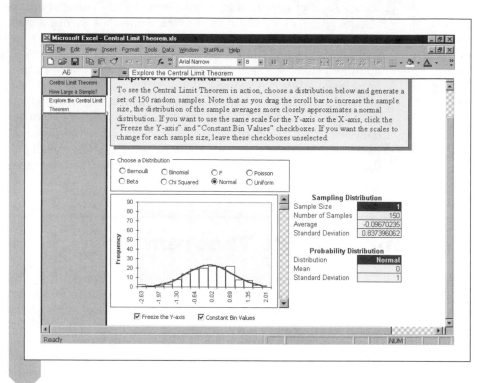

The worksheet lets you generate 150 random samples from one of eight common probability distributions with up to 16 observations per sample. The worksheet also calculates and displays the distribution of the sample averages. You can change the sample size by dragging the scroll bar up or down. The worksheet opens, displaying the distribution of the sample average for the standard normal distribution. To see how the worksheet works, move the scroll bar, increasing the sample size from 1 to 16.

To change the sample size:

1 Drag the scroll bar down. The sample size increases from 1 to 16.

As you drag the scroll bar down, the histogram displays the change in the sample size, and the standard deviation decreases from 1 to about 0.25.

2 Drag the scroll bar back up to return the sample size to 1.

Note that if you want to view the histogram with different bin values under different sample sizes, you can deselect the Constant Bin Values checkbox. When the bin values are the same, you can compare the spread of the data from one histogram to another because the same bin values are used for all charts. Deselecting the Constant Bin Values checkbox fits the bin values to the data and gives you more detail, but it's more difficult to compare histograms from different sample sizes. You can also "freeze" the y-axis to retain the y-axis scale from one sample size to another, making it easier to compare one chart with another. To scale the y-axis to the data, unselect the Freeze the Y-Axis checkbox.

Now that you've viewed the sampling distribution for the standard normal, you'll choose another distribution from the list. Another commonly used probability distribution is the uniform distribution. In the **uniform distribution**, probability values are uniform across a continuous range of values. The probability density function for the uniform distribution is

Uniform Probability Density Function

$$f(x) = \frac{1}{b - a}, a < b$$

where b is the upper boundary and a is the lower boundary of the distribution. Figure 5-25 displays the Uniform distribution where $a = -2$ and $b = 2$.

Figure 5-25
The uniform probability distribution

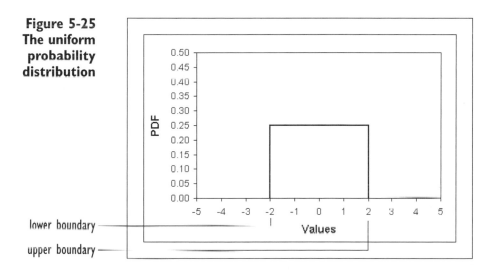

The mean, μ, and standard deviation, σ, for the uniform distribution are

$$\mu = \frac{b + a}{2} \qquad \sigma = \frac{b - a}{\sqrt{12}}$$

Thus if $a = -2$ and $b = 2$, $\mu = 0$ and $\sigma = 4/\sqrt{12} = 1.1547$.

Having learned something about the uniform distribution, let's observe the sampling distribution of the sample average.

To generate the sampling distribution of the uniform distribution:

1 Click the **Uniform** option button on the Central Limit Theorem worksheet.

2 Enter **–2** in the Minimum box and **2** in the Maximum box. See Figure 5-26.

Figure 5-26
Setting the
parameter
values for
the uniform
distribution

Uniform Distribution Parameters	
Minimum	-2
Maximum	2
OK	Cancel

3 Click the **OK** button.

Excel generates 150 random samples for the uniform distribution. The initial sample size is 1, which is equivalent to generating 150 different observations from the uniform distribution. The initial average values should be close to 0 and 1.15.

4 Drag the scrollbar down to increase the sample size from 1 to 16. Figure 5-27 shows the sampling distribution for the average under different sample sizes (your charts and values will be slightly different). You may want to unfreeze and freeze the *y*-axis in order to display the histograms in a more detailed scale.

Figure 5-27
Sampling
distribution
of the
average of
uniform
distribu-
tions for
different
sample sizes

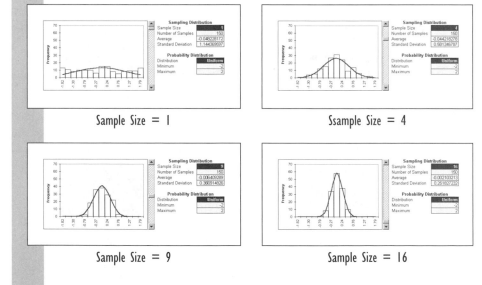

Sample Size = 1   Ssample Size = 4

Sample Size = 9   Sample Size = 16

5 Try some of the other distributions in the list under various sample sizes. When you're finished working with the workbook, close it. You do not have to save your changes.

A few points about the Central Limit Theorem should be considered. First, the theorem applies only to probability distributions that have a finite mean and standard deviation. Second, the sample size and the properties of the original distribution govern the degree to which the sampling distribution approximates the normal distribution. For large sample sizes, the approximation can be very good, whereas for smaller samples, the approximation might not be good at all. If the probability distribution is very skewed, the larger sample size might be necessary. If the distribution is symmetric, the sample size need not be very large. How large is large? If the original distribution is symmetric and already close in shape to a normal distribution, a sample size of 15 or 20 should be large enough. For a highly skewed distribution, a sample size of 40 or 50 might be required. Usually, the Central Limit Theorem can be safely applied if the sample size is 30 or more.

The Central Limit Theorem is probably the most important theorem in statistics. With this theorem, statisticians can make reasonable inferences about the sample mean without having to know about the underlying probability distribution. You'll learn how to make some of these inferences in the next chapter.

Exercises

1. Explore the following statistical concepts.
   a. Define the term *random variable*.
   b. How is a random variable different from an observation?
   c. What is the distinction between $\bar{x}$ and μ?

2. A sample of the top 50 women-owned businesses in Wisconsin is undertaken. Does this constitute a random sample? Explain your reasoning. Can you make any inferences about women-owned businesses on the basis of this sample?

3. The administration counts the number of low-birth-weight babies born each week in a particular hospital. Assume, for the sake of simplicity, that the rate of low birth-weight births is constant from week to week.
   a. What is the probability distribution of this value?
   b. If the average number of low-birth-weight babies is 5, what is the probability that no low-birth-weight babies will be born in a single week?

c. The administration counts the low-birth-weight babies every week and then calculates the average count for the entire year. What is the approximate distribution of the average?

4. A flipping coin follows a probability distribution called the **Bernoulli distribution**. There are two outcomes for the Bernoulli, 0 or 1. The probability function for the Bernoulli is simply

> Bernoulli Distribution
>
> $P(Y = 1) = p$, $P(Y = 0) = 1 - p$ $0 \leq p \leq 1$

The mean value in the Bernoulli distribution is p and the standard deviation is $\sqrt{p(1 - p)}$. In the example of flipping a fair coin, the value of p is 0.5.

a. If you toss a coin, recording 1 for every head and 0 for every tail, what is the average value you would expect to toss? What is the standard deviation of those values?

b. If you toss the coin 100 times, what is the expected average value? What is the approximate standard error?

c. A friend hands you a coin. You toss it 100 times and obtain 35 heads. Is this evidence that the coin is not a fair coin? Explain why or why not. *Hint:* use the Central Limit Theorem.

5. The **Binomial distribution** is used for performing multiple trials of a Bernoulli random variable. For example, instead of flipping a single coin, the Binomial distribution predicts the probable outcome of flipping multiple coins. The probability function for the Binomial distribution is

Binomial Distribution

$$P(Y = y) = \binom{n}{y} p^y (1 - p)^{n-y} \quad y = 0, 1, 2, \ldots, n$$

$$\text{where } \binom{n}{y} = \frac{n!}{y!(n - y)!}$$

where p is the probability of a head or tail and n is the number of replications. The mean of the Binomial distribution is np. The standard deviation is $\sqrt{np(1 - p)}$. For example, if 100 coins are flipped, with $p = 0.5$, then the expected number of heads would be $100 \times 0.5 = 50$. The standard deviation is $\sqrt{0.5 \cdot 0.5 \cdot 100} = 5$.

a. If you toss a fair coin 25 times, what is the expected number of heads? What is the standard deviation?

b. You toss a hundred coins 25 times. The average number of heads from those samples is 45. Is this evidence that the coins are biased? Explain why or why not. *Hint:* use the Central Limit Theorem.

6. The mean of a Poisson distribution is λ, and the standard deviation is $\sqrt{\lambda}$.

a. The number of accidents at a factory per year follows a Poisson distribution with a mean value of 10. If you calculate the average number of accidents over the past 25 years at the factory, what would be the value of the standard error of the average?

b. If you didn't know the value of λ and, after examining 25 years' worth of data, you calculated a sample average of 10, within what range of values would you be 95% confident that the true value of λ lies? For what range of values would you be 99% confident that it contains the true value of λ? *Hint:* use the Central Limit Theorem.

7. Excel includes the function NORMSDIST(Y), which calculates the probability of a standard normal random variable having a value $\leq y$.

a. Use this function to find the probability of a standard normal random variable having a value of 0.5 or less.

b. What is the probability of a standard normal random variable having a value exactly equal to 0.5?

c. The NORMDIST Excel function finds the probability for random variables coming from any normal distribution. Using this function, find the probability of a random variable coming from a normal distribution with $\mu = 10$ and $\sigma = 2$, having a value less than 5.

8. Open the BASE.XLS workbook. The mean career batting average is 0.263 and the standard deviation is 0.02338.

a. Assuming that the batting averages are normally distributed, use Excel's NORMDIST function to find the probability that a player will bat 0.300 or better (that is, 1 − probability that a player will bat less than 0.300).

b. How many players actually batted 0.300 or better? Compare this to the expected number.

c. Create a histogram with a normal curve of the baseball salary data.

d. Create a normal probability plot of the salary data. Does the data appear to be normally distributed?

e. Calculate the expected maximum salary based on the normal scores (you can find the maximum normal score by holding the mouse pointer over the highest point on the normal probability plot). How does this compare to the observed maximum salary?

f. Is the distribution of the salary data positively skewed, negatively skewed, or symmetric?

g. Save your work as E5BASE.XLS.

9. The HOMEDATA.XLS workbook contains a sample of 117 housing prices for Albuquerque, New Mexico.

a. Open the workbook and create a histogram (with a normal curve) and a normal probability plot of the housing prices. Does the data appear to follow a normal distribution?

b. Calculate the average housing price and the standard deviation of the housing price. Because this is a sample of all of the house prices in Albuquerque, the average serves as an estimate of the mean house price. Use the values you calculate to determine a range of values that you are 95% confident contains the true mean value.

c. Create a new column containing the log of the home price values. Create a histogram with a normal curve and a normal probability plot. Do the transformed values appear more normally distributed than the untransformed values?

d. Save your workbook as E5HOMEDATA.XLS.

10. The dispersion of shots used in shooting at the target in the Random Samples workbook follows a bivariate normal distribution (a combination of two normal distributions, one in the vertical direction and one in the horizontal direction). The value of σ for each level of accuracy is

Accuracy	Standard Deviation
Highest	0.1
Good	0.2
Moderate	0.4
Poor	0.6
Lowest	1.0

Open the Random Samples workbook and create a distribution of shots around the target with good accuracy.

a. Explain why the predicted percentages have the values they have.

b. For a marksman with the lowest accuracy, how many shots would the marksman have to take before he could assume with 95% confidence that the average horizontal location of his shots was within 0.2 unit of the bull's eye?

c. How many shots would a marksman with the highest accuracy have to take before achieving similar confidence in the average of his shots?

11. Open a blank workbook and, using the Create Random Data command from StatPlus, create 9 columns of 100 rows of Poisson random values with $\lambda = 0.25$.

a. Create a tenth column in your workbook containing the average values from the first 9 columns.

b. Create a histogram of the first column of random Poisson values with a superimposed normal curve. Create a second histogram of the column averages with a superimposed normal curve. Compare the two curves.

c. Calculate the average and standard deviation for the first and tenth columns. How do these calculated values compare to the value of λ? See Exercise 6 for more information on the Poisson distribution.

d. Save your workbook as E5POISSON.XLS. Close the workbook.

12. Repeat Exercise 11 with 9 columns of 100 rows of binomial random values where the number of trials is 16 and the value of p is 0.25. Save your workbook as E5BINOM.XLS.

13. Repeat Exercise 11 with 9 columns of 100 rows of Bernoulli random values where $p = 0.25$. Save your workbook as E5BERN.XLS.

14. Repeat Exercise 11 with 9 columns of 100 rows of uniform random values where the lower boundary is 0 and the upper boundary is 100. Save the workbook as E5UNI.XLS.

15. *True or false*: According to the Central Limit Theorem, as the size of the sample increases, the distribution of the observations approximates a normal distribution. Defend your answer.

16. You want to collect a sample of values from a uniform distribution where μ is unknown but $\sigma = 10$.
    a. How large a sample would you need to estimate the value of μ within 2 units with a confidence of 95%?
    b. How large a sample would you need to estimate the value of μ within 2 units with a confidence of 99%?
    c. If the sample size is 25 and μ is 50, what is the probability that the sample average will have a value of 48 or less?

17. At the 1996 Summer Olympic games in Atlanta, Linford Christie of Great Britain was disqualified from the finals of the men's 100-meter race because of a false start. Christie did not react before the starting gun sounded, but he did react in less than 0.1 second. According to the rules, anyone who reacts in less than a tenth of a second *must* have false-started by anticipating the race's start. Christie bitterly protested the ruling, claiming that he had just reacted very quickly. Using the reaction times from the first heat of the men's 100-meter race, try to weigh the merits of Christie's claim versus the argument of the race officials that no one can react as fast as Christie did without anticipating the starting gun.
    a. Open the RACE.XLS workbook and create a histogram of the reaction times. Where would a value of 0.1 second fall on the chart?
    b. Calculate the mean and standard deviation of the first heat's reaction times. Use these values in Excel's NORMDIST function and calculate the probability that a value of 0.1 or less could be observed.
    c. Create a normal probability plot of the reaction times. Do the data appear to follow the normal distribution?
    d. Save your workbook as E5RACE.XLS.
    e. Write a summary of your findings. Include in your summary a discussion of the difficulties in determining whether Christie anticipated the starter's gun. Are the data appropriate for this type of analysis? What are some limitations of the data? What kind of data would give you better information regarding a runner's reaction times to the starter's gun (specifically, runners taking part in the finals of an Olympic event)?

Chapter 6

Statistical Inference

Objectives

In this chapter you will learn to:

- Create confidence intervals

- Apply a hypothesis test

- Use the *t*-distribution in a hypothesis test

- Perform a one-sample and a two-sample *t*-test

- Analyze data using nonparametric approaches

T he concepts you learned in Chapter 5 provide the basis for the subject of this chapter, statistical inference. Two of the main tools of statistical inference are confidence intervals and hypothesis tests. In this chapter, you'll apply these tools to reach conclusions about your data. You'll be introduced to a new distribution, the t-distribution, and you'll see how to use it in performing statistical inference. You'll also learn about nonparametric tests that make fewer assumptions about the distribution of your data.

Confidence Intervals

In the previous chapter, you learned two very important facts about distributions and samples:

1. A sample average will approximately follow a normal distribution with mean μ and standard deviation $\sigma/\sqrt{n}$, where μ is the mean of the probability distribution the sample is drawn from, σ is the standard deviation of the probability distribution, and n is the size of the sample. Another way of writing this is

$$x \sim N\left(\mu, \sigma/\sqrt{n}\right)$$

2. In a normal distribution, about 95% of the time, the values fall within 2 standard deviations of the mean.

From these two facts, we can reach general conclusions about how precisely the sample average estimates the value of μ. For example, if σ = 10 and our sample size is 25, the sample average will approximately follow a normal distribution with mean μ and standard deviation 2, so 95% of the time, the sample average will fall within 4 units of μ. This means that if the sample average is 20, we could construct a **confidence interval** from about 16 to 24 that should, with 95% confidence, "capture" the value of μ. If we want this confidence interval to be smaller, we simply increase the sample size. A sample of 100 observations would result in a 95% confidence interval for μ ranging from about 18 to 22.

The use of the "2 standard deviations" rule is an approximation. What if we wanted a more exact estimate of the 95% confidence interval, or what if we wanted to construct other confidence intervals, such as a 99% confidence interval? How would we go about doing that?

z-Test Statistic and z-Values

In order to derive a more general expression of the confidence interval, we first have to express the sample average in terms of a standard normal distribution. We can do this by subtracting the value of μ and dividing by the standard error.

This value will then follow a standard normal distribution; that is,

$$\frac{\bar{x} - \mu}{\sigma/\sqrt{n}} \sim N(0,1)$$

This value is called a **z-test statistic**. We then need to compare the z-test statistic to z-values. A **z-value**, usually written as z_p, is the point z on a standard normal curve such that for random variable Z, $P(Z \le z_p) = p$. For example, $z_{0.95} = 1.645$ because 95% of the values on a standard normal curve are less than 1.645. See Figure 6-1.

Figure 6-1
The z-value

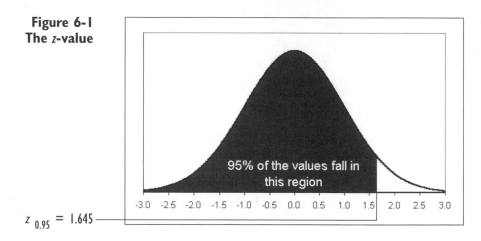

$z_{0.95} = 1.645$

Figure 6-1 shows a one-sided z-value, but for confidence intervals, we're more interested in a two-sided z-value, where p is the probability of the value falling in the center of the distribution, and α (which equals $1 - p$) is the probability of its falling in one of the two tails. For a two-sided range of size p, these z-values are $-z_{1-\alpha/2}$ and $z_{1-\alpha/2}$. In other words, for a random variable Z, $P(-z_{1-\alpha/2} < Z < z_{1-\alpha/2}) = 1 - \alpha = p$. If we want to find the central 95% of the standard normal curve, $p = 0.95$, $\alpha = 0.05$, and $z_{1-0.05/2} = z_{0.975} = 1.96$. This means that 95% of the values on a standard normal curve lie between -1.96 and 1.96. See Figure 6-2.

**Figure 6-2
Two sided
z-values**

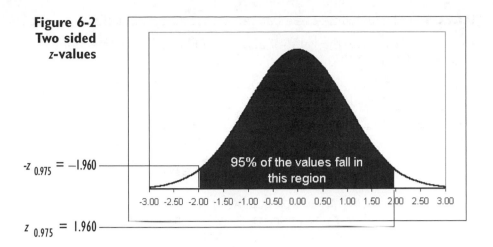

We can use a two-sided z-value to construct a general expression for the confidence interval. The more general expression is:

$$P\left(\bar{x} - z_{1-a/2}\frac{\sigma}{\sqrt{n}} < \mu < \bar{x} + z_{1-a/2}\frac{\sigma}{\sqrt{n}}\right) = 1 - a$$

Thus the upper and lower confidence limits for μ are $x \pm z_{1-a/2}\sigma/\sqrt{n}$. For example, if $\alpha = 0.05$, then $z_{1-0.05/2} = 1.96$ and the 95% confidence limits are $\bar{x} + 1.96 \times \sigma/\sqrt{n}$, which is pretty close to our rule-of-thumb estimate of ± 2 standard errors from the sample average. Table 6-1 shows confidence intervals of various sizes using this approach.

Table 6-1 Confidence Intervals

$1 - \alpha$	$z_{1-\alpha/2}$	**Confidence Band**
0.800	1.282	$\bar{x} \pm 1.282 \times \sigma/\sqrt{n}$
0.900	1.645	$\bar{x} \pm 1.645 \times \sigma/\sqrt{n}$
0.950	1.960	$\bar{x} \pm 1.960 \times \sigma/\sqrt{n}$
0.990	2.576	$\bar{x} \pm 2.576 \times \sigma/\sqrt{n}$
0.999	3.290	$\bar{x} \pm 3.290 \times \sigma/\sqrt{n}$

For example, if you want to construct a confidence interval around the sample average that will capture the value of μ 99.9% of the time, calculate the sample average ± 3.3 times the standard error. Admittedly, this will tend to be a very large interval.

- To calculate the value of $z_{1-\alpha/2}$ with Excel, use the function NORMSINV(x), where $x = 1 - \alpha/2$.

- To find the probability associated with a z-test statistic, use the function NORMSDIST(z), where z is the z-test statistic.

Calculating the Confidence Interval with Excel

You can use Excel's functions to calculate a confidence interval if you know the standard deviation of the underlying probability distribution. For example, suppose you are conducting a survey on the cost of a medical procedure as part of research on health care reform. The cost of the procedure follows the normal distribution, where $\sigma = 1000$. After sampling 50 different hospitals at random, you calculate the average cost to be $5,500. What is the 90% confidence interval for the value of μ — the mean cost of all hospitals? (That is, how far above and below $5,500 must you go to say, "I'm 90% confident that the mean cost of this procedure lies in this range"?)

To calculate the 90% confidence interval:

1 Start Excel and open a blank workbook.

2 Type **Average** in cell A1, **Std. Error** in cell B1, **Lower** in cell C1, and **Upper** in D1.

3 Click cell **A2** and type **5500** (the observed sample average.)

4 Type **=1000/sqrt(50)** in cell B2. This is the standard error of the sample average.

5 Type **=A2-B2*NORMSINV(0.95)** in cell C2. Because we are interested in the 90% confidence interval, $1 - \alpha/2 = 1 - 0.10/2 = 0.95$, and we use the NORMSINV(0.95) function.

6 Type **=A2+B2*NORMSINV(0.95) in cell D2**. Figure 6-3 shows the resulting 90% confidence interval.

**Figure 6-3
Using Excel
to calculate
a confidence
interval**

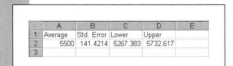

7 Close your workbook. You do not have to save the changes.

Excel returns a 90% confidence band ranging from $5,267.38 to $5,732.62. If you were trying to estimate the mean cost of this procedure for all hospitals, you could state that you were 90% confident that the cost was not less than $5,267.38 or more than $5,732.62.

Interpreting the Confidence Interval

It's important that you understand what is meant by statistical confidence. When you calculated the confidence interval for the cost of the hospital procedure, you were *not* stating that the probability of the mean cost falling between $5,267.38 and $5,732.62 was 0.90. That would incorrectly imply that the range you calculated and the mean cost are random variables. They're not. After drawing a specific sample and from that sample calculating the confidence interval, we're no longer working with random variables but with actual observations. The mean cost is also some fixed (but unknown) number and is not random. The term *confidence* refers to our confidence in the procedure we used to calculate the range. The term *90% confident* means that we are confident that our procedure will capture the value of μ 90% of the times it is used.

CONCEPT TUTORIALS:
The Confidence Interval

To get a visual picture of the confidence interval in action, you can use the Explore workbook named Confidence Intervals to read about, and work with, a confidence interval.

To use the Confidence Intervals workbook:

1 Open the file **Confidence Intervals**, located in the Explore folder of your Student disk. Enable any macros in the workbook.

2 Move through the sheets in this workbook, viewing the material on z-values and confidence intervals.

3 Click **Explore the Confidence Interval** from the Table of Contents column. See Figure 6-4.

**Figure 6-4
The
Confidence
Intervals
workbook**

true mean value

confidence
intervals that don't
capture the true
mean

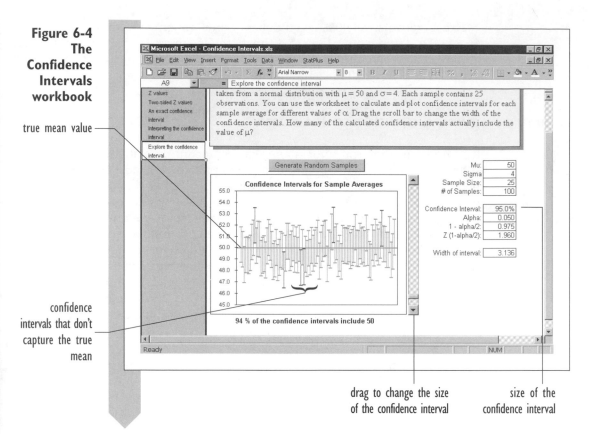

drag to change the size
of the confidence interval

size of the
confidence interval

The worksheet in Figure 6-4 shows 100 simulated confidence intervals taken from a normal distribution with $\mu = 50$ and $\sigma = 4$. Each of the 100 samples contains 25 observations, so the standard error of each sample average is 0.8.

If a confidence interval captures the true value of μ, it shows up on the chart as a vertical green line. If a confidence interval does not include μ, it shows up as a vertical red line. You would expect that 95 of the 100 samples would include the value of μ in their confidence intervals. Because there is some random variation, Figure 6-4 shows that only 94% of the sample confidence intervals include the value of μ. Using this worksheet, you can generate a new random sample or change the width of the confidence band. Try this now by reducing the confidence interval from 95% to 75%.

To reduce the size of the confidence interval:

Drag the vertical scroll bar up until the value **75.0%** appears in the highlighted Confidence Interval box. See Figure 6-5.

**Figure 6-5
75%
confidence
intervals**

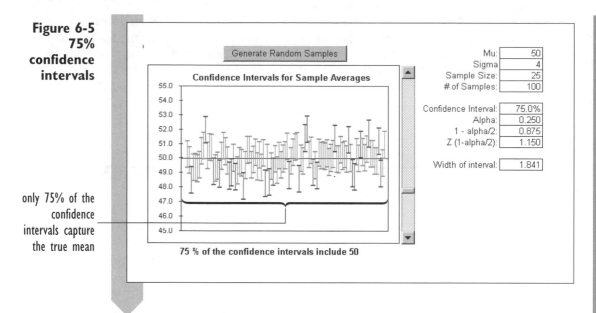

only 75% of the
confidence
intervals capture
the true mean

By reducing the confidence interval to 75%, you've reduced the width of the confidence band, but at the cost of many more red lines appearing on your chart. If you were relying upon this confidence interval to capture the value of μ, you would run a great risk of making an error. Let's go the other way and increase the confidence interval.

To increase the size of the confidence interval:

Drag the vertical scroll bar down until the value **99.0%** appears in the Confidence Interval box.

All of the confidence intervals now capture the value of μ, but the size of the confidence bands has greatly increased. As you can see, there is a trade-off in using the confidence interval. Selecting too small a value could result in missing the value of μ. Selecting a larger value will almost certainly capture μ, but at the expense of having a range of values too broad to be useful. Statisticians have generally favored the 95% confidence interval as a compromise between these two positions.

An important lesson to learn from this simulation is to not take the sample average at face value. Confidence intervals help you quantify how precise the sample average estimates the value of μ. The next time you hear in the news that a study has shown that a drug causes a mean decrease in blood pressure, or that polls predict a certain election result, you should ask, "And what is the confidence interval?"

If you want to generate a new set of random samples, you can click the Generate Random Samples button in the Confidence Intervals workbook. Continue exploring the workbook until you understand the relationship among the confidence interval, the sample average, and the value of μ. Close the workbook when you're finished. You do not have to save your changes.

Hypothesis Testing

Confidence intervals are one way of performing statistical inference; another way is hypothesis testing. In a **hypothesis test**, you formulate a theory about the phenomenon you're studying and examine whether that explanation is supported by the statistical evidence. In statistics, we formulate a hypothesis first, then collect data, and then perform a statistical test. The order is important. If we formulate our hypothesis *after* collecting the data, we run the risk of having a biased test, because our hypothesis might be designed to fit the data. To guard against a biased test, the hypothesis should be tested on a new set of data. Figure 6-6 displays a classical approach to developing and testing a theory.

Figure 6-6
Testing a theory

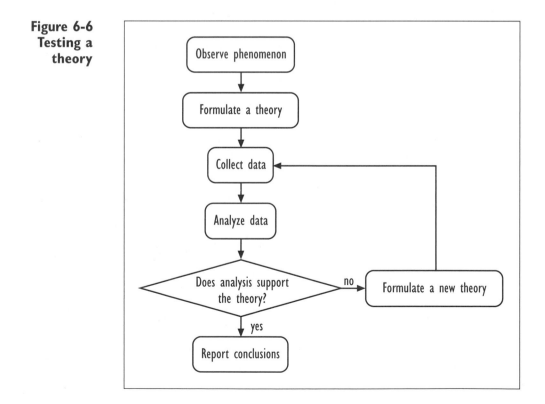

There are four elements in a hypothesis test:

1. A null hypothesis, H_0
2. An alternate hypothesis, H_a
3. A test statistic
4. A rejection region

The **null hypothesis**, usually labeled $\mathbf{H_0}$, represents the default or status quo theory about the phenomenon that you're studying. You accept the null hypothesis as true unless you have convincing evidence to the contrary. The **alternative hypothesis**, or $\mathbf{H_a}$, represents an alternate theory that is automatically accepted as true if the null hypothesis is rejected. Often the alternative hypothesis is the hypothesis you want to accept. For example, a new medication is being studied that claims to reduce blood pressure. The null hypothesis is that the medication *does not* affect the patient's blood pressure. The alternative hypothesis is that the medication *does* affect the patient's blood pressure (in either a positive or a negative direction.)

The **test statistic** is a statistic calculated from the data that you use to decide whether to reject or accept the null hypothesis. The **rejection region** specifies the set of values of the test statistic under which you'll reject the null hypothesis (and accept the alternative.)

Types of Error

We can never be sure that our conclusions are free from error, but we can try to reduce the probability of error. In hypothesis testing, we can make two types of errors:

1. **Type I error**: Rejecting the null hypothesis when the null hypothesis is actually true
2. **Type II error**: Failing to reject the null hypothesis when the alternative hypothesis is actually true

The probability of Type I error is denoted by the Greek letter α, and the probability of Type II error is identified by the Greek letter β.

Generally, statisticians are more concerned with the probability of Type I error, because rejecting the null hypothesis often results in some fundamental change in the status quo. In the blood pressure medication example, incorrectly accepting the alternative hypothesis could result in prescribing an ineffective drug to thousands of people. Statisticians will set a limit, called the **significance level**, that is the highest probability of Type I error allowed. An accepted value for the significance level is 0.05, which means that if the probability of Type I error is higher than 0.05, the null hypothesis will not be rejected.

Reducing Type II error becomes important in the design of experiments, where the statistician wants to ensure that the study will detect an effect if a true difference exists. An analysis of the probability of Type II error can aid the statistician in determining how many subjects to have in the study.

An Example of Hypothesis Testing

Let's put these abstract ideas into a concrete example. You work at a plant that manufactures resistors. Previous studies have shown that the number of defective resistors in a batch follows a normal distribution with a mean of 50 and a standard deviation of 15. A new process has been proposed that will reduce the number of defective resistors, saving the plant money. You put the process in place and create a sample of 25 batches. The average number of defects in a batch is 45. Does this prove that the new process reduces the number of defective resistors, or is 45 simply a random aberration, and the process does not make any difference at all?

Here are our hypotheses:

H_0: There is no change in the mean number of defective resistors under the new process.

H_a: The mean number of defective resistors *has* changed.

or, equivalently,

H_0: The mean number of defective resistors in the new process is 50.

H_a: The mean number of defective resistors is *not* 50.

Acceptance and Rejection Regions

To decide between these two hypotheses, we'll assume that the null hypothesis is true. Let μ_0 be the mean under the null hypothesis. This means that under the null hypothesis,

$$P\left(-z_{1-a/2} < \frac{\bar{x} - \mu_0}{\sigma/\sqrt{n}} < z_{1-a/2}\right) = 1 - a$$

By multiplying by the standard error and adding μ_0 to each term in the inequality, we get

$$P\left(\mu_0 - z_{1-a/2}\frac{\sigma}{\sqrt{n}} < \bar{x} < \mu_0 + z_{1-a/2}\frac{\sigma}{\sqrt{n}}\right) = 1 - a$$

This means that the sample average should be in the range $\mu_0 \pm z_{1-\alpha/2} \sigma / \sqrt{n}$ with probability $1 - \alpha$, *if the null hypothesis is true*. Now let α be our significance level, so that if the sample average lies outside this range, we'll reject the null hypothesis and accept the alternative. These outside values would constitute the rejection region, mentioned earlier. The values within the range constitute the **acceptance region**, under which we'll accept the null hypothesis. The upper and lower boundaries of the acceptance region are known as **critical values**, because they are critical in deciding whether to accept or reject the null hypothesis.

Let's apply this formula to our example: $\mu_0 = 50$, $\sigma = 15$, $n = 25$, and we'll set $\alpha = 0.05$ so that the probability of Type I error is 5%. The acceptance region is therefore

$$= 50 \pm 1.96 \times 15 / \sqrt{25}$$

$$= 50 \pm 5.88$$

$$= (44.12, 55.88)$$

Any value that is less than 44.12 or greater than 55.88 will cause us to reject the null hypothesis. Because 45 falls in the acceptance region, we accept the null hypothesis and do not conclude that the new process decreases the number of defective resistors in a batch.

p-Values

The *p-value* is the probability of a value as extreme as the observed value. We can calculate that by examining the z-test statistic. The z-test statistic is

$$z = \frac{\bar{x} - \mu_0}{\sigma / \sqrt{n}}$$

$$= \frac{45 - 50}{15 / \sqrt{25}}$$

$$= -1.67$$

The probability of a standard normal value of less than -1.67 is 0.0478. To calculate the p-value, we need to take into account the terms of the alternate hypothesis. In this case, the alternative hypothesis was that the new process made no difference (either positive or negative) in the number of defects. Thus, we need to calculate the probability of an extreme value 1.67 units from 0 in *either* direction. Because the standard normal distribution is symmetric, the probability of a value being < -1.67 is equal to the probability of a value being > 1.67, so we can simply double the probability, resulting in a p-value of $2 \times 0.0478 = 0.0956$.

This was an example of a **two-tailed test**, in which we assume that extreme values can come in either direction. We can also construct a **one-tailed test**, in which we consider differences in only one direction. A one-tailed test could have these hypotheses:

H_0: The mean number of defective resistors in the new process is 50.

H_a: The mean number of defective resistors is < 50.

We use this hypothesis if something in the new process would *absolutely* rule out the possibility of an increase in the number of defective resistors. If that were the case, we would not need to double the probability, and the *p*-value would be 0.0478, so the sample average lies outside the acceptance region if α=0.05. We would call this result statistically significant and would reject the null hypothesis, accepting the hypothesis that the new process reduces the number of defective resistors.

It sounds like we've got something for nothing, but we haven't. We've attained significant results at the cost of assuming something that we hadn't assumed before. Because it's easier to achieve "significant" results in one-tailed tests, they should be used with extreme caution and only when warranted by the situation. You should always state your alternative hypothesis *before* doing your analysis (rather than deciding on a one-tailed test after seeing the results with the two-tailed test).

EXCEL TIPS

- To calculate the *p*-value with Excel, first calculate the value of the *z*-test statistic using the Excel function z=(AVERAGE(*data range*)-μ_0)/(σ/SQRT(*n*)), where *data range* is the range of cells in your worksheet containing the sample values, μ_0 is the mean under the null hypothesis, σ is the standard deviation of the probability distribution, and *n* is the sample size.

- For a one-tailed test where *z* is negative, the *p*-value = NORMSDIST(z)

- For a one-tailed test where *z* is positive, the *p*-value = 1-NORMSDIST(z)

- For a two-tailed test where *z* is negative, the *p*-value = 2*NORMSDIST(z)

- For a two-tailed test where *z* is positive, the *p*-value = 2*(1-NORMSDIST(z))

CONCEPT TUTORIALS:
Hypothesis Testing

You can get a visual picture of the principles of hypothesis testing by opening the Hypothesis Testing workbook.

To use the Hypothesis Testing workbook:

1 Open the file **Hypothesis Testing**, located in the Explore folder of your Student disk. Enable any macros in the workbook.

2 Move through the workbook, reviewing the material on hypothesis testing.

3 Click **Explore Hypothesis Testing** from the Table of Contents column. See Figure 6-7

**Figure 6-7
The
Hypothesis
Testing
workbook**

click highlighted
boxes to change
conditions

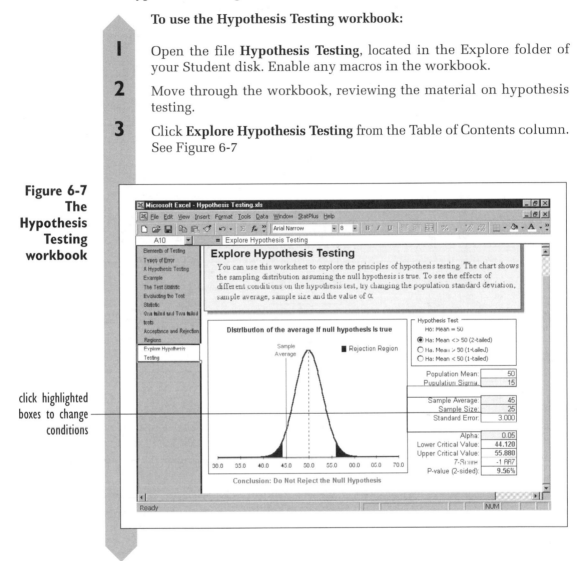

The workbook shows the sampling distribution under the null hypothesis, where it is assumed that $\mu = 50$. The rejection region is displayed on the chart in black. This workbook allows you to vary four values — σ, $\bar{x}$, the sample size, and α — to see their effect on the hypothesis test. You can also choose whether to perform a one-tailed or a two-tailed test. The results of your choices are automatically displayed on the workbook and in the chart. By working with different values of these factors, you can get a clearer picture of how hypothesis testing works.

For example, what impact would doubling the sample size have on the hypothesis test, assuming all other factors remained the same? Let's find out.

To increase the sample size:

I Click the **Sample Size** box, and change the sample size from 25 to **50**. See Figure 6-8.

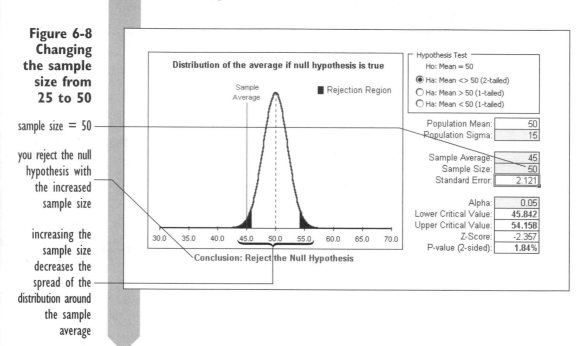

Figure 6-8 Changing the sample size from 25 to 50

sample size = 50

you reject the null hypothesis with the increased sample size

increasing the sample size decreases the spread of the distribution around the sample average

The width of the acceptance region shrinks from 11.76 (44.12 to 55.88) with a sample size of 25 to 8.32 (45.84 to 54.16) with a sample size of 50. The observed sample average lies within the rejection region, so you reject the null hypothesis with a *p*-value of 1.84%.

Now let's see what happens when you increase the value of σ from 15 to 20.

To increase the value of σ:

I Click the **Population Sigma** box, and change the value from 15 to **20**.

Because the value of σ has increased, the value of the standard error has increased too, from 2.121 to 2.828. The lower critical value has fallen to 44.456 and the *p*-value has increased to 7.71%, so you do not reject the null hypothesis. The variability of the data is one of the most important factors in hypothesis testing; much of statistical analysis is concerned with reducing or explaining variability.

Finally, let's find out what our conclusions would be if we used a one-tailed test, where H_a is the hypothesis that the mean < 50.

To change to a one-tailed test:

2 Click the **Ha: Mean < 50 (1-tailed)** option button.

The chart changes to a one-tailed test. See Figure 6-9.

**Figure 6-9
Switching
from a
one-tailed
test to a
two-tailed
test**

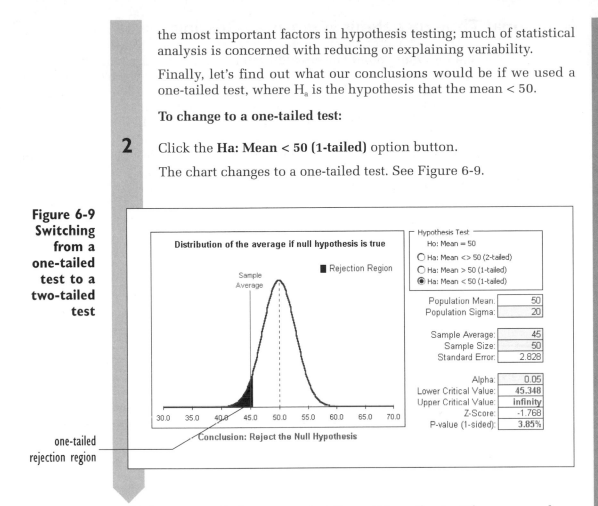

one-tailed
rejection region

Conclusion: Reject the Null Hypothesis

Under a one-tailed test, we reject the null hypothesis. Of course, we have to be very careful with this approach, because we are changing our hypothesis *after* seeing the data. If this were an actual situation, changing the hypothesis like this would be inappropriate. The better course would be to draw another random sample of 25 batches and test the new hypothesis on that set of data (and only if we have compelling reasons for doing a one-tailed test.)

Try other combinations of hypothesis and parameter values to see how they affect the hypothesis test. Close the workbook when you're finished. You do not have to save your changes.

Additional Thoughts about Hypothesis Testing

One important point you should keep in mind when hypothesis testing is that accepting the null hypothesis does *not* mean that the null hypothesis is true. Rather, you are stating that there is insufficient reason to reject it. The distinction is subtle but important. To state that accepting the null hypothesis means that $\mu = 50$ excludes the possibility that μ actually equals 49 or 49.9 or 49.99. But you didn't test any of these possibilities. What you *did* test was whether the data are incompatible with the assumption that $\mu = 50$. You found that in some cases, they are not.

You've looked at two approaches to statistical inference: the confidence interval and the hypothesis test. For a particular value of α, the width of the confidence interval around the sample average is equal to the width of the two-sided acceptance region around μ_0. This means that the following two statements imply each other:

1. The value μ_0, lies outside the $(1 - \alpha)$ confidence interval around $\bar{x}$.
2. Reject the null hypothesis that $\mu = \mu_0$ at the α significance level.

The *t*-Distribution

Up to now, you've been assuming that the value of σ is known. What if you didn't know the value of σ? One solution is to substitute the standard deviation of the sample values, s, for σ in the hypothesis-testing equations. However there are problems with this approach. If s underestimates σ, then you'll overestimate the significance of the results, perhaps causing you to reject the null hypothesis falsely. Or if s overestimates the value of σ, you could accept the null hypothesis when the null hypothesis isn't true.

In the early years of this century, William Gosset, working at the Guinness brewery in Ireland, became worried about the uncertainty caused by substituting s for σ. He believed that the resulting error could be especially bad for small sample sizes. What Gosset discovered was that when you substitute s for σ, the ratio

$$\frac{\bar{x} - \mu}{s/\sqrt{n}}$$

does *not* follow the standard normal distribution; rather, it follows a distribution called the **t-distribution**.

The *t*-distribution is a probability distribution centered around zero and characterized by a single parameter called the **degrees of freedom**, which is equal to the sample size, n, minus 1. For example, if the sample size is 20, the degrees of freedom equal 19. The *t*-distribution is similar to a standard normal distribution except that it has heavier tails. As the sample size increases, the *t*-distribution approaches the standard normal, but for smaller sample sizes there can be big differences.

CONCEPT TUTORIALS:
The *t*-Distribution

To see how the *t*-distribution differs from the standard normal, open the Distributions workbook.

To explore the *t*-distribution:

1 Open the file **Distributions**, located in the Explore folder of your Student disk.

2 Click **t** from the Table of Contents column. Review the material and scroll to the bottom of the worksheet. See Figure 6-10.

Figure 6-10
The
***t*-distribution**

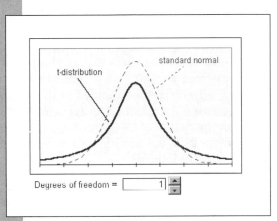

3 Click the Spin Arrow buttons located next to the Degrees of Freedom box to change the degrees of freedom for the *t*-distribution.

The workbook opens, displaying the *t*-distribution with 1 degree of freedom. Notice that the *t*-distribution has heavier tails than the superimposed standard normal distribution. As you increase the degrees of freedom, the *t*-distribution more closely approximates the standard normal.

Continue exploring the *t*-distribution by changing the degrees of freedom and observing the changing curve. Close the workbook when you're finished. You do not have to save your changes.

Working with the *t*-Statistic

Excel provides several functions to work with the *t*-distribution. Two of these are displayed in Table 6-2.

Table 6-2 Two *t*-distribution functions in Excel

Function	Description
TDIST(*t*, *df*, *tails*)	Returns the *p*-value for the *t*-distribution for a given value of *t*, degrees of freedom *df*, and *tails* = 1 (one-tailed) or *tails* = 2 (two-tailed)
TINV(*p*, *df*)	Returns the two-tailed *t*-value from the *t*-distribution with degrees of freedom *df* for a *p*-value = *p*. For a one-tailed *t*-value, replace *p* with 2*p.

Let's use Excel to apply the *t*-distribution to a problem involving textbook prices. The college administration claims that students should not expect to spend more than an average of $200 each semester for books. A student associated with the school newspaper decides to investigate this claim and interviews 25 randomly selected students. The average spent by the 25 students is $220, and the standard deviation of these purchases is $50. Is this significant evidence that the statement from the administration is wrong?

First, let's construct our hypothesis:

H_0: The average cost of textbooks is $200.

H_a: The average cost of textbooks is not equal to $200.

Now we'll construct the ***t*-statistic, t_{n-1}**:

$$t_{n-1} = \frac{\bar{x} - \mu_0}{s/\sqrt{n}} = \frac{220 - 200}{50/\sqrt{25}} = \frac{20}{10} = 2$$

To test the null hypothesis with Excel:

1 Open a new blank workbook.

2 In cell A1, type **=TDIST(2,24,2)** and press **Enter.**

In this example, 2 is the value of the *t* statistic, 24 is the degrees of freedom, and we enter 2 because this is a two-tailed test.

The TDIST function returns a *p*-value of 0.05694, so we do not reject the null hypothesis at the 5% level. Thus we conclude that there is not sufficient evidence that the college administration is underestimating the price of textbooks. If we had used the *z*-test statistic rather than the *t*-statistic in this example, the *p*-value would have been 0.0455, and we would have erroneously rejected the null hypothesis.

Constructing a *t* Confidence Interval

Still we have a sample average that doesn't completely match the administration's claim. Let's construct a 95% confidence interval for the mean value to see in what range of values the true mean might lie. Because we don't know the value of σ, we can't use the confidence interval equation discussed earlier—we'll have to use one based on the *t*-distribution. The equation for the **t confidence interval** is

$$\left(\bar{x} - t_{1-\alpha/2,\, n-1}\frac{s}{\sqrt{n}} \, , \, \bar{x} + t_{1-\alpha/2,\, n-1}\frac{s}{\sqrt{n}} \right)$$

Here, $t_{1-\alpha/2,\, n-1}$ is the point on the *t*-distribution with $n-1$ degrees of freedom, such that the probability of a *t* random variable being less than it is $1 - \alpha/2$. To calculate this value in Excel, you use the TINV function. However, in the TINV function, you enter the value of α, not $1 - \alpha/2$. For example, to calculate the value of $t_{1-\alpha/2,\, n-1}$, enter the function TINV(α,n - 1). Use this information to construct a 95% confidence interval.

To construct a 95% confidence interval for the price of textbooks:

1 In cell A2, type **=220-TINV(0.05,24)*50/SQRT(25)** and press **Tab**.

2 In cell B2, type **=220+TINV(0.05,24)*50/SQRT(25)** and press **Enter**.

The 95% confidence interval is (199.36 , 240.64). We do not expect the mean price of textbooks to be much less than $200, nor should it be much greater than $240. By comparison, the confidence interval based on the standard normal distribution is (200.40 , 239.60), so the confidence intervals are very close in size.

The Robustness of *t*

When you use the *t*-distribution to analyze your data, you're assuming that the data follow a normal distribution. What are the consequences if this turns out not to be the case? The *t*-distribution has a property called **robustness**, which means that even if the assumption of normality is moderately violated, the *p*-values returned by the *t*-statistic will still be pretty accurate. As long as the distribution of the data does not violate the assumption of normality in an extreme way, you can use the *t*-distribution with confidence.

Applying the *t*-Test to Paired Data

The *t*-distribution becomes useful in analyzing **paired data**, where observations come in natural pairs and you wish to explore the difference between the two pairs. For example, a doctor might measure the effect of a drug by measuring the physiological state of patients before and after administering the drug. Each patient in this study has two observations, and the observations are paired with each other. To determine the drug's effectiveness, the doctor looks at the difference between the before and after readings.

The Excel workbook WLABOR.XLS contains data on the percentage of women in the work force in 1968 and 1972 taken from a sample of 19 cities. The workbook contains the following variables:

Table 6-3 Data on percentage of women in the work force

Range Name	Range	Description
City	A2:A20	The name of the city
Year_68	B2:B20	The percent of women in the work force in 1968
Year_72	C2:C20	The percent of women in the work force in 1972
Diff	D2:D20	The change in percentage from 1968 to 1972 for each city

To open the WLabor.xls workbook:

1 Open the **WLabor** workbook from the folder containing your student files.

2 Click **File > Save As** and save the workbook as **WLabor2**. The workbook appears as shown in Figure 6-11.

Figure 6-11
The WLabor workbook

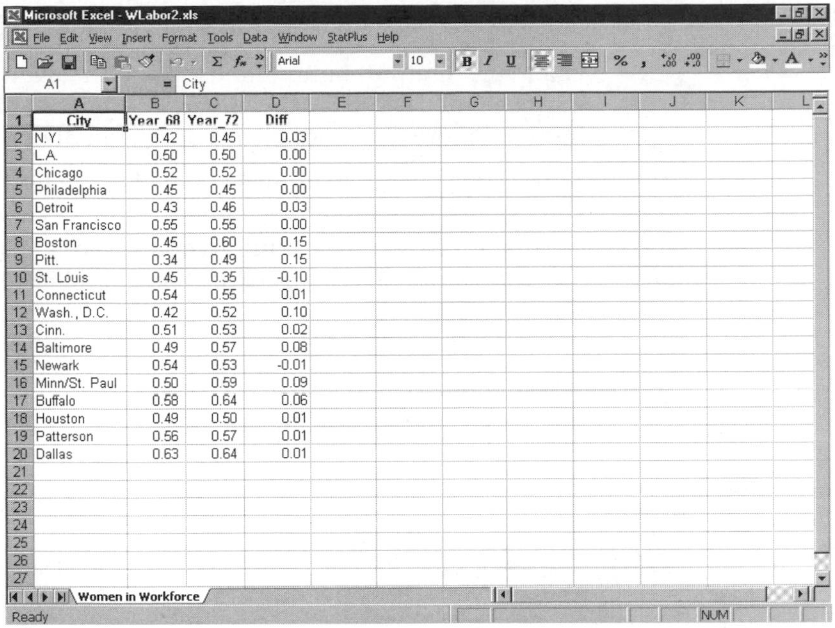

Microsoft Excel - WLabor2.xls

A1 = City

	A	B	C	D
1	City	Year_68	Year_72	Diff
2	N.Y.	0.42	0.45	0.03
3	L.A.	0.50	0.50	0.00
4	Chicago	0.52	0.52	0.00
5	Philadelphia	0.45	0.45	0.00
6	Detroit	0.43	0.46	0.03
7	San Francisco	0.55	0.55	0.00
8	Boston	0.45	0.60	0.15
9	Pitt.	0.34	0.49	0.15
10	St. Louis	0.45	0.35	-0.10
11	Connecticut	0.54	0.55	0.01
12	Wash., D.C.	0.42	0.52	0.10
13	Cinn.	0.51	0.53	0.02
14	Baltimore	0.49	0.57	0.08
15	Newark	0.54	0.53	-0.01
16	Minn/St. Paul	0.50	0.59	0.09
17	Buffalo	0.58	0.64	0.06
18	Houston	0.49	0.50	0.01
19	Patterson	0.56	0.57	0.01
20	Dallas	0.63	0.64	0.01

Women in Workforce

Ready — NUM

There are two observations from each city, and the observations constitute paired data. You've been asked to determine whether this sample of 19 cities demonstrates a statistically significant increase in the percentage of women in the work force. Let μ be the mean change in the percent of women in the work force. You have two hypotheses:

H_0: $\mu_0 = 0$ (There is no change in the percentage from 1968 to 1972.)

H_a: $\mu_0 \neq 0$ (There is some change, but we're not assuming the direction of the change.)

You can use StatPlus to test these hypotheses and create a t confidence interval for the change in percent from 1968 to 1972. Do this now by analyzing the Diff variable to test whether the average difference is significantly different from zero.

To test whether there is a significant difference:

1 Click **StatPlus > One Sample Tests > 1 Sample t-test** from the Excel menu.

We use the one-sample t-test because we are essentially looking at one sample of data—the sample of paired differences.

2 Verify that the **1-sample t-test** option button is selected.

3 Click the **Input** button and then click the **Use Range Names** option button. Select **Diff** from the list of range names and click the **OK** button.

4 Click the **Output** button and select the **New Worksheet** option button. Enter **t-test** for the new worksheet name and click the **OK** button. Figure 6-12 shows the completed dialog box.

Note that we could have also selected the Paired t-test (two columns) and then selected the Year_68 and Year_72 columns. Note also that the dialog box will test a null hypothesis of $\mu = 0$ versus the alternate hypothesis of "not equal to" 0. You can change these values if you wish to test other hypotheses.

Figure 6-12
The
One-Sample
or Paired
t-Test
dialog box

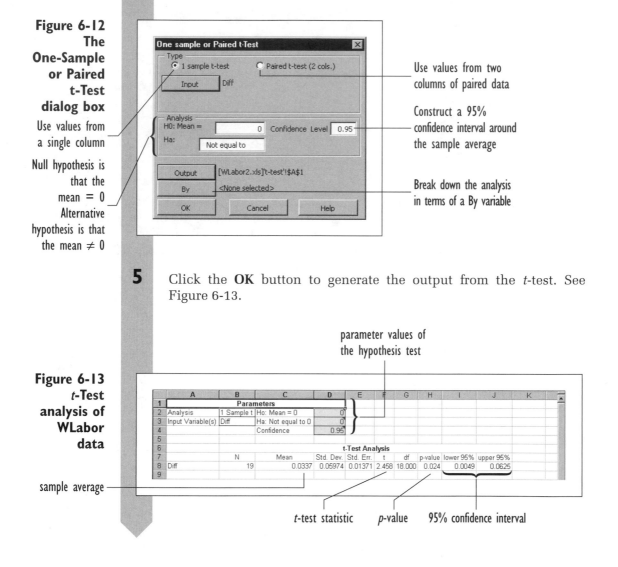

Use values from
a single column

Null hypothesis is
that the
mean = 0
Alternative
hypothesis is that
the mean ≠ 0

5 Click the **OK** button to generate the output from the *t*-test. See Figure 6-13.

Figure 6-13
t-Test
analysis of
WLabor
data

Based on our analysis, there is an average increase in the percent of women in the labor force of 3.37 percentage points between 1968 and 1972. This is statistically significant with a p-value of 0.024, so we reject the null hypothesis and accept the alternative. There has been a significant change in women's participation in the work force in those 4 years. The 95% confidence interval for this estimate ranges from 0.49 percentage point up to 6.25 percentage points.

A government spokesman, viewing some other data on this topic, claims that the percentage of women in the work force has increased 5 points from 1968 to 1972. Does your data conflict with his statement? Let's test the hypothesis.

$$H_0: \mu = 0.05$$
$$H_a: \mu \neq 0.05$$

Rather than rerunning the StatPlus command, we can simply enter the new hypothesis directly into the t-test worksheet.

To test the new hypothesis:

Click cell **D2**, change the value from 0 to **0.05**, and press **Enter**.

The p-value changes from 0.024 to 0.249. A p-value of 0.249 is not small enough to reject the null hypothesis. We conclude that our data does not conflict with the government statement in a significant way.

You can also change other values in the hypothesis test. You can switch to a one-sided test by changing the value of cell D3 to either –1 or 1. You can also change the size of the confidence interval by changing the value in cell D4.

EXCEL TIPS

- You can also perform a paired t-test of your data using the Analysis ToolPak, supplied by Excel. To perform a paired t-test, load the Analysis ToolPak and then choose "Data Analysis" from the Tools menu. In the Analysis Tools list box, click "t-Test: Paired Two Sample for Means" and specify the two columns containing the paired data. This command does not calculate the confidence interval for you, so you have to calculate that using the formulas supplied in this chapter.

You should not accept your analysis at face value without further investigating the assumptions of the t-test. One of these assumptions is that the data follows a normal distribution. The t-test is robust, but that doesn't mean you shouldn't explore the possibility that the data is seriously non-normal. Do this now by creating a histogram and normal probability plot of the data in the Diff column.

To create a histogram of the difference data:

1 Click **StatPlus > Single Variable Charts > Histograms** from the Excel menu bar.

2 Click the **Data Values** button and select **Diff** from the list of range names in the workbook.

3 Click the **Normal Curve** checkbox to add a normal curve to the histogram.

4 Click the **Output** button and then click the **Cell** option button and select cell **A10** in the *t*-test worksheet.

5 Click the **OK** button twice to create the histogram. See Figure 6-14.

**Figure 6-14
Histogram
of difference
data**

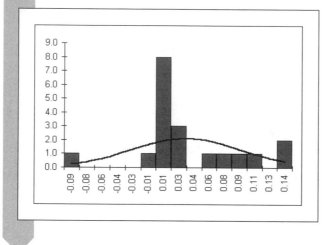

The observed difference values don't appear to follow the normal curve particularly well. Let's see whether the normal probability plot provides more information.

To create a normal probability plot of the difference data:

1 Click **StatPlus > Single Variable Charts > Normal P-plots** from the Excel menu bar.

2 Click the **Data Values** button and select **Diff** from the list of range names in the workbook.

3 Click the **Output** button and then click the **Cell** option button and select cell **A26** in the *t*-test worksheet.

4 Click the **OK** button twice to create the normal probability plot. See Figure 6-15.

Figure 6-15
Normal
probability
plot of
difference
data

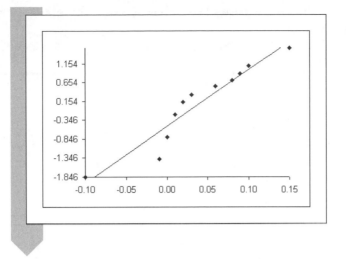

From the two plots, there is enough graphical evidence to make us worry that the data does not follow the normal distribution. This is a problem because now we can't feel completely comfortable about the p-values the t-test gave us. Because the assumption of normality may have been violated, we can't be sure that the p-value is accurate.

Applying a Nonparametric Test to Paired Data

The t-test is an example of a **parametric test** because you compare the data values to a distribution whose shape can be quantified on the basis of the value of one or more parameters. As you've seen, the t-distribution's shape is determined by the value of the degrees-of-freedom parameter. A **nonparametric** test makes fewer (and simpler) assumptions about the distribution of the data. Most nonparametric tests are based on ranks and not the actual data values (this frees them from making specific assumptions about the data.) The study of nonparametric statistics can fill an entire textbook. We'll just cover the high points and show how to apply a nonparametric test to your data.

The Wilcoxon Signed Rank Test

The nonparametric counterpart to the t-test is the Wilcoxon Signed Rank test. In the **Wilcoxon Signed Rank test**, we rank the absolute values of the original data from smallest to largest, and then each rank is multiplied by the sign of the original value (−1, 0, or 1.) In case of a tie, we assign an average rank to the tied values. Table 6-4 shows the values of a variable, along with the values of the signed ranks.

Table 6-4 Signed Ranks

Variable Values	Signed Ranks
18	7.0
4	2.0
15	6.0
−5	−3.5
−2	−1.0
10	5.0
5	3.5

There are seven values in this data set, so the ranks go from 1 (for lowest in absolute value) up to 7 (for the highest in absolute value.) The lowest in absolute value is −2, so that observation gets the rank 1 and then is multiplied by the sign of the observation to get the sign rank value −1. The value 4 gets the sign rank value 2 and so forth. Two observations are tied with absolute values equal to −5 and 5. They should get ranks 3 and 4 in our data set, but because they're tied, they both get an average rank of 3.5 (or −3.5.)

Next we calculate the sum of the signed ranks. If most of the values were positive, this would be a large positive number. If most of the values were negative, this would be a large negative number. The sum of the signed ranks in our example equals: $7 + 2 + 6 − 3.5 − 1 + 5 + 3.5 = 19$.

The only assumption we make with the Wilcoxon Signed Rank test is that the distribution of the values is symmetric around the median. If under the null hypothesis, we assume that the median = 0, this would imply that we should have as many negative ranks as positive ranks and that the sum of the signed ranks should be 0. Using probability theory, we can then determine how probable it is for a collection of 7 observations to have a total signed rank of 19 or more, if the null hypothesis is true. Without going into the details of how to make this calculation, the p-value in this particular case is 0.133, so we would not reject the null hypothesis. In addition to calculating p-values, you can also calculate confidence intervals using the Wilcoxon test statistic.

One advantage in using ranks instead of the actual values is that the hypothesis test is much less sensitive to the effect of outliers. Also, nonparametric procedures can be applied to situations involving ordinal data, such as surveys in which subjects rank their preferences. The downside of nonparametric tests is that they are not as efficient as parametric tests when the data *is* normally distributed. This means that for normal data you need a larger sample size in order to detect statistically significant effects (5% larger when the Wilcoxon Signed Rank test is used in place of the t-test.) Of course, if the data is not normally distributed, you can often detect statistically significant effects with smaller sample sizes using nonparametric procedures. The nonparametric test can be *more* efficient in those cases.

The bottom line is that if there is some question about whether to use a parametric or a nonparametric test, do the analysis both ways.

Excel does not include any nonparametric tests, but you can use StatPlus to generate test results using the Wilcoxon Signed Rank test. Apply this test now to the work force data. Your hypotheses are

$$H_0: \text{Median difference} = 0$$
$$H_a: \text{Median difference} \neq 0$$

To analyze the difference data using the Wilcoxon Signed Rank test:

1 Click **StatPlus > One Sample Tests > 1 Sample Wilcoxon Sign Rank test** from the Excel menu.

2 Verify that the **1-sample W-test** option button is selected.

3 Click the **Input** button, select **Diff** from the list of range names, and click the **OK** button.

4 Click the **Output** button and select the **New Worksheet** option button. Enter **W-test** for the new worksheet name and click the **OK** button.

5 Click the **OK** button to generate the output from the Wilcoxon Signed Rank test. See Figure 6-16.

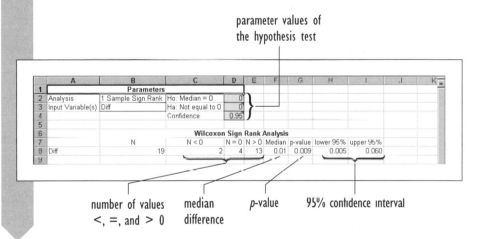

Figure 6-16
Wilcoxon Signed Rank test analysis of WLabor data

The results of our analysis using the Wilcoxon Signed Rank test are similar to the results with the t-test. We still reject the null hypothesis—this time with a stronger p-value of 0.009. The 95% confidence interval is pretty close to what we calculated before, (0.005, 0.060). Moreover, we learn that the number of cities whose percentage of women in the work force increased from 1968 to 1972 was 13 and that only 2 cities showed a decrease in percentage points (4 cities were unchanged.) This information strengthens our earlier conclusion that there was a significant increase in women in the work force during those 4 years.

EXCEL TIPS

- Excel doesn't include a command to perform the signed rank test, but you can approximate the *p*-values and confidence intervals of the signed rank test by calculating the sign ranks of your data and then performing a one-sample *t*-test on those values.

- To calculate the ranks of your data, use Excel's RANK function. If your data contains ties, you should use the RANKTIED function (*StatPlus required*).

- To calculate the signed ranks of your data, use the SIGNRANK function (*StatPlus required*).

The Sign Test

Another nonparametric test that makes even fewer assumptions than the Wilcoxon Signed Rank test is the Sign test. In the **Sign test**, we ignore the values entirely and simply count the number of positive values and the number of negative values. We then test whether there are more positive (or negative) values than there should be. The test is similar to what we might use to test whether a two-sided coin is fair.

The Sign test is usually less efficient (requires a larger sample size) than either the *t*-test or the signed rank test, except in cases where the data comes from a heavy tailed distribution. In those cases, the Sign test may be more effective than either the *t*-test or the signed rank test.

Let's apply the Sign test to our data set. Our hypotheses are

H_0: Probability of a negative value = probability of a positive value

H_a: Probability of a negative value ≠ probability of a positive value

To analyze the difference data using the Sign test:

1 Click **StatPlus > One Sample Tests > 1 Sample Sign test** from the Excel menu.

2 Verify that the **1-sample s-test** option button is selected.

3 Click the **Input** button, select **Diff** from the list of range names, and click the **OK** button.

4 Click the **Output** button and select the **New Worksheet** option button. Enter **s-test** for the new worksheet name and click the **OK** button.

5 Click the **OK** button to generate the output from the Sign test. See Figure 6-17.

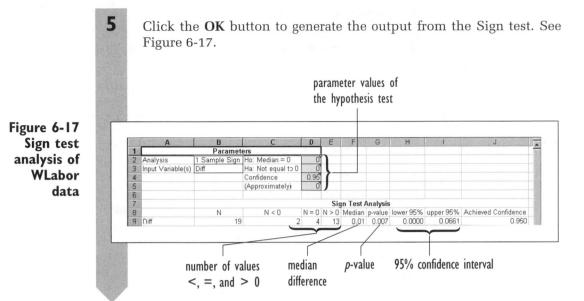

Figure 6-17 Sign test analysis of WLabor data

Even under the Sign test, we reject the null hypothesis and accept the alternative, concluding that the percent of women in the work force has significantly increased. The *p*-value for the Sign test is 0.007. We can also construct a 95% confidence interval with the Sign test, but because of the nature of the test, we can only approximate the confidence interval at this level. The output in Figure 6-17 shows the approximate 95% confidence interval of the change in the percentage of women in the workforce to be (0.00, 0.66). We can also find exact confidence intervals under the Sign test that are closest to 95% without going over or going under 95%.

To find the exact confidence intervals under the Sign test:

1 Click cell **D5** and type **−1**.

The output changes to give you the exact confidence interval that is *at most* 95%. In this case that is a 93.6% confidence interval, and it ranges from 0.00 to 0.060.

2 Click cell **D5** and type **1**.

The output changes and displays the exact confidence interval that is *at least* 95%. Excel displays the 98.1% confidence interval: (0.00, 0.80).

You've completed your research with the WLabor workbook. You've concluded that there is sufficient evidence in this sample of 19 cities to conclude that there has been an increase in the percentage of women participating in the work force during the 4-year period from 1968 to 1972.

To complete your work:

Save your changes to the WLabor workbook and close the file.

The Two-Sample *t*-Test

In the one-sample or paired *t*-test, you compared the sample average to a fixed value specified in the null hypothesis. In a **two-sample *t*-test**, you compare the averages from two independent samples to determine whether a significant difference exists between the samples. For example, one sample might contain the salaries of the male professors at a university, a second sample the female professors' salaries. You would test to see whether there is a statistically significant difference between the two sample averages.

To compare the sample averages, you have a choice of two *t*-tests. One test statistic, called the **unpooled two-sample *t*-statistic**, has the form

$$t = \frac{(\overline{x}_1 - \overline{x}_2) - (\mu_1 - \mu_2)}{\sqrt{\dfrac{s_1^2}{n_1} + \dfrac{s_2^2}{n_2}}}$$

where $\overline{x}_1$ and $\overline{x}_2$ are the sample averages for the first and second samples, s_1 and s_2 are the sample standard deviations, n_1 and n_2 are the sample sizes, and μ_1 and μ_2 are the means of the two distributions.

This form of the *t*-statistic assumes that the samples come from distributions with different standard deviations, having values of σ_1 and σ_2. It may also be the case that both distributions share a common standard deviation, σ. If that is the case, we can construct a *t*-statistic by *pooling* the estimates of the standard deviation from the two samples into a single estimate, which we'll label as *s*. The value of *s* is

$$s = \sqrt{\frac{(n_1 - 1)s_1^2 + (n_2 - 1)s_2^2}{n_1 + n_2 - 2}}$$

The **pooled two-sample *t*-statistic** would then be equal to

$$t = \frac{(\overline{x}_1 - \overline{x}_2) - (\mu_1 - \mu_2)}{s\sqrt{\dfrac{1}{n_1} + \dfrac{1}{n_2}}}$$

Comparing the Pooled and Unpooled Test Statistics

There are important differences between the two test statistics. The unpooled statistic, although we refer to it as a *t*-test, *does not* strictly follow a *t*-distribution. However, we can closely approximate the correct *p*-values for this statistic by assuming it does and then compare the test statistic to a *t*-distribution with degrees of freedom equal to

$$df = \frac{\left(\dfrac{s_1^2}{n_1} + \dfrac{s_2^2}{n_2}\right)^2}{\left[\dfrac{\left(\dfrac{s_1^2}{n_1}\right)^2}{n_1 - 1} + \dfrac{\left(\dfrac{s_2^2}{n_2}\right)^2}{n_2 - 1}\right]}$$

Here s_1 and s_2 are the standard deviations of the values in the first and second samples. The degrees of freedom for this statistic often result in a fractional value. In actual practice, you'll probably never have to make this calculation yourself; your statistics package will do it for you.

For the pooled statistic, the situation is much easier. The pooled *t*-statistic *does* follow a *t*-distribution with degrees of freedom equal to

$$df = n_1 + n_2 - 2$$

If the standard deviations are different and you apply the pooled *t*-statistic to the data, you run the risk of reporting an erroneous *p*-value. To guard against this problem, it may be best to perform both a pooled and an unpooled test and then compare the results. If they agree, report the pooled *t*, because this test statistic is more widely known. Use the unpooled *t* if the two tests disagree. You should also examine the standard deviations of the two samples and determine whether they're close in value.

Working with the Two-Sample *t*-Statistic

To see how the two-sample *t*-test works, let's consider two groups of students: One group has learned to write with a word processor on a Macintosh computer, and the other has learned on a PC. There are 25 students in each group. At the end of the session, each student writes an essay that is graded on a 100-point scale. The average grade for the Macintosh students is 75 with a standard deviation of 8. The average for the PC students is 80 with a standard deviation of 6. Could the difference in sample averages be attributed to differences between learning on a Macintosh and learning on a PC? Our hypotheses are

$$H_0: \mu_1 - \mu_2 = 0$$
$$H_a: \mu_1 - \mu_2 \neq 0$$

where μ_1 is the assumed mean of the first distribution and μ_2 is the mean of the second distribution. Notice that we are not making any assumptions about what the actual values of μ_1 and μ_2 are; we are interested only in the difference between them. Because the standard deviations of the two samples are close in value, we'll use a pooled t-test to test the null hypothesis. First, we must calculate the value of the pooled standard deviation, s:

$$
\begin{aligned}
s &= \sqrt{\frac{(n_1 - 1)s_1^2 + (n_2 - 1)s_2^2}{n_1 + n_2 = 2}} \\
&= \sqrt{\frac{(25 - 1)8^2 + (25 - 1)6^2}{25 + 25 - 2}} \\
&= 7.07
\end{aligned}
$$

Thus, the t-statistic equals

$$
\begin{aligned}
t &= \frac{(\bar{x}_1 - \bar{x}_2) - (\mu_1 - \mu_2)}{s\sqrt{\dfrac{1}{n_1} + \dfrac{1}{n_2}}} \\
&= \frac{(75 - 80) - 0}{7.07\sqrt{\dfrac{1}{25} + \dfrac{1}{25}}} \\
&= -2.5
\end{aligned}
$$

which should follow a t-distribution with 48 degrees of freedom. Is this a significant value?

To evaluate the t-statistic:

Open a blank workbook. In cell A1 type **=TDIST(2.5,48,2)** and press **Enter**.

Note that we enter the value 2.5 rather than −2.5 because the TDIST function works with positive values. We enter the value 2 as the third parameter because this is a two-sided test.

Excel returns a p-value of 0.016 and the null hypothesis is rejected. We conclude that the evidence supports the conclusion that students learn to write better on PCs.

This example is artificial, but it is based on research by Dr. Marcia Peoples Halio of the University of Delaware (*Academic Computing*, January 1990, pp. 16–19, 45). She found that when students had a choice of computer, the students who chose the PC had much better writing scores than the students who chose the Macintosh. Note that the students chose which computer to use, and these were therefore not random samples. Because the Macintosh has been advertised as the computer for those who know nothing about computers, it's possible that more confident students chose the PC.

Applying the *t*-Test to Two-Sample Data

The Excel workbook NURSEHOME.XLS contains data on a random sample of nursing homes collected by the Department of Health and Social Services in New Mexico. The following variables were collected:

Table 6-5 Nursing Home data

Range Name	Range	Description
Beds	A2:A53	The number of beds in the home
Medical_Days	B2:B53	Annual medical in-patient days (hundreds)
Total_Days	C2:C53	Annual total patient days (hundreds)
Revenue	D2:D53	Annual total patient care revenue ($hundreds)
Salaries	E2:E53	Annual nursing salaries ($hundreds)
Expenses	F2:F53	Annual facility expenses ($hundreds)
Location	G2:G53	Rural and non-rural homes

To open the NurseHome.xls workbook:

1 Open the **NurseHome** workbook from the folder containing your student files.

2 Click **File > Save As** and save the workbook as **NurseHome2**. The workbook appears as shown in Figure 6-18.

Figure 6-18 The NurseHome workbook

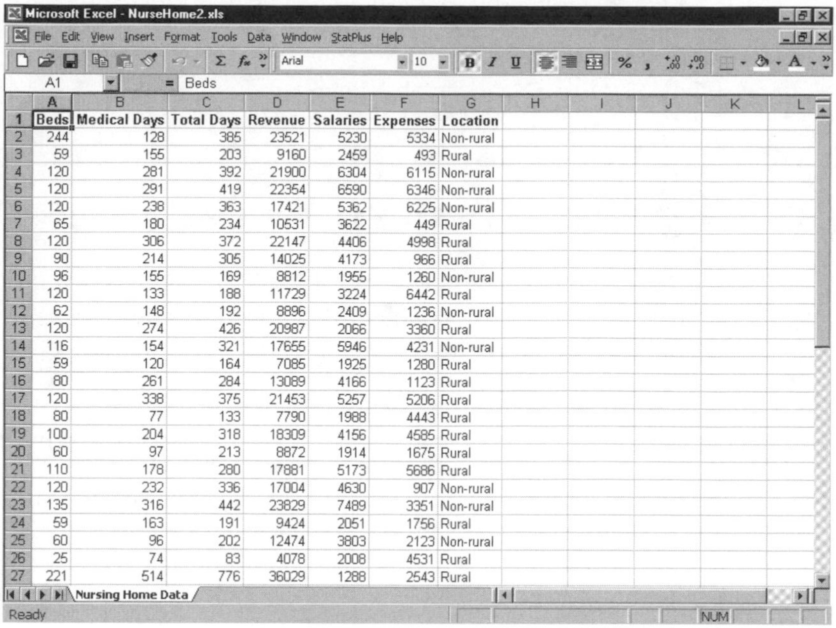

Beds	Medical Days	Total Days	Revenue	Salaries	Expenses	Location
244	128	385	23521	5230	5334	Non-rural
59	155	203	9160	2459	493	Rural
120	281	392	21900	6304	6115	Non-rural
120	291	419	22354	6590	6346	Non-rural
120	238	363	17421	5362	6225	Non-rural
65	180	234	10531	3622	449	Rural
120	306	372	22147	4406	4998	Rural
90	214	305	14025	4173	966	Rural
96	155	169	8812	1955	1260	Non-rural
120	133	188	11729	3224	6442	Rural
62	148	192	8896	2409	1236	Non-rural
120	274	426	20987	2066	3360	Rural
116	154	321	17655	5946	4231	Non-rural
59	120	164	7085	1925	1280	Rural
80	261	284	13089	4166	1123	Rural
120	338	375	21453	5257	5206	Rural
80	77	133	7790	1988	4443	Rural
100	204	318	18309	4156	4585	Rural
60	97	213	8872	1914	1675	Rural
110	178	280	17881	5173	5686	Rural
120	232	336	17004	4630	907	Non-rural
135	316	442	23829	7489	3351	Non-rural
59	163	191	9424	2051	1756	Rural
60	96	202	12474	3803	2123	Non-rural
25	74	83	4078	2008	4531	Rural
221	514	776	36029	1288	2543	Rural

You've been asked to examine the data and determine whether there is a difference in facility usage between the rural and non-rural homes. Specifically, you're to compare the total number of patient days per year for both types of facilities. Your conclusions will be used in forming the department's budget for the next year. Here are your hypotheses:

H_0: Average total patient days in the rural nursing homes = average total patient days in non-rural homes

H_a: Average total patient days in the rural nursing homes ≠ average total patient days in non-rural homes

To test the null hypothesis, you can use the StatPlus Two-Sample *t*-test command. We'll assume that there is equal variance in the two samples; however, we'll reexamine this assumption as we go along.

To perform a two sample *t*-test on the nursing home data:

1. Click **StatPlus > Two Sample Tests > 2-Sample t-test** from the menu bar.
2. Verify that the **Use Column of Category Values** option button is selected.

Your data can be organized in one of two ways: (1) with two separate columns for each sample or (2) with one column of data values and one column of category values. The NurseHome data is organized in the second way, with the Location column indicating whether a particular nursing home is rural or non-rural.

3   Click the **Data Values** button and select **Total_Days** from the list of range names.

4   Click the **Categories** button and select **Location** from the range names list.

5   Click the **Output** button and direct the output to a new worksheet named **t-test**.

Figure 6-19 shows the completed dialog box.

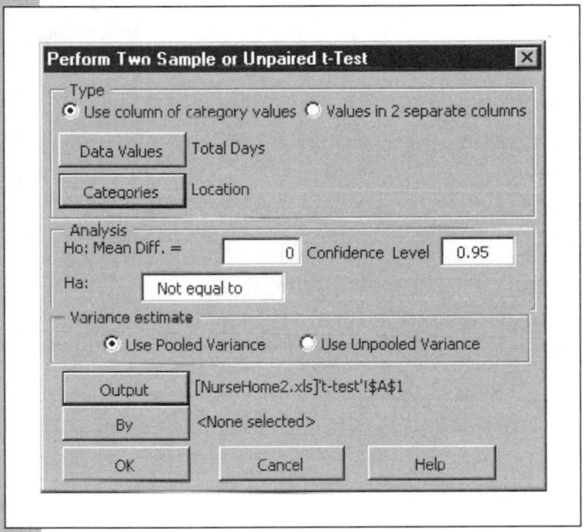

**Figure 6-19
The
Two-Sample
t-Test dialog
box**

6   Click the **OK** button to start the two-sample *t*-test. Figure 6-20 shows the completed output.

Figure 6-20 Pooled two-sample t-test output for the NurseHome data

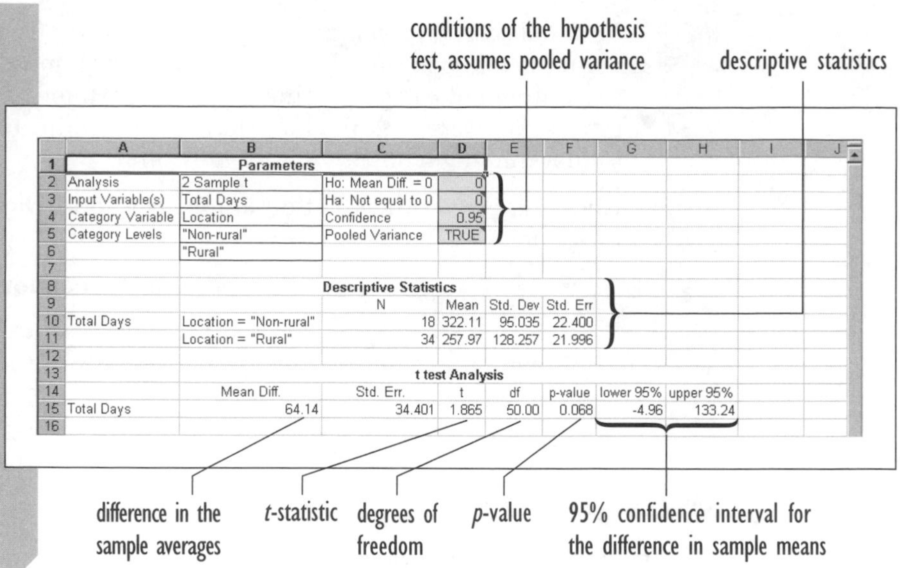

conditions of the hypothesis test, assumes pooled variance

descriptive statistics

	A	B	C	D	E	F	G	H	I	J
1		Parameters								
2	Analysis	2 Sample t	Ho: Mean Diff. = 0	0						
3	Input Variable(s)	Total Days	Ha: Not equal to 0	0						
4	Category Variable	Location	Confidence	0.95						
5	Category Levels	"Non-rural"	Pooled Variance	TRUE						
6		"Rural"								
7										
8			Descriptive Statistics							
9			N		Mean	Std. Dev	Std. Err			
10	Total Days	Location = "Non-rural"	18	322.11	95.035	22.400				
11		Location = "Rural"	34	257.97	128.257	21.996				
12										
13			t test Analysis							
14		Mean Diff.	Std. Err.	t	df	p-value	lower 95%	upper 95%		
15	Total Days	64.14	34.401	1.865	50.00	0.068	-4.96	133.24		
16										

difference in the sample averages | t-statistic | degrees of freedom | p-value | 95% confidence interval for the difference in sample means

According to the analysis, the average number of patient days for non-rural hospitals is 322.11 per year. Because this data is in terms of hundreds, this represents an average of 32,211 patient days. By comparison, the average number of patient days for rural hospitals is 25,797, for a difference of 6,414 patient days. The data suggests that rural hospitals have a higher mean number of patient days, but the p-value for the t-test is 0.068, which would cause us not to reject the null hypothesis. The 95% confidence interval for the difference in patient days ranges from −4.96 to 133.24 (−496 to 13,324 in absolute numbers.)

However, the standard deviation of the value in the non-rural sample is 95.035 compared to 128.257 in the rural sample. This might lead us to doubt that a common standard deviation applies to both samples. Perhaps we should use the t-test without pooling the standard deviation estimates. To do this, we could run the two-sample t-test command again or simply edit the worksheet containing the output.

To change the output to display the results of an unpooled test:

1   Click cell **D5** in the **t-test** worksheet.

Cell D5 contains the value TRUE if a pooled test is used and FALSE for an unpooled test. To switch from one to the other, change the value of D5.

2   Replace TRUE with **FALSE** in cell D5.

The output changes to display the results of an unpooled test. See Figure 6-21.

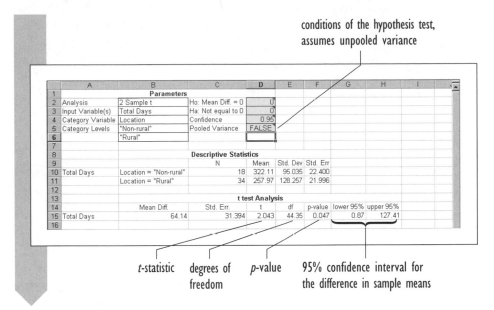

conditions of the hypothesis test, assumes unpooled variance

Figure 6-21 Unpooled two-sample *t*-test output for the NurseHome data

	A	B	C	D	E	F	G	H	I
1		**Parameters**							
2	Analysis	2 Sample t	Ho: Mean Diff. = 0	0					
3	Input Variable(s)	Total Days	Ha: Not equal to 0	0					
4	Category Variable	Location	Confidence	0.95					
5	Category Levels	"Non-rural"	Pooled Variance	FALSE					
6		"Rural"							
7									
8			**Descriptive Statistics**						
9			N	Mean	Std. Dev	Std. Err			
10	Total Days	Location = "Non-rural"	18	322.11	95.035	22.400			
11		Location = "Rural"	34	257.97	128.257	21.996			
12									
13			**t test Analysis**						
14		Mean Diff.	Std. Err.	t	df	p-value	lower 95%	upper 95%	
15	Total Days	64.14	31.394	2.043	44.35	0.047	0.87	127.41	
16									

t-statistic   degrees of freedom   *p*-value   95% confidence interval for the difference in sample means

With the unpooled test, the *p*-value changes to 0.047, leading us to reject the null hypothesis and accept the alternative—non-rural nursing homes are used to a greater degree than rural homes. The 95% confidence interval for the difference between the two samples ranges from 0.87 to 127.41, this translates to a difference of 87 to 12,741 patient days between the two locations.

We have a problem. Depending on which test we use, we reach a different conclusion. The difference between the two tests depends on the assumption we make about the distributions of the rural and non-rural samples. To get a clearer picture, we should create a histogram of the two samples.

To create histograms of the two samples:

1 Click **StatPlus > Multi-variable Charts > Multiple Histograms** from the Excel menu.

2 Verify that the **Use a column of category levels** option button is selected.

3 Click the **Data Values** button and select **Total_Days** from the list of range names.

4 Click the **Categories** button and select the range name, **Location**.

5 Click the **Output** button, click the **Cell** option button and select cell **A17** on the t-test worksheet.

6 Click the **OK** button twice to generate the two histograms. See Figure 6-22.

Figure 6-22
Distribution
of the
Total_Days
variable for
the rural
and
non-rural
areas

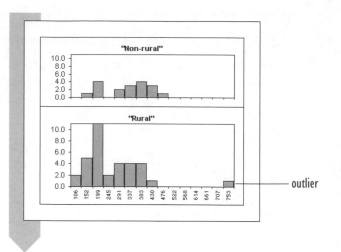

Figure 6-22 makes it clear why the standard deviation for the rural nursing homes is higher than that for the non-rural homes—there is an outlier in the rural home sample. One way of reducing the effect of this outlier on our hypothesis test is to switch to a nonparametric procedure, which is less influenced by the presence of outliers.

EXCEL TIPS

- You can also perform a two-sample *t*-test of your data using the Analysis ToolPak, supplied by Excel. To perform a two-sample *t*-test, load the Analysis ToolPak and then choose "Data Analysis" from the Tools menu.

- For a pooled test, click "t-Test: Two Sample Assuming Equal Variances" and specify the two columns containing the paired data.

- For an unpooled test, click "t-Test: Two Sample Assuming Unequal Variances."

- The Analysis ToolPak does not include confidence intervals for the two-sample *t*.

Applying a Nonparametric Test to Two-Sample Data

The two sample nonparametric test is the **Mann-Whitney test**. In the Mann-Whitney test we rank all of the values from smallest to largest and then sum the ranks in each sample. Unlike the Wilcoxon test, we do not rank the absolute data values or multiply the ranks by the sign of the original data. Table 6-6 shows an example of two sample data along with the calculated ranks.

Table 6-6 Two-Sample data

Sample-1 Values	Ranks	Sample-2 Values	Ranks
22	12.0	−3	3.0
16	11.0	−1	4.0
1	5.0	2	6.0
−4	1.5	8	9.0
7	8.0	−4	1.5
3	7.0		
9	10.0		

Note that we don't need to have equal sample sizes. Our null hypothesis is that both samples have the same median value. In this example, the sum of the Sample-1 ranks is 54.5, and the sum of the Sample-2 ranks is 23.5. We can use probability theory to determine the probability of the first sample having a rank sum of 54.5 or greater if the null hypothesis were true. In this case, that p-value would be 0.176, which would not support rejecting the null hypothesis.

When using the Mann-Whitney test, we also need to calculate the median difference between the two samples. This is done by calculating the difference for each pair of observations taken from Sample 1 and Sample 2 and then determining the median of those differences. For the data in Table 6-6, there are 35 pairs, starting with the difference between 22 and −3 (the first observations in the samples) and going down to the difference between 9 and −4 (the last observations). The median of these 35 differences is 7. By comparison, the difference of the sample averages is 7.31, so the median difference is pretty close. When the sample sizes get large, these calculations cannot be easily done by hand.

The Mann-Whitney test makes only four assumptions:

1. Both samples are random samples taken from their respective probability distributions.
2. The samples are independent of each other.
3. The measurement scale is at least ordinal.
4. The two distributions have the same shape.

The Mann-Whitney test relies on ranks, which lessens the effect of outliers. The downside is that under certain situations, the Mann-Whitney test will not be as efficient as the two-sample t-test in detecting differences between samples. Also, some researchers may not be familiar or comfortable with this nonparametric approach.

Let's apply the Mann-Whitney test to the nursing home data. Our hypotheses are

H_0: Median total patient days in the rural nursing homes = median total patient days in the non-rural homes

H_a: Median total patient days in the rural nursing homes ≠ median total patient days in the non-rural homes

To apply the Mann-Whitney test to the nursing home data:

1 Click **StatPlus > Two Sample Tests > 2 Sample Mann-Whitney Rank test** from the menu bar.

2 Verify that the **Use Column of Category Values** option button is selected.

3 As before, you can have your data in two separate columns (1 for each sample) or in a single column alongside a column of category levels.

4 Click the **Data Values** button and select **Total_Days** from the list of range names.

5 Click the **Categories** button and select the range name **Location**.

6 Click the **Output** button and send your output to a new worksheet named **MW-test**.

7 Click **OK** to generate the Mann-Whitney analysis for the nursing home data. See Figure 6-23.

Figure 6-23
Mann-
Whitney
Rank test
analysis of
the nursing
home data

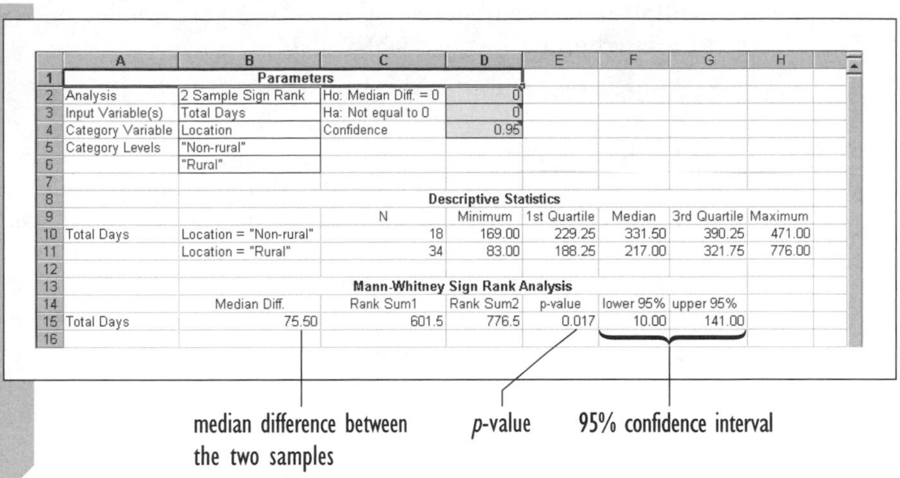

	A	B	C	D	E	F	G	H
1			Parameters					
2	Analysis	2 Sample Sign Rank	Ho: Median Diff. = 0	0				
3	Input Variable(s)	Total Days	Ha: Not equal to 0	0				
4	Category Variable	Location	Confidence	0.95				
5	Category Levels	"Non-rural"						
6		"Rural"						
7								
8			Descriptive Statistics					
9			N	Minimum	1st Quartile	Median	3rd Quartile	Maximum
10	Total Days	Location = "Non-rural"	18	169.00	229.25	331.50	390.25	471.00
11		Location = "Rural"	34	83.00	188.25	217.00	321.75	776.00
12								
13			Mann-Whitney Sign Rank Analysis					
14		Median Diff.	Rank Sum1	Rank Sum2	p-value	lower 95%	upper 95%	
15	Total Days	75.50	601.5	776.5	0.017	10.00	141.00	
16								

median difference between   *p*-value   95% confidence interval
the two samples

From the output in Figure 6-23, we note that the median number of patient days in non-rural hospitals is 331.50, or 33,150 in absolute numbers. The median number of patient days in rural hospitals is 217.00, or 21,700. We calculate the median difference between these two samples to be 75.50, or 7,550 patient days. This difference is statistically significant with a *p*-value of 0.017. Our final decision: We decide to reject the null hypothesis and accept the alternative, concluding that non-rural nursing homes are much more heavily utilized in terms of total patient days than their rural counterparts.

The 95% confidence interval gives a range of values for this difference. We conclude that the median difference is not less than 1,000 total patient days per year and not more than 14,100 patient days.

To complete your work:

Save your changes to the **NurseHome2** workbook and close the file.

Final Thoughts about Statistical Inference

The previous example displays some of the challenges and dangers in doing statistical inference. It is tempting to see a *p*-value or a confidence interval as the authoritative answer to your research. However, to use the tools of statistical inference properly, you should always be aware of the limitations of your statistical tests. Here are some general rules you should follow when performing statistical inference:

1. State your hypotheses clearly and, if possible, *before* collecting and analyzing your data.
2. Understand the nature and limitations of the statistical tests you use. Be aware of any assumptions that the test makes about the nature of your data. Try to verify that these assumptions are met (or at least that there is no evidence that they are being violated.)
3. Graph your data; it will help you more easily detect any departures from the assumptions of your statistical test. Calculate descriptive statistics of your data for the same reason.
4. If appropriate, perform more than one kind of statistical test. A different test, such as a nonparametric one, may provide important insight into your data.
5. Your goal is *not* to reject the null hypothesis. A study that fails to reject the null hypothesis is not a failure, nor is a low *p*-value a sign of success (especially if you're rejecting the null hypothesis in error.) Your goal should be to determine what, if any, conclusions you can reach about your data in a fair and impartial way and then to ascertain how reliable those conclusions are.

Exercises

1. *True or false (and why)*: "A 95% confidence interval that covers the range $(-5, 5)$ tells you that the probability is 95% that μ will have a value between -5 and 5.

2. *True or false (and why)*: "Accepting the null hypothesis means that the null hypothesis is true."

3. *True or false (and why)*: "Rejecting the null hypothesis means that the null hypothesis is false."

4. Consider a sample of 25 normally distributed observations with a sample average of 50.
   a. Calculate the 95% confidence interval if $\sigma = 20$.
   b. Calculate the 95% confidence interval if σ is unknown but if the sample standard deviation $= 20$.

5. The nationwide mean price for a 3-year-old Honda Civic is $8,500 with a known standard deviation of $600. You check the newspaper and find nine 3-year-old Civics in San Francisco selling for an average price of

$9,000. You wonder whether the cost of Civics in San Francisco is higher than for the rest of the nation.

a. State your question about the price of Civics in terms of a null and an alternate hypothesis. What are you assuming about the distribution of Civic prices?

b. Will the alternative hypothesis be one- or two-sided? Defend your answer.

c. Test your null hypothesis. Do you accept or reject it and at what p-value? Construct a 95% confidence interval for Civic prices in San Francisco.

d. Redo your analysis, but this time assume that the sample size is 10 with a sample average of $9,000 and a sample standard deviation of $600. Assume that you don't know the value of the nationwide standard deviation.

6. In tests of stereo speakers, ten American-made speakers had an average performance rating of 90 with a standard deviation of 5. Five imported speakers had an average rating of 85 with a standard deviation of 4.

a. Write a null and an alternative hypothesis comparing the two types of speakers.

b. Test the null hypothesis. What is the p-value?

c. If you decide to change the significance level to 10%, does your conclusion change?

7. Derive the formula for the t confidence interval based on the definition of the t-statistic shown earlier in this chapter.

8. Reopen the NURSEHOME.XLS workbook. Explore whether there is a significant difference between rural and non-rural homes in terms of size (as expressed by the number of beds in the homes).

a. Write down a set of hypotheses for exploring the question of whether the numbers of beds in rural and non-rural homes differ.

b. Apply a two-sample t-test to the data. Report your results assuming a pooled estimate of the standard deviation and assuming an unpooled estimate. What are the p-value and confidence interval under each assumption?

c. Plot the distribution of the Beds variable for the different locations. Do the plot and the descriptive statistics give you any help in determining whether to use a pooled or a non-pooled test? Explain your answer.

d. Apply the Mann-Whitney test to the data. What are your hypotheses? What is your conclusion? Include information on the p-value and confidence interval in your discussion.

e. Summarize your results. Do your conclusions differ based on which test you apply? Which statistical test would you report and why?

f. Save your workbook as E6NURSEHOME.XLS.

9. Reopen the NURSEHOME.XLS workbook again. This time you've been asked to explore whether rural homes are used at a lower rate than non-rural homes after adjusting for the differing size of the homes.
   a. Create a new variable named "Days_Beds" equal to the ratio of the total number of patient days to the number of beds in the home. Format the data to display three decimal places.
   b. Compare the average value of the Days_Beds variable for rural and non-rural homes. What are your hypotheses? What test or tests will you use to evaluate your null hypothesis?
   c. Plot the distribution of the Days_Beds variable for the two locations.
   d. Summarize your results, including any descriptive statistics, p-values, and confidence intervals you created during your analysis. Is there evidence to suggest that rural homes are being utilized at a lower rate?
   e. Save your workbook as E6NURSEHOME2.XLS.

10. Draft numbers from the Vietnam War have been saved in the DRAFT.XLS workbook. (See the Chapter 4 exercises for a discussion of the draft lottery.) It's been claimed that people whose birthday fell in the second half of the year had lower draft numbers and therefore were more likely to be drafted. Open the DRAFT.XLS workbook and explore this claim.
    a. Write down the null and alternative hypotheses for your study.
    b. Create a two-sample t-test to analyze your hypotheses. Do you use a pooled or an unpooled test? Which type of test does the distribution of the data support?
    c. Create a histogram of the distribution of draft numbers broken down by whether the number was assigned in the first half of the year or the second. What probability distribution does the data resemble? What property of the t-statistic allows you still to apply the t-test to your data?
    d. Calculate a 95% confidence interval for the average draft number for people born in the first half of the year and then for people born in the second half of the year. (*Hint*: You can do this using StatPlus's 1-sample t-test command, specifying the Half variable as the BY variable.)
    e. Report your results. What is the mean difference in draft number between people born in the first half of the year and those born in the second half? Is this a significant difference? What are the practical ramifications of your conclusions?
    f. Save your workbook as E6DRAFT.XLS.

11. The JRCOL.XLS workbook contains salary information for faculty at a college. The female faculty members claim that they are underpaid relative to their male counterparts. Open the JRCOL.XLS workbook and investigate their claim.
    a. Write down your null and alternative hypotheses. What is the significance level for this test?
    b. Perform a two-sample t-test on the salary data broken down by gender. Does it make any difference whether you perform a pooled or an

unpooled test? Does the data suggest that there is a salary difference between male and female faculty members? Create histograms of the distribution of salary data for male and female instructors.

c. Redo the two-sample t-test, this time breaking down the analysis by the Rank_Hired variable. Are there significant differences between salaries for the various employee ranks? (*Note*: Some combinations of gender and rank hired will have sample sizes of 0. This will result in Excel displaying a #VALUE! result in the workbook. You can ignore these employee ranks because there is no data to investigate.)

d. Write your conclusions. Is there evidence that the college has underpaid its female faculty? If so, does this difference exist for all teaching ranks? Why does this study not prove sexual discrimination? What factors have been ignored?

e. Save your workbook as E6JRCOL.XLS.

12. The BIGTEN.XLS workbook has graduation information on Big Ten schools. (See Chapter 3 for a discussion of this data set.) Open the BIGTEN.XLS workbook and explore whether there is a difference in the graduation rates between white male athletes and white female athletes.

a. State your null and alternative hypotheses.

b. Perform a paired t-test of the white male and white female athlete graduation rates. Is there statistically significant evidence of a difference in the graduation rates? What is a 95% confidence interval for the difference? What is the 90% confidence interval?

c. Redo the analysis using the Wilcoxon Signed Rank test and the Sign test.

d. Why is this an example of paired data?

e. Summarize your conclusions. Can you apply your results to universities in general? Defend your answer.

f. Save your workbook as E6BIGTEN.XLS.

13. The MORTGAGE.XLS workbook contains information on refusal rates from 20 lending institutions broken down by race and income status. It has been claimed that lending institutions have significantly higher refusal rates for minorities. Open this workbook and test that claim.

a. State your null and alternative hypotheses.

b. Apply a paired t-test to the refusal rates for minority and white applicants. What is the 95% confidence interval for the difference in refusal rates? What is the p-value for the test?

c. Create a histogram and normal probability plot of the difference in refusal rate. Does the data appear normal?

d. Redo your analysis using the Wilcoxon Signed Rank test. How do your results compare to the paired t-test?

e. Redo questions a through d using the refusal rates for high-income whites and minorities. How do the results of the two analyses compare, especially in terms of the confidence interval for the difference in refusal rate? Is there evidence to suggest that there is no refusal rate gap for higher-income minorities?

f. Summarize your conclusions and save the workbook as E6MORT-GAGE.XLS.

14. The TEACHER.XLS workbook stores average teacher salary, public school spending per pupil, and the ratio of teacher to pupil spending for 1985, broken down by state and region. Open the workbook and do the following:

    a. Construct a 95% *t*-confidence interval for each of the numeric variables, broken down by region.

    b. Construct a 95% Wilcoxon Signed Rank confidence interval for the numeric variables, by region.

    c. Save your workbook as E6TEACHER.XLS.

15. The POLU.XLS workbook contains data on the number of unhealthful pollution days for 14 U.S. cities from 1980 to 1989. Analyze what impact environmental regulations have had on pollution.

    a. State your null and alternative hypotheses for this analysis.

    b. Open the POLU.XLS workbook and perform a one-sample *t*-test on the difference between the number of pollution days in 1980 and the average number of pollution days between 1985 and 1989 (use the Diff80 variable). What is the 95% confidence interval for the difference in pollution days? Do you accept or reject the null hypothesis and at what *p*-value?

    c. Create a normal probability plot of the difference data. Does the data seem normally distributed? Are there any problems in the distribution of the data that may cause problems with the one-sample *t*-test?

    d. Perform a Wilcoxon Signed Rank test on the difference data. How do your results compare with the one-sample *t*? What would account for this? Which test would you use in a report and why?

    e. Summarize your findings. Save your workbook as E6POLU.XLS.

16. In a NASA-funded study, seven men and eight women spent 24 days in seclusion to study the effects of gravity on circulation. Without gravity, there is a loss of blood from the legs to the upper part of the body. The study started with a 9-day control period in which the subjects were allowed to walk around. Then followed a 10-day bed-rest period in which the subjects' feet were somewhat elevated to simulate weightlessness in space. The study ended with a 5-day recovery period in which the subjects again were allowed to walk around. Every few days, the researchers measured the electrical resistance at the calf, which increases when there is a blood loss. The electrical resistance gives an indirect measure of the blood loss and indicates how the subject's body responds to the conditions. Data for this study have been stored in the SPACE.XLS workbook. You've been asked to examine whether the male subjects and the female subjects differed in how they

responded to the study. You're to perform your analysis for each of the days in the study. Open the SPACE.XLS workbook to start your investigation.

a. State your null and alternative hypotheses.

b. Perform a two-sample *t*-test comparing the value of the Resistance variable between the male and female subjects, broken down by day. You do not have to summarize your results across days.

c. On what day or days is there a significant difference between the two groups? Do your results change if you use an unpooled rather than a pooled estimate of the standard deviation?

d. Create a scatterplot of Resistance versus Days. Break the scatterplot down by gender using the StatPlus command shown in Chapter 3. Describe the effect displayed in the scatterplot (you may want to change the scatterplot scales to view the data better).

e. Redo your analysis using the Mann-Whitney test. Do your conclusions from part b change any with the nonparametric procedure?

f. Summarize your findings. Explain how (if at all) the male and female subjects differed in their response to the study. Include in your discussion the various parts of the study (control period, bed rest period, etc.) and how the patients responded during those specific intervals. Include any pertinent statistics.

g. Save your workbook as E6SPACE.XLS.

17. The MATH.XLS workbook contains data from a study analyzing two methods of teaching mathematics. Students were randomly assigned to two groups: a control group that was taught in the usual way and an experimental group that was regularly assigned homework and given frequent quizzes. Students in the experimental group were allowed to retake their exams to raise their grades (though a different exam was given for the retake). The final exam scores of the two groups were recorded. Open the MATH.XLS workbook and investigate whether there is compelling evidence that students in the experimental group had higher scores than those in the control group.

a. State your null and alternative hypotheses. Is this a one-sided test or a two-sided test? Why?

b. Perform a two-sample *t*-test on the final exam score. Use a pooled estimate of the standard deviation. What is the 95% confidence interval for the difference in scores? What is the *p*-value? Do you accept or reject the null hypothesis? Do your conclusions change if you use an unpooled test?

c. Chart the distribution of the final exam scores for the two groups. What do the charts tell about the distributions? Do the charts cast any doubt on your conclusions in part b? Why?

d. Do a second analysis of the data using the Mann-Whitney Rank test. How do these results compare to the two-sample *t*? Are your conclusions the same?

e. Summarize your findings and report your conclusion. Is there a significant change in the exam scores under the experimental approach?

f. Save your workbook as E6MATH.XLS.

18. The VOTING.XLS workbook contains the percentage of the presidential vote that the Democratic candidate received in 1980 and 1984, broken down by state and region. You've been asked to investigate the difference between the 1980 and 1984 voting patterns.
    a. Open the workbook and calculate the paired *t*-test for the voting percentage, broken down by area. Summarize your findings for all areas as well.
    b. For which areas was there a significant change in the voting percentage? For which areas was there no significant change? What was the overall change in the voting percentage across all areas?
    c. Summarize your findings, including your descriptive statistics, *p*-values, and confidence intervals.
    d. Save your workbook as E6VOTING.XLS.

19. Open the workbook CALCFM.XLS. The workbook shows the first-semester calculus scores for male and female students. Analyze the data set to determine whether there is a significant difference between the two groups.
    a. State your null and alternative hypotheses.
    b. Perform a two-sample *t* test on the data, using a pooled estimate of the variance. What is the 95% confidence interval of the difference between the two groups? Do you reject or accept the null hypothesis? At what *p*-value?
    c. Chart the distribution of exam scores for the two groups. Do the distributions appear normal? What property of exam scores makes it unlikely that these exam scores follow the normal distribution? (*Hint*: Test scores are usually constrained to fall between 0 and 100.) What property of the *t*-distribution might allow you to use the *t*-test anyway?
    d. Summarize your findings and state your conclusions.
    e. Save the workbook as E6CALCFM.XLS.

20. The RACEPAIR.XLS workbook contains information on reaction times and race times, recorded by sprinters running in the first three rounds of the 100-meter dash at the 1996 Summer Olympics. You're asked to determine whether there is evidence that the sprinter's reaction time (the time it takes for the sprinter to leave the starting block at the sound of the gun) changes as he advances in the competition. Open the RACEPAIR.XLS workbook.
    a. Use the paired *t*-test and analyze the differences between the following variables: React 1 vs. React 2, React 1 vs. React 3, and React 2 vs. React 3. Calculate the 95% confidence interval for each difference pair, and test for statistical significance at the 5% level. Are there any pairs of rounds in which there is a significant difference in the average reaction time?

b. Create three new columns in the Reaction Times worksheet displaying the three paired differences, and then create three normal probability plots of those differences. Does the distribution of the paired differences follow a normal distribution?

c. Redo the analysis, this time using the Wilcoxon Sign Rank test. Do your conclusions change when you use this test?

d. Summarize your results and give your conclusions. Have you accepted the null hypothesis, or is there evidence that reaction times do change from one round to another?

e. Save your workbook as E6RACEPAIR.XLS.

21. Reopen the RACEPAIR workbook from the previous exercise. This time, analyze the race times from the three rounds of the 100-meter dash.

a. Perform a paired *t*-test of the race times (use the Race1, Race2, and Race3 variables), comparing the differences between Round 1 and Round 2, and then Round 2 and Round 3. Is there significant evidence that the race times decrease as the runner advances in the competition? Calculate the 95% confidence interval for the change in race time.

b. Summarize your results, including any descriptive statistics and *p*-values. Does your evidence suggest any difference in the competition level as a runner goes from Round 1 to Round 2 as compared to going from Round 2 to Round 3?

c. Save your workbook as E6RACEPAIR2.XLS.

Chapter 7

Tables

Objectives

In this chapter you will learn to:

- Create pivot tables of a single categorical variable

- Create pie charts and bar charts

- Relate two categorical variables with a two-way table

- Apply the chi-square test to a two-way table

- Compute expected values of a two-way table

- Combine or eliminate small categories to get valid tests

- Test for association between ordinal variables

- Create a custom sort order for your workbook

In this chapter you'll learn how to work with categorical data in the form of tables and ordinal variables. You'll learn how to use Excel's PivotTable feature to create tables, and you'll explore how to analyze the data in the table using StatPlus.

Pivot Tables

In the previous chapter you used t-tests and nonparametric tests to analyze continuous variables. You can also apply hypothesis tests to categorical and ordinal data. This type of data is most commonly seen in surveys, which record counts broken down by categories. For example, you create a table of instructors broken down by title (assistant professor, associate professor, or full professor) and gender (male or female). Are there significantly more male full professors than females with that title? How many female professors would you expect given the data? An analysis of categorical data addresses questions of this type.

To illustrate how to work with categorical variables, let's look at data from a survey of professors who teach statistics courses. The SURVEY.XLS data set includes 392 responses to questions about whether the course requires calculus, whether statistical software is used, how the software is obtained by students, what kind of computer is used, and so on. The workbook contains the following variables:

Table 7-1 Survey of Statistics Professors data

Range Name	Range	Description
Computer	A2:A393	Computer used in the course
Dept	B2:B393	Department
Available	C2:C393	Type of computer system available to the student
Interest	D2:D393	The amount of interest in a supplementary statistics text
Calculus	E2:E393	The extent to which calculus is required for the course
Uses_Software	F2:F393	Whether the course uses software or not
Enroll_A	G2:G393	Categorical variable indicating semester course enrollment in the instructor's course. For example, 001-050 means that from 1 to 50 students are enrolled each semester.
Enroll_B	H2:H393	Categorical variable indicating annual course enrollment
Max_Cost	I2:I393	Maximum cost for a supplementary computer text

To open Survey.xls:

1 Start Excel if necessary and maximize the Excel window.

2 Open **Survey.xls** from your Student disk.

3 Save the workbook as **Survey2.xls**. The workbook appears as shown in Figure 7-1.

Figure 7-1
Survey data

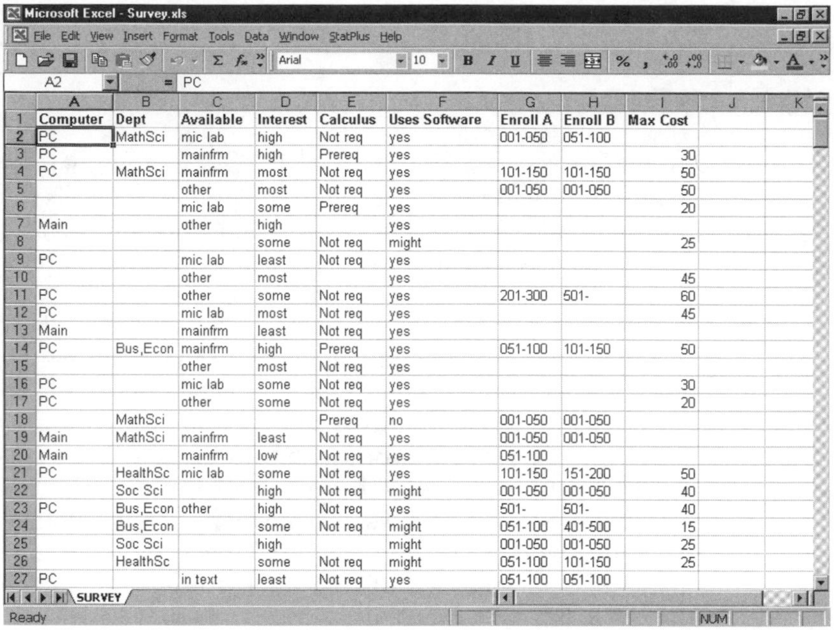

	A	B	C	D	E	F	G	H	I	J	K
1	Computer	Dept	Available	Interest	Calculus	Uses Software	Enroll A	Enroll B	Max Cost		
2	PC	MathSci	mic lab	high	Not req	yes	001-050	051-100			
3	PC		mainfrm	high	Prereq	yes			30		
4	PC	MathSci	mainfrm	most	Not req	yes	101-150	101-150	50		
5			other	most	Not req	yes	001-050	001-050	50		
6			mic lab	some	Prereq	yes			20		
7	Main		other	high		yes					
8				some	Not req	might			25		
9	PC		mic lab	least	Not req	yes					
10			other	most		yes			45		
11	PC		other	some	Not req	yes	201-300	501-	60		
12	PC		mic lab	most	Not req	yes			45		
13	Main		mainfrm	least	Not req	yes					
14	PC	Bus,Econ	mainfrm	high	Prereq	yes	051-100	101-150	50		
15			other	most	Not req	yes					
16	PC		mic lab	some	Not req	yes			30		
17	PC		other	some	Not req	yes			20		
18		MathSci			Prereq	no	001-050	001-050			
19	Main	MathSci	mainfrm	least	Not req	yes	001-050	001-050			
20	Main		mainfrm	low	Not req	yes	051-100				
21	PC	HealthSc	mic lab	some	Not req	yes	101-150	151-200	50		
22		Soc Sci		high	Not req	might	001-050	001-050	40		
23	PC	Bus,Econ	other	high	Not req	yes	501-	501-	40		
24		Bus,Econ		some	Not req	might	051-100	401-500	15		
25		Soc Sci		high		might	001-050	001-050	25		
26		HealthSc		some	Not req	might	051-100	101-150	25		
27	PC		in text	least	Not req	yes	051-100	051-100			

You've been asked to determine what computers are used most often in statistics instruction. The Computer variable contains this information. The answers fall into four categories: two kinds of microcomputers (Macintoshes and PCs), minicomputers, and mainframes. (It makes sense to put the IBM PC and all of its clones in one category, because they all use the same type of software.)

Using the PivotTable Wizard

In Excel, you obtain category counts by generating a **pivot table**, a worksheet table that summarizes data from the source data list (in this case the survey data). Excel pivot tables are interactive—that is, you can update them whenever you change the source data list, and you can switch row and column headings to view the data in different ways (hence the term *pivot*).

Try creating a pivot table that summarizes the computer usage data. Excel provides a PivotTable Wizard, similar to the ChartWizard you've used to make charts, that makes creating a pivot table easy.

To start the PivotTable wizard:

1 Click **Data > PivotTable and PivotChart Report** to open the PivotTable Wizard - Step 1 of 4 dialog box.

2 Verify that the **Microsoft Excel List** or **Database option button** is selected and that the **PivotTable** option button is selected.

3 Click **Next**.

Now the PivotTable wizard requests the worksheet range that contains the data you want to use. If the data in your worksheet are arranged as a list, this range appears in the Range box.

4 Verify that the range **$A$1:$I$393** is entered in the Range box; then click **Next**.

Now you control the layout of the pivot table. A pivot table has four areas. The **Row area** determines the categories that will appear in each row of the table. Similarly, the **Column area** controls the categories for each table column. The **Data area** determines the values that will appear at each intersection of row and column categories. Finally, the **Page area** is used to create different tables for different categories of a categorical variable.

To specify the pivot table layout:

1 Click the **Layout** button.

Excel displays a graphical layout of the pivot table. You can place different variables in the table by dragging a button on the right to the right pivot table diagram.

2 Drag the **Computer** button to the Row section of the table.

3 This tells Excel to use different computer types (PC, Main, Mini, and Mac) as the row labels in the table.

4 Drag the **Computer** button again to the Data area of the table.

The label changes to "Count of Computer," indicating that the pivot table will show the counts of each type of computer. Your dialog box should appear as shown in Figure 7-2.

**Figure 7-2
Pivot table
layout**

column area ——

page area ——

row area ——

data area ——

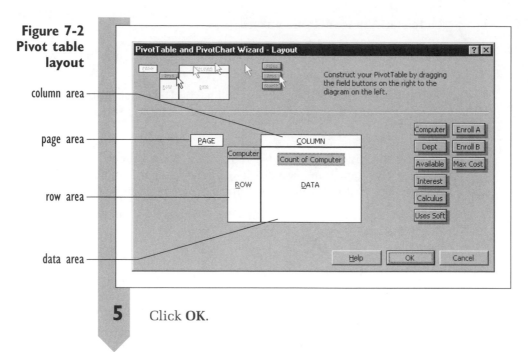

5 Click **OK**.

Next you specify some options for the pivot table's appearance.

To control the pivot table's appearance:

1 Click the **Options** button.

2 Click the **Grand totals for rows** checkbox to deselect it.

The grand total checkboxes allow you to calculate the sum of the counts across the different category levels. In this example, you don't have a category variable in the column area, so you don't need to calculate the grand total across each row.

3 Type **0** in the **For empty cells, show** box.

4 This means that if there are no observations for a particular cell in the table, Excel will show a count of 0.

The PivotTable Options dialog box appears as shown in Figure 7-3.

Figure 7-3
Pivot table
options

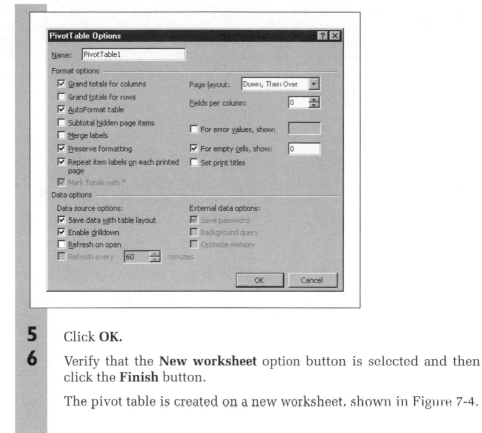

5 Click **OK.**

6 Verify that the **New worksheet** option button is selected and then click the **Finish** button.

The pivot table is created on a new worksheet, shown in Figure 7-4.

Figure 7-4
Pivot table
displaying
the types of
computers
used in the
Survey
workbook

Computer
drop-down box

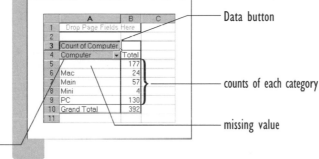

The first column of the pivot table contains the four categories in the Computer column (Mac, Main, Mini, and PC), plus a blank cell for missing values (cell A3). The second column shows the number of statistics professors who use computers of each type in their classes. (Excel might also display a PivotTable toolbar. You will not use the toolbar in this chapter, but you can learn more about it using Excel's on-line Help.) You discover that 130 of the 392 professors used only PCs, 57 used only mainframes, 24 used only Macintoshes, 4 used only minicomputers, and 177 did not respond or the data was not recorded. The number of nonresponders might be large because many classes used more than one computer type, and nothing was entered for those classes in the Computer column. If you were to include classes that used more than one computer type, you would still find the PC far in front. More classes use the PC than all other types of computers put together.

Removing Categories from a Pivot Table

The number of nonresponders doesn't interest you right now, so you decide to remove that category from the pivot table. Pivot tables include drop-down list boxes that you can use to specify which categories are displayed in the table.

To remove a category from the pivot table:

1 Click the **Computer** drop-down box on the pivot table.

2 Deselect the blank checkbox in the list of categories. See Figure 7-5.

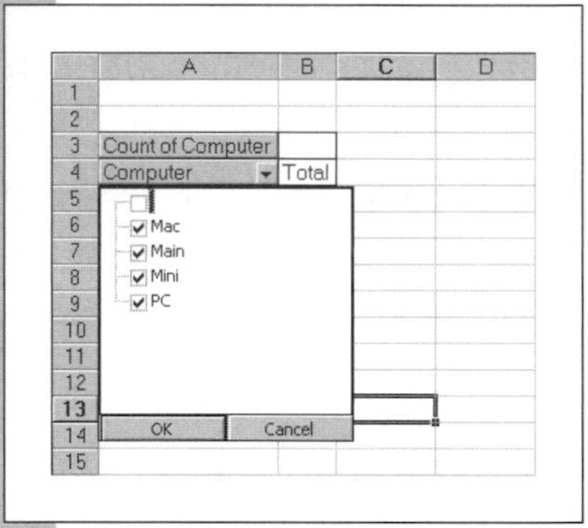

**Figure 7-5
Removing
the blank
category
from the
pivot table**

3 Click the **OK** button.

4 The pivot table changes and no longer displays the nonresponders category. See Figure 7-6.

Figure 7-6
The pivot
table with
the blank
category
removed

Figure 7-6 The pivot table with the blank category removed

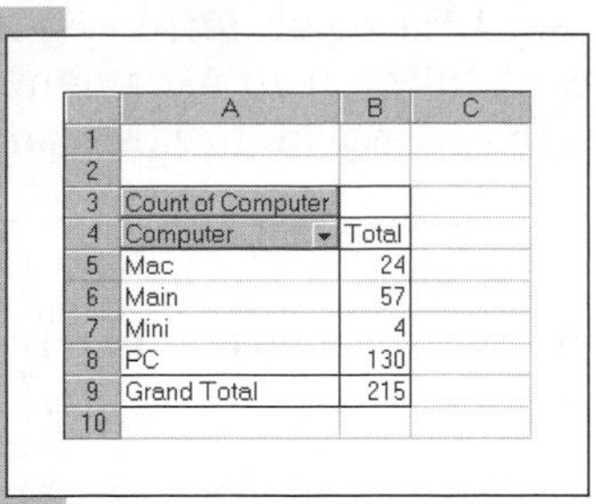

	A	B	C
1			
2			
3	Count of Computer		
4	Computer ▼	Total	
5	Mac	24	
6	Main	57	
7	Mini	4	
8	PC	130	
9	Grand Total	215	
10			

You can now see that 215 professors who responded to the survey listed only one computer. Of these, 130 listed the PC. If you wanted to continue the analysis, you could remove other categories from the table, perhaps reformatting the table to show only PCs and Macs.

Changing the Values Displayed by the Pivot Table

By default, the pivot table displays the count for each cell in the table. You can choose a variety of other types of values to display, including sums, maximums, minimums, averages, and percentages. When you choose to display percentages, you can show the percentage of all of the cells in the table or the percentage within each row or column. Try this now by changing the table to show the percentage for each category.

To display percentages of the values within the columns:

1 Right-click any of the count values in column B, and then click **Field Settings** from the menu.

2 Click the **Options >>** button to display the Show Data as list box.

3 Click **% of column** in the Show Data As list box; then click **OK**.

The pivot table is modified to show values as percentages of the total rather than as counts. See Figure 7-7.

Figure 7-7
Pivot table
showing
percentages

Figure 7-7 Pivot table showing percentages

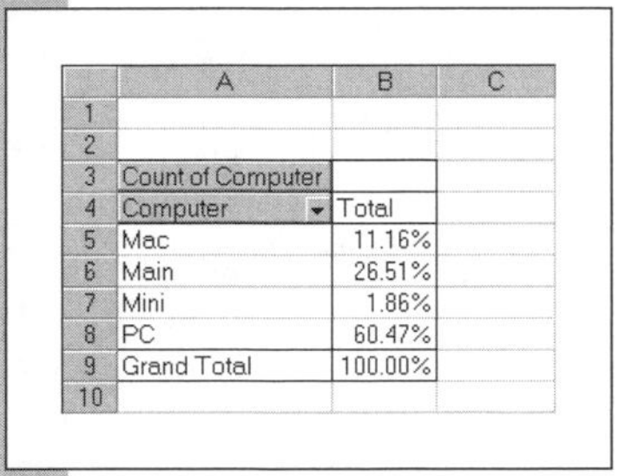

	A	B	C
1			
2			
3	Count of Computer		
4	Computer ▼	Total	
5	Mac	11.16%	
6	Main	26.51%	
7	Mini	1.86%	
8	PC	60.47%	
9	Grand Total	100.00%	
10			

You can see at a glance that 60.47% of professors who listed only one computer use the PC, followed by 26.51% who use the mainframe, 11.16% who use the Macintosh, and less than 2% who use the minicomputer.

To view the counts of the responses again:

1 Right-click any of the percentages in column B; then click **Field Settings** in the shortcut menu.

2 If necessary, click the **Options >>** button to display the Show Data as list box.

3 Click **Normal** in the Show Data as list box; then click **OK**.

Displaying Categorical Data in a Bar Chart

You can quickly display your pivot table data in a pivot chart. The default pivot chart is a bar chart, in which the length of the bar is proportional to the number of counts in each cell.

To create a bar chart for computer usage:

1 Right-click anywhere within the pivot table and click **PivotChart** from the menu.

Excel produces a bar chart on a separate sheet as shown in Figure 7-8.

**Figure 7-8
Bar chart of
computer
usage data**

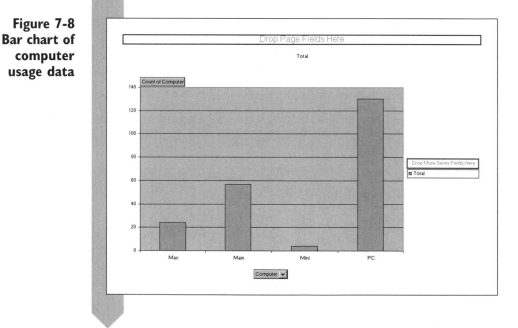

The pivot chart works the same as the pivot table. For example, if you click the Computer drop-down arrow, you can add categories to or remove them from the chart. The pivot chart is also linked to the pivot table, so any changes you make to layout or formatting of the chart are automatically reflected in the pivot table.

What can we learn from the bar chart? Obviously, that PCs are most often used in these courses. We can quickly see that PCs are used about twice as often as the nearest competitor—mainframes. Bar charts are often used by statisticians when the need is to show the relative sizes of the groups. What is not clear from the chart is the size of each group compared to the whole. For example, do PCs comprise more than half of the total computer use? It's not so easy to determine that kind of information from the bar chart. To deal with that problem, we can have Excel add the counts to the chart.

To add counts to the bar chart:

1 Click **Chart > Chart Options**.

2 Click the **Data Labels** tab.

3 Click the **Show value** option button and then click **OK**.

Excel updates the chart, which now shows the counts for each bar. See Figure 7-9.

Figure 7-9
Bar chart
with counts

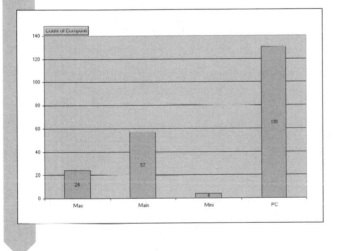

From this display we can see that 130 courses use PCs, and a little arithmetic tells us that this is more than half of all of the courses in the sample.

Displaying Categorical Data in a Pie Chart

Another way of comparing the size of individual groups to the whole is the pie chart. A **pie chart** displays a circle (like a pie), in which each pie slice represents a different category. Let's see how the pie chart displays the computer data. Rather than recreating the chart from scratch, we'll simply change the chart type of the current graph.

To display a pie chart:

1 Click **Chart > Chart Type**.

2 Click **Pie** from the list of Chart types and click the **OK** button.

Excel displays the pie chart in Figure 7-10.

Figure 7-10
Pie chart
with counts

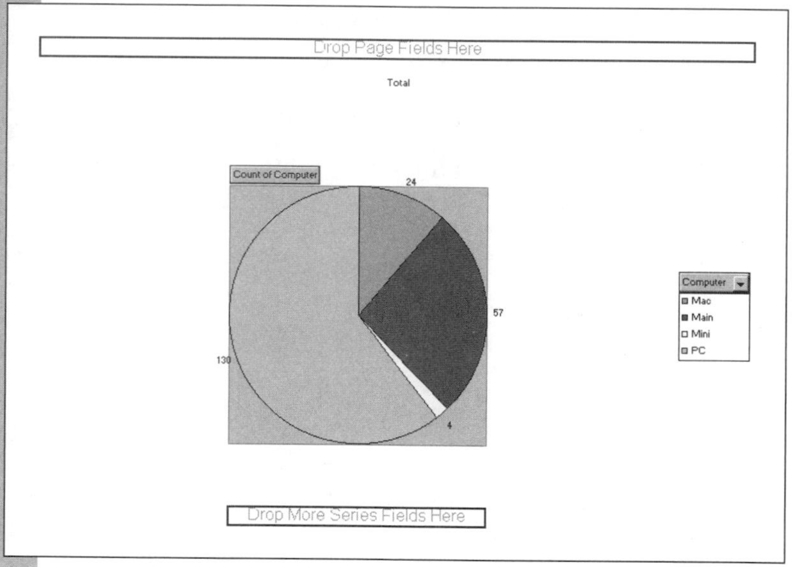

From the pie chart we can easily see that PCs account for more than half of the computers used—something we confirmed in the bar chart only by adding counts to the chart. The graph retained these count labels when we converted to the pie chart, but we can change that option as well. We can display the percentage of each slice, and we can add a label to the slice identifying which category it represents.

To change the pie chart's display options:

1 Right-click the pie chart and click **Format Data Series** from the menu.

2 Click the **Data Labels** tab.

3 Within the Data Labels dialog sheet, click the **Show label and percent** option button.

4 Click **OK**.

The chart changes as shown in Figure 7-11, displaying the percentages and labels for each pie slice.

**Figure 7-11
Pie chart
with
percents
and labels**

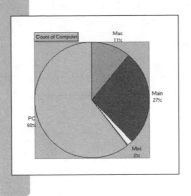

The pie chart is an effective tool for displaying categorical data, though it tends to be used more often in business reports than in statistical analyses.

EXCEL TIPS

- To view the source data for any particular cell in a pivot table, double-click the cell. Excel will open a new worksheet containing the observations from the original data source that comprise that cell.

- You can specify data sources other than the current workbook for your pivot table data, such as databases and external files.

- If the values in the data source change, the pivot table updates automatically to reflect those changes.

- You can create your own customized functions for the pivot table. To do so, right-click any cell in the table and choose "Formulas" and "Calculated Field" from the menu.

Two-Way Tables

What about the relationship between two categorical variables? You might want to see how computer use varies by department. Do different departments tend to choose different types of computers? Does one department tend to use Macintoshes, whereas another tends to use PCs? Excel's pivot table feature can compile a table of counts for computer use by department. Departments fall into four categories:

- Bus,Econ for business and economics
- HealthSc for health sciences
- MathSci for mathematics, statistics, physical sciences, and engineering
- Soc Sci for psychology and other social sciences

To create a pivot table for computer by department:

1 Return to the Survey worksheet.

2 Click **Data > PivotTable and PivotChart Report**.

Because you are using the same data source as was used for the first pivot table, you don't need to reselect it.

3 Click the **Another PivotTable or PivotChart** option button and click **Next**.

Excel displays a list of pivot tables in the workbook. There is only one other pivot table, so it is automatically selected.

4 Click **Next**.

5 Click the **Layout** button.

6 Drag the **Computer** button to the Row area and then again to the Data area.

7 Drag the **Dept** button to the Column area. See Figure 7-12.

**Figure 7-12
Creating a
two-way
pivot table**

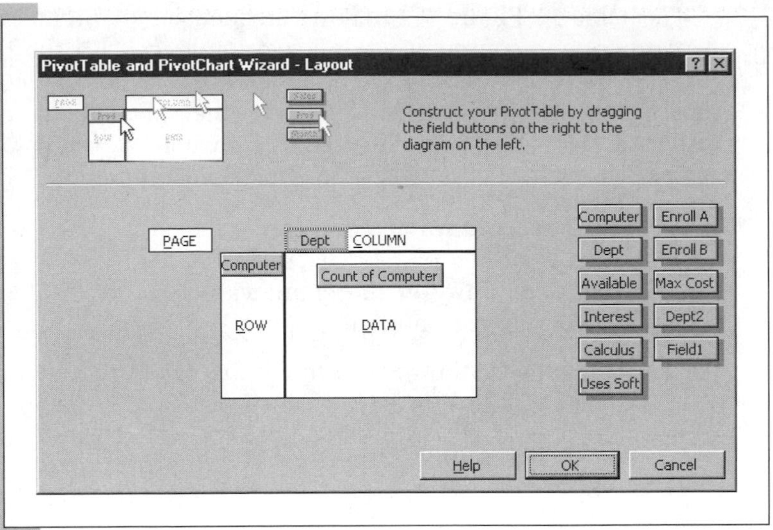

8   Click **OK**.

9   Click the **Options** button.

10   Type **0** in the **For Empty cells, show** box and click **OK**.

11   Click **Finish** to start creating the pivot table.

When you created the first pivot table, you hid the blank category levels for the table. Do this as well for the two-way table.

To hide the missing values:

1   Click the **Computer** drop-down list arrow in the pivot table and deselect the blank checkbox. Click **OK**.

2   Click the **Dept** drop-down list arrow and deselect the blank checkbox. Click **OK**.

Figure 7-13 shows the completed pivot table.

**Figure 7-13
Two-way
table of
computer
versus
department**

	A	B	C	D	E	F	G
1							
2							
3	Count of Computer	Dept					
4	Computer	Bus.Econ	HealthSc	MathSci	Soc Sci	Grand Total	
5	Mac	4	0	12	6	22	
6	Main	12	3	22	13	50	
7	Mini	1	0	2	1	4	
8	PC	49	8	29	28	114	
9	Grand Total	66	11	65	48	190	
10							

The table in Figure 7-13 shows frequencies of different combinations of department and computer used. For example, cell E5, the intersection of Soc Sci and Mac, shows that six professors in the Social Science departments reported the Macintosh as the choice for their classes. There are a total of 190 responses. How do these values compare when viewed as percentages within each department? Let's modify the pivot table to find out.

To show column percentages:

1 Right-click any of the count values in the table; then click **Field Settings** in the shortcut menu.

2 Click the **Options >>** button to display the Show Data as list box if it is not visible.

3 Click **% of column** in the Show Data as list box; then click **OK**. See Figure 7-14.

Figure 7-14
Two-way
table of
percentages

	A	B	C	D	E	F	G
1							
2							
3	Count of Computer	Dept					
4	Computer	Bus,Econ	HealthSc	MathSci	Soc Sci	Grand Total	
5	Mac	6.06%	0.00%	18.46%	12.50%	11.58%	
6	Main	18.18%	27.27%	33.85%	27.08%	26.32%	
7	Mini	1.52%	0.00%	3.08%	2.08%	2.11%	
8	PC	74.24%	72.73%	44.62%	58.33%	60.00%	
9	Grand Total	100.00%	100.00%	100.00%	100.00%	100.00%	
10							

The PC percentages are high for Bus,Econ and HealthSc, and, relatively speaking, the Mac percentage is high for MathSci. That is, 74.24% of Bus,Econ classes use the PC (cell B8), but only 6.06% of Bus,Econ classes use the Macintosh (cell B5). For MathSci, only 44.62% use the PC (cell D8), but 18.46% use the Macintosh (cell D5). You are looking for relative differences here: The MathSci 18.46% is not high when compared to the PC percentage, but it is high when compared to the 6.06% Macintosh usage for Bus,Econ. It would be instructive to redo the table to show the Macintosh usage as a percentage not of the total computer use but of the PC use. This might do a better job of highlighting the differences between departments in the ratio of Macintosh use to PC use.

To calculate the percentage of each cell relative to the number of PCs used:

1 Right-click any of the percentages in the table; then click **Field Settings** in the shortcut menu.

2 Click **% Of** in the Show Data as list box (the third option in the list).

3 Click **Computer** in the Base Field list box if it's not already selected.

4 Click **PC** in the Base Item list box. See Figure 7-15.

Figure 7-15
PivotTable
Field dialog
box with %
Of option
selected

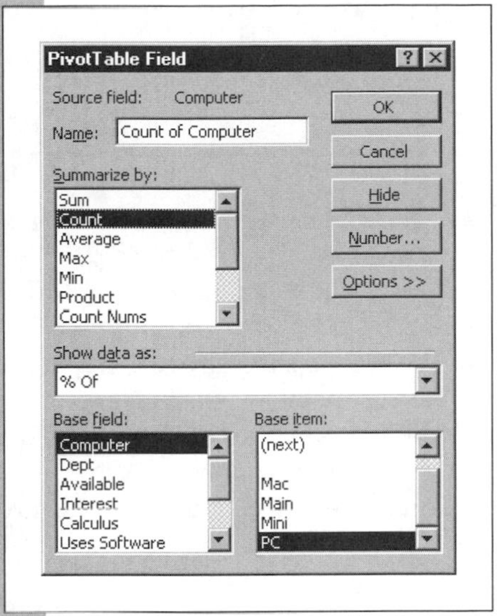

5 Click **OK**.

Figure 7-16 shows the pivot table, where each cell is expressed as a percentage of the number of courses that use PC's within each department.

Figure 7-16
Pivot table
values
expressed
as a
percentage
of PC usage

	A	B	C	D	E	F	G
1							
2							
3	Count of Computer	Dept					
4	Computer	Bus,Econ	HealthSc	MathSci	Soc Sci	Grand Total	
5	Mac	8.16%	0.00%	41.38%	21.43%	19.30%	
6	Main	24.49%	37.50%	75.86%	46.43%	43.86%	
7	Mini	2.04%	0.00%	6.90%	3.57%	3.51%	
8	PC	100.00%	100.00%	100.00%	100.00%	100.00%	
9	Grand Total						
10							

The Bus,Econ departments used Macintoshes at a rate of 8.16% of the PC usage rate (cell B5), whereas the MathSci departments used Macs at a rate of 41.38% of the rate of PC usage (cell D5). Overall, the Macs were used almost $\frac{1}{5}$ as often as the PC. Perhaps you anticipated that business and economics departments would be inclined to use the PC because of its prevalence in industry. It's also probably not a surprise that the Macintosh is popular with mathematicians.

To reformat the table to show counts again:

1 Right-click any of the percentages in the pivot table; then click **Field Settings** in the shortcut menu.

2 Click **Normal** in the Show Data as list box; then click OK.

Computing Expected Counts

If the pattern of computer use were the same in each department, we would expect to find the column percentages (shown in Figure 7-14) to be about the same for each department. We would then say that department and choice of computer are **independent** of each other, so that the pattern of usage does not depend on the department. It's the same for all of them. On the other hand, if there *is* a difference in how the computers are used between departments, we would say that department and computer use are **related**. We cannot say anything about what types of computers are being used without knowing which department is being examined.

You've seen that there might be a relationship between computer use and department. There is some evidence, for example, that the way Macs are being used differs between one department and another. Is this difference significant? We could formulate the following hypotheses:

H_0: The pattern of computer usage is the same in all departments

H_a: The pattern of computer usage is related to the department

How can you test the null hypothesis? Essentially, you want a test statistic that will examine the pattern of computer use across departments and then compare it to what we would expect to see if computer use and department type were independent variables.

How do you compute the expected counts? Under the null hypothesis, the percentage of each computer used should be the same in each department. Our best estimate of these percentages comes from the percentage of the grand total, shown in column F of Figure 7-14. Thus we expect that about 11.58% of the courses use Macs, 26.32% use mainframes, and so forth. To express this value in terms of counts, we multiply the expected percentage by the total number of courses in each department. For example, there are 65 courses in the MathSci departments, and if 11.58% of these were using Macintoshes, this would be 65 × 0.1158, or 7.53, departments. Note that the actual observed value is 12 (cell B5 in Figure 7-13), so more Macintoshes are being used in these departments than expected. This is equivalent to the formula

$$\text{Expected Count} = \frac{(\text{Row Total}) \times (\text{Column Total})}{\text{Total Observations}}$$

Thus the expected count of Macintoshes for the MathSci department could also be calculated as follows:

$$\text{Expected Count} = \frac{22 \times 65}{190} = 7.53$$

To create a table of expected counts, you can either use Excel to perform the manual calculations or use the StatPlus add-in to create the table for you.

To create a table of expected counts:

1 Click **StatPlus > Table Statistics** from the Excel menu.

2 Select the range **A4:E8**.

3 Click the **Output** button; then click the **New Worksheet** option button and type the worksheet name, **Computer Department Table**. Click **OK**.

4 Click the **OK** button to start generating the table of expected counts. See Figure 7-17.

Figure 7-17
Tables of observed and expected counts

Observed Counts	Bus,Econ	HealthSc	MathSci	Soc Sci
Mac	4	0	12	6
Main	12	3	22	13
Mini	1	0	2	1
PC	49	8	29	28

Expected Counts	Bus,Econ	HealthSc	MathSci	Soc Sci
Mac	7.64	1.27	7.53	5.56
Main	17.37	2.89	17.11	12.63
Mini	1.39	0.23	1.37	1.01
PC	39.60	6.60	39.00	28.80

The command generates some output in addition to the table of expected counts shown in Figure 7-17, which we'll discuss later. The values in the Expected Counts table are the counts we would expect to see if computer usage were independent of department.

The Pearson Chi-Square Statistic

With our tables of observed counts and expected counts, we need to calculate a single test statistic that will summarize the amount of difference between the two tables. In 1900, the statistician Karl Pearson devised such a test statistic, called the **Pearson chi-square**. The formula for the Pearson chi-square is

$$\text{Pearson Chi-Square} = \sum_{\text{all cells}} \frac{(\text{Observed Count} - \text{Expected Count})^2}{\text{Expected Count}}$$

If the frequencies all agreed with their expected values, this total would be 0. If there were a substantial difference between the observed and expected counts, this value would be large. For the data in Figure 7-13, this value is

$$\text{Pearson Chi-Square} = \frac{(4 - 7.64)^2}{7.64} + \frac{(0 - 1.27)^2}{1.27} + \frac{(12 - 7.53)^2}{7.53} + \cdots + \frac{(28 - 28.8)^2}{28.8} = 14.525$$

Is this value large or small? Pearson discovered that when the null hypothesis is true, values of this test statistic approximately follow a distribution called the χ^2-distribution (pronounced "chi-squared"). Therefore, one needs to compare the observed value of the Pearson chi-square with the χ^2-distribution to decide whether the value is large enough to warrant rejection of the null hypothesis.

CONCEPT TUTORIALS:
The χ^2-Distribution

To understand the χ^2-distribution better, use the instructional workbook for Distributions.

To use the Distribution workbook:

1 Open the file **Distributions**, located in the Explore folder of your Student disk. Enable the macros in the workbook.

2 Click **Chi-squared** from the Table of Contents column. Review the material and scroll to the bottom of the worksheet. See Figure 7-18.

**Figure 7-18
Chi-square
distribution
worksheet**

Unlike the normal distribution and t-distribution, the χ^2-distribution is limited to values ≥ 0. However, like the t-distribution, the χ^2-distribution involves a single parameter—the degrees of freedom. When the degrees of freedom are low, the distribution is highly skewed. As the degrees of freedom increase, the weight of the distribution shifts farther to the right and becomes less skewed. To see how the shape of the distribution changes, try changing the degrees of freedom in the worksheet.

To increase the degrees of freedom for the χ^2-distribution:

1 Click the **Degrees of freedom** spin arrow and increase the degrees of freedom to **9**.

The distribution changes shape as shown in Figure 7-19.

**Figure 7-19
Chi-square
distribution
with
9 degrees
of freedom**

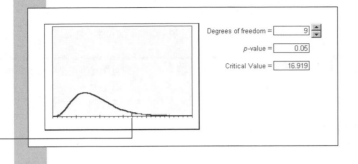

critical boundary

Like the normal and *t*-distributions, the χ^2-distribution has a critical boundary for rejecting the null hypothesis, but unlike those distributions, it's a one-sided boundary. There are a few situations where one might use upper and lower critical boundaries.

The critical boundary is shown in your chart with a vertical red line. Currently, the critical boundary is set for $\alpha = 0.05$. In Figure 7-19, this is equal to 16.919. You can change the value of α in this worksheet to see the critical boundary for other *p*-values.

To change the critical boundary:

1 Click the *p*-value box, type **0.10**, and press **Enter**.

The critical boundary changes, moving back to 14.684.

Experiment with other values for the degrees of freedom and the critical boundary.

When you're finished with the worksheet:

1 Close the Distributions workbook. Do not save any changes.

2 Return to the Survey2 workbook, displaying the Computer Department Table worksheet.

The degrees of freedom for the Pearson chi-square are determined by the numbers of rows and columns in the table. If there are *r* rows and *c* columns, the number of degrees of freedom are $(r-1) \times (c-1)$. For our table of computer use by department, there are 4 rows and 4 columns, and the number of degrees of freedom for the Pearson chi-square statistic is $(4-1) \times (4-1)$ or 9.

Where does the formula for degrees of freedom come from? The Pearson chi-square is based on the difference between the observed and expected counts. Note that the sum of these differences is 0 for each row and column in the table. For example, in the first row of the table, the expected and observed counts are as follows:

Table 7-2 Counts for computer use by department

Observed	Expected	Difference
4	7.64	−3.64
0	1.27	−1.27
12	7.53	4.47
6	5.56	0.44
	Sum	0.00

Because this sum is 0, the last difference can be calculated on the basis of the previous three, and there are only three cells that are free to vary in value. Applied to the whole table, this means that if we know 9 of the 16 differences, then we can calculate the values of the remaining 7 differences. Hence the number of degrees of freedom is 9.

Working with the χ^2-Distribution in Excel

Now that we know the value of the test statistic and the degrees of freedom, we are ready to test the null hypothesis. Excel includes several functions to help you work with the χ^2-distribution. Table 7-3 shows some of these.

Table 7-3 Excel functions for χ^2-distribution

Function	Description
CHIDIST(x, df)	Returns the p-value for the χ^2-distribution for a given value of x and degrees of freedom, df.
CHIINV(p, df)	Returns the χ^2 value from the χ^2-distribution with degrees of freedom, df, and p-value, p.
CHITEST(observed, expected)	Calculates the Pearson chi-square, where observed is a range containing the observed counts and expected is a range containing the expected counts.
PEARSONCHISQ(observed)	Calculates the Pearson chi-square, where observed is the range containing the observed counts. StatPlus required.
PEARSONP(observed)	Calculates the p-value of the Pearson chi-square, where observed is the range containing the observed counts. StatPlus required.

The output you generated earlier displays (among other things) the value for the Pearson chi-square statistic. The χ^2-value is 14.525 with a p-value of 0.105. Because this probability is not less than 0.05, there is not a significant departure from independence at the 5% level, and you accept the null hypothesis that the computer used and the department are independent of each other.

Breaking Down the Chi-Square Statistic

The value of the Pearson chi-square statistic is built up from every cell in the table. You can get an idea of which cells contributed the most to the total value by observing the table of standardized residuals. The value of the **standardized residual** is

$$\text{Standardized Residual} = \frac{\text{Observed Count} - \text{Expected Count}}{\sqrt{\text{Expected Count}}}$$

Note that the standardized residual is equal to the square root of each component of the Pearson chi-square statistic. Figure 7-20 displays the standardized residuals for the Computer Use by Department table.

**Figure 7-20
Table of
standardized
residuals**

Std. Residuals	Bus,Econ	HealthSc	MathSci	Soc Sci
Mac	-1.32	-1.13	1.63	0.19
Main	-1.29	0.06	1.18	0.10
Mini	-0.33	-0.48	0.54	-0.01
PC	1.49	0.54	-1.60	-0.15

Other Table Statistics

A common mistake is to use the value of χ^2 to measure the degree of association between the two categorical variables. However, the χ^2, along with the p-value, measures only the *significance* of the association. This is because the value of χ^2 is partly dependent on sample size and the size of the table. For example, in a 3×3 table, a χ^2-value of 10 is significant with a p-value of 0.04, but the same value in a 4×4 table is *not* significant with a p-value of 0.35.

A **measure of association**, on the other hand, gives a value to the association between the row and column variables that is not dependent on the sample size and the size of the table. Generally, the higher the measure of association, the stronger the association between the two categorical variables.

Figure 7-21 shows other test statistics and measures of association created by the StatPlus Table Statistics command.

Figure 7-21 Other table statistics

Test Statistics	Value	df	p-value
Pearson Chi-Square	14.525	9	0.105
Continuity Adjusted Chi-Square	10.806	9	0.289
Likelihood Ratio Chi-Square	16.125	9	0.064

Measures of Association	Value	Std. Error	p-value
Phi	0.276		
Contigency	0.266		
Cramer's V	0.160		
Goodman-Kruskal Gamma	-0.244	0.095	0.011
Kendalls tau-b	-0.154	0.062	0.012
Stuart's tau-c	-0.128	0.051	0.013
Somer's D (C|R)	-0.172	0.069	0.012
Somer's D (R|C)	-0.138	0.055	0.013
Warning: More than 1/5 of Fitted Cells are Sparse			

Table 7-4 summarizes these statistics and their uses.

Table 7-4 StatPlus Table Statistics

Statistic	Description
Pearson Chi-Square	Calculates the difference between the observed and expected counts. Approximately follows a χ^2-distribution with $(r-1) \times (c-1)$ degrees of freedom, where r is the number of rows in the table and c is the number of table columns.
Continuity Adjusted Chi-Square	Similar to the Pearson chi-square, except that it adjusts the χ^2-value for the continuity of the χ^2-distribution.
Likelihood Ratio Chi-Square	It approximately follows a χ^2-distribution with $(r-1) \times (c-1)$ degrees of freedom.
Phi	Measures the association between the row and column variables, varying from -1 to 1. A value near 0 indicates no association.
Contingency	A measure of association ranging from 0 (no association) to a maximum of 1 (high association). The upper bound may be less than 1, depending on the values of the row and column totals.
Cramer's V	A variation of the Contingency measure, modifying the statistic so that the upper bound is always 1.
Goodman-Kruskal Gamma	A measure of association used when the row and column values are ordinal variables. Gamma ranges from -1 to 1. A negative value indicates negative association, a positive value indicates positive association, and 0 indicates no association between the variables.

(continued)

Kendall's tau-b	Similar to Gamma, except that tau-b includes a correction for ties. Used only for ordinal variables.
Stuart's tau-c	Similar to tau-b, except that it includes a correction for table size. Used only for ordinal variables.
Somers' D	A modification of the tau-b statistic. Somers' D is used for ordinal variables in which one variable is used to predict the value of the other variable. Somers' (C|R) is used when the column variable is used to predict the value of the row variable. Somers' (R|C) is used when the row variable is used to predict the value of the column variable.

Because the χ^2-distribution is a continuous distribution and counts represent discrete values, some statisticians are concerned that the Pearson chi-square statistic is not appropriate. They recommend using the Continuity Adjusted chi-square statistic instead. We feel that the Pearson chi-square statistic is more accurate and can be used without adjustment.

Among the other statistics in Table 7-4, the Likelihood Ratio chi-square statistic is usually close to the Pearson chi-square statistic. Many statisticians prefer using the Likelihood Ratio chi-square because it is used in log-linear modeling—a topic beyond the scope of this book.

None of the three test statistics shown in Figure 7-21 is significant at the 5% level (though the p-values do differ greatly). The association between Computer Use and Department ranges from 0.276 (for Phi) to 0.160 (for Cramer's V). None of the associations is large in size. The final four measures of association (Gamma, tau-b, tau-c and Somers' D) are used for ordinal data and are not appropriate for nominal data.

Validity of the Chi-Square Test with Small Frequencies

Another thing to notice in Figure 7-21 is the warning printed after the table of statistics: "Warning: More than $\frac{1}{5}$ of Fitted Cells are Sparse." "Sparse" means that there are cells whose expected value is less than 5. Refer to Figure 7-17. In the table of expected counts, 6 of the 16 cells (37.5%) have values below 5. Because this is greater than 20%, the warning message is displayed.

The problem is that it might not be valid to use the Pearson chi-square test on a table with a large number of sparse cells. The test requires large samples, and this means that cells with small counts can be a problem. You might get by with as many as one-fifth of the expected counts under 5, but if it's more than that, the *p*-value returned by the Pearson chi-square might lead you erroneously to reject or accept the null hypothesis.

Eliminating Sparse Data

What can you do to create a valid test? When there are low counts in a cell, you can either pool columns or rows together to increase the cell counts or remove rows or columns from the table altogether.

Three of the sparse cells are found in the health science column. You could try to pool the values in this column with another column. It's hard to decide which other column to pool them with. Are health science courses similar to math science courses? Or would the social sciences be a better fit? It may be safer simply to eliminate the column, though you would have to make a note of this fact when you report your results.

Four of the sparse cells are found in the Mini computer row (clearly, mini-computers are not heavily used in these courses). Minicomputers are closely related to mainframes, so you could argue for combining them in a single column.

Let's recreate the table reducing the number of sparse cells by removing the health science category and pooling the minicomputers with the mainframes (we could also restructure the current table instead of recreating it from scratch, but this way we'll have both tables in the workbook).

To recreate the Computer Use by Department table:

1 Click **Data > PivotTable and PivotChart Report** from the Excel menu.

2 Click the **Another PivotTable or PivotChart** option button and click the **Next** button twice.

3 Click the **Layout** button and drag the **Computer** button to the Row area and then again to the Data area. Drag the **Dept** button to the Column area.

4 Click **OK**.

5 Click the **Options** button.

6 Type **0** in the **For Empty cells, show** box and click **OK**.

7 Click **Finish** to start creating the pivot table.

8 Remove the blank categories from the rows and columns of the table.

9 Remove **HealthSc** from the Dept drop-down list box.

Of the 12 cells, 4, or 33%, have counts of less than 5, whereas 5 of 12, or about 42%, have values of 6 or less. There is still a problem with sparse cells. To compensate, group the mainframes and minicomputers together.

Grouping Categories

To pool the cell counts of rows or columns, you combine the various categories into groups. Creating a group does not affect the underlying data.

To combine the mainframe and minicomputer categories:

1 Select cells **A6:A7** (the row labels for Main and Mini)

2 Right-click the selection, point to **Group and Outline**, and then click **Group** from the menu.

The row labels in column A are shifted to column B, and Excel adds a new row level in column A in the pivot table (Computer2), calling the grouping of Mains and Minis "Group 1." Let's give these more descriptive names.

3 Click cell **A4**, type **Computer Groups**, and press **Enter**.

4 Click cell **A6**, type **Main/Mini**, and press **Enter**.

Now remove the ungrouped row fields from the table.

5 Click cell **B4**, the Computer field button, drag it off below the table, and when the pointer appears with a red "X" next to it, release the mouse button. Excel removes the ungrouped Computer field from the pivot table. See Figure 7-22.

**Figure 7-22
Pivot table with grouped categories**

Main/Mini is combination of counts for mainframes and minicomputers

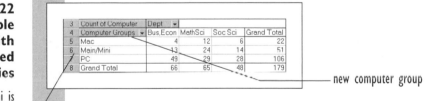

new computer group

Notice that now only 1 of the 9 cells (11%) has a count below 5; 2 of the 9 cells (22%) have values of 6 or less. Restructuring the table has reduced the sparseness problem, but what has it done to the probability value of the Pearson chi-square test?

To check the table statistics on the revised table:

1 Click **StatPlus > Table Statistics**.

2 Select the range **A4:D7**.

3 Click the **Output** button and send the output to the new worksheet, **Computer Department Table 2**.

4 Click the **OK** button to start generating the new batch of table statistics. See Figure 7-23.

Figure 7-23
Table
statistics
for the
new table

Table Statistics			
Observed Counts	Bus,Econ	MathSci	Soc Sci
Mac	4	12	6
Main/Mini	13	24	14
PC	49	29	28
Expected Counts	Bus,Econ	MathSci	Soc Sci
Mac	8.11	7.99	5.90
Main/Mini	18.80	18.52	13.68
PC	39.08	38.49	28.42
Std. Residuals	Bus,Econ	MathSci	Soc Sci
Mac	−1.44	1.42	0.04
Main/Mini	−1.34	1.27	0.09
PC	1.59	−1.53	−0.08
Test Statistics	Value	df	p-value
Pearson Chi-Square	12.384	4	0.015
Continuity Adjusted Chi-Square	10.356	4	0.035
Likelihood Ratio Chi-Square	12.706	4	0.013
Measures of Association	Value	Std. Error	p-value
Phi	0.263		
Contigency	0.254		
Cramer's V	0.186		
Goodman-Kruskal Gamma	−0.234	0.102	0.022
Kendalls tau-b	−0.145	0.065	0.025
Stuart's tau-c	−0.132	0.059	0.025
Somer's D (C|R)	−0.159	0.071	0.025
Somer's D (R|C)	−0.133	0.060	0.026

The Pearson chi-square test statistic p-value has changed greatly, from about 0.105 to 0.015. Because the p-value 0.015 is less than 0.05, the Pearson statistic is significant at the 5% level. Thus, you would reject the null hypothesis of independence and accept the alternative hypothesis that there is a relationship between the department and the type of computer used.

The table of standardized residuals gives us a hint about the nature of that relationship. Notice that we have low residuals for the social science courses, indicating that the observed values closely agree with the expected values, calculated on the basis of the null hypothesis. The differences are mainly for the business/economics courses and the mathematics/science courses. The business/economics courses use PCs more often than would be predicted by the null hypothesis (note the positive residual value of 1.59). On the other hand, mathematics/science courses used Macintoshes, mainframes, and minicomputers more often than expected (the residuals are 1.42 and 1.27).

There could be many different reasons for this. One possible explanation is that when the survey was taken, more business software had been developed for PCs than for Macintoshes, mainframes, and minicomputers.

Tables with Ordinal Variables

The two-way table you just produced was for two nominal variables. Now let's look at categorical variables that have inherent order. For ordinal variables, there are more powerful tests than the Pearson chi-square, which often fails to give significance for ordered variables.

As an example, consider the Calculus and Enroll B variables. The Calculus variable tells the extent to which calculus is required for a given statistics class (Not req or Prereq). From this point of view, Calculus is an ordinal variable, although it would also be possible to think of it as a nominal variable. When a variable takes on only two values, there really is no distinction between nominal and ordinal, because any two values can be regarded as ordered. The other variable, Enroll B, is a categorical variable that contains measures of the size of the annual course enrollment. In the survey, instructors were asked to check one of eight categories (0–50, 51–100, 101–150, 151–200, 201–300, 301–400, 401–500, and 501–) for the number of students in the course. You might expect that classes requiring calculus would have smaller enrollments.

Testing for a Relationship Between Two Ordinal Variables

We want to test whether there is a relationship between a class requiring calculus as a prerequisite and the size of the class. Our hypotheses are

H_0: The pattern of enrollment is the same regardless of a calculus prerequisite

H_a: The pattern of enrollment is related to a calculus prerequisite

To test the null hypothesis, first form a two-way table for categorical variables, Calculus and Enroll B.

To form the table:

1   Click **Data > PivotTable and PivotChart Report** from the Excel menu.

2   Click the **Another PivotTable or PivotChart** option button and click the **Next** button twice.

3   Click the **Layout** button and drag the **Enroll B** button to the Row area and then again to the Data area. Drag the **Calculus** button to the Column area.

4 Click **OK**.

5 Click the **Options** button.

6 Type **0** in the **For Empty cells, show** box and click **OK**.

7 Verify that the **New Worksheet** option button is selected and click **Finish** to start creating the pivot table.

8 Remove the blank categories from the rows and columns of the table. See Figure 7-24.

Figure 7-24
Table of
enrollment
versus
calculus
prerequisite

Count of Enroll B	Calculus ▾		
Enroll B ▾	Not req	Prereq	Grand Total
001-050	54	20	74
051-100	62	15	77
101-150	41	6	47
151-200	36	4	40
201-300	26	3	29
301-400	20	1	21
401-500	7	2	9
501-	38	6	44
Grand Total	284	57	341

In the table you just created, Excel automatically arranges the Enroll B levels in a combination of numeric and alphabetic order (alphanumeric order), so be careful with category names. For example, if "051–100" were written as "51–100" instead, Excel would place it near the bottom of the table because 5 comes after 1, 2, 3, and 4.

Remember that most of the expected values should exceed 5, so there is cause for concern about the sparsity in the second column between 201 and 500, but because only 4 of the 16 cells have expected values less than 5, the situation is not terrible. Nevertheless, let's combine enrollment levels from 200 to 500 of the pivot table.

To combine levels, use the same procedure you did with the computer table:

1 Highlight **A9:A11**, the enrollment row labels for the categories from 200 through 500.

2 Right-click the selection to display the shortcut menu; then click **Group and Outline > Group**.

3 Click cell **A4**, type **Enrollment**, and press **Enter**.

4 Click cell **A9**, type **201-500**, and press **Enter**.

5 Click cell **B4** and drag it off the table. See Figure 7-25.

Figure 7-25
Enrollment
table with
sparse cell
pooled

pooled values —

Figure 7-25 Enrollment table with sparse cell pooled

Count of Enroll B	Calculus		
Enrollment	Not req	Prereq	Grand Total
001-050	54	20	74
051-100	62	15	77
101-150	41	6	47
151-200	36	4	40
201-500	53	6	59
501-	38	6	44
Grand Total	284	57	341

Now generate the table statistics for the new table.

To generate table statistics:

1 Click **StatPlus > Table Statistics**.

2 Select the range **A4:C10**.

3 Click the **Output** button and send the table to a new worksheet named **Enrollment Statistics**.

4 Click the **OK** button.

The statistics for this table are as shown in Figure 7-26.

**Figure 7-26
Statistics for
the
Enrollment
table**
statistics that
don't assume —
ordinal data

statistics that
assume ordinal —
data

Test Statistics	Value	df	p-value
Pearson Chi-Square	10.013	5	0.075
Continuity Adjusted Chi-Square	7.818	5	0.167
Likelihood Ratio Chi-Square	9.762	5	0.082

Measures of Association	Value	Std. Error	p-value
Phi	0.171		
Contigency	0.169		
Cramer's V	0.171		
Goodman-Kruskal Gamma	0.280	0.101	0.006
Kendalls tau-b	-0.133	0.049	0.006
Stuart's tau-c	-0.127	0.048	0.007
Somer's D (C|R)	-0.077	0.029	0.007
Somer's D (R|C)	-0.229	0.083	0.006

It's interesting to note that those statistics that do not assume that the data is ordinal (the Pearson chi-square, the Continuity Adjusted X^2, and the Likelihood Ratio X^2) all fail to reject the null hypothesis at the 0.05 level. On the other hand, the statistics that take advantage of the fact that we're using ordinal data (the Goodman-Kruskal Gamma, Kendall's tau-b, Stuart tau-c, and Somers' D) all reject the null hypothesis. This illustrates an important point: Always use the statistics test that best matches the characteristics of your data. Relying on the ordinal tests, we reject the null hypothesis, accepting the alternative hypothesis that the pattern of enrollment differs on the basis of whether calculus is a prerequisite.

To explore how that difference manifests itself, let's examine the table of expected values and standardized residuals in Figure 7-27.

**Figure 7-27
Expected
counts and
standardized
residuals for
the
Enrollment
table**

Observed Counts	Not req	Prereq
001-050	54	20
051-100	62	15
101-150	41	6
151-200	36	4
201-500	53	6
501-	38	6

Expected Counts	Not req	Prereq
001-050	61.63	12.37
051-100	64.13	12.87
101-150	39.14	7.86
151-200	33.31	6.69
201-500	49.14	9.86
501-	36.65	7.35

Std. Residuals	Not req	Prereq
001-050	-0.97	2.17
051-100	-0.27	0.59
101-150	0.30	-0.66
151-200	0.47	-1.04
201-500	0.55	-1.23
501-	0.22	-0.50

From the tables, we see that the null hypothesis underpredicts the number of courses with class sizes in the 1–50 range that require knowledge of calculus. The null hypothesis predicts that 12.37 classes fit this classification, and there were 20 of them. As the size of the classes increases, the null hypothesis increasingly overpredicts the number of courses that require calculus. For example, if the null hypothesis were true, we would expect to see almost 10 courses with 201–500 students each require knowledge of calculus. The observed number from the survey was 6. From this we conclude that class size and a calculus prerequisite are not independent and that the courses that require knowledge of calculus are more likely to be smaller.

Custom Sort Order

With ordinal data you want the values to appear in the proper order when created by the pivot table. If order is alphabetic or the variable itself is numeric, this is not a problem. However, what if the variable being considered has a definite order, but this order is neither alphabetic nor numeric? Consider the Interest variable from the Survey workbook, which measures the degree of interest in a supplementary statistics text. The values of this variable have a definite order (least, low, some, high, most), but this order is not alphabetic or numeric. You could create a numeric variable based on the values of Interest, such as 1 = least, 2 = low, 3 = some, 4 = high, and 5 = most. Another approach is to create a custom sort order, which lets you define a sort order for a variable.

You can define any number of custom sort orders. Excel already has some built in for your use, such as months of the year (Jan, Feb, Mar, . . . , Dec), so if your data has a variable with month values, you can sort the data list by months (you can also do this with pivot tables). Try creating a custom sort order for the values of the Interest variable.

To create a custom sort order for the values of the Interest variable:

1   Click **Tools > Options**.

2   Click the **Custom Lists** tab.

3   Click the **List entries** list box.

4   Type **least** and press **Enter**.

5   Type **low** and press **Enter**.

6   Type **some** and press **Enter**.

7   Type **high** and press **Enter**.

8   Type **most** and click **Add**.

9   Your Custom Lists tab of the Options dialog box should look like Figure 7-28.

**Figure 7-28
Custom sort
order**

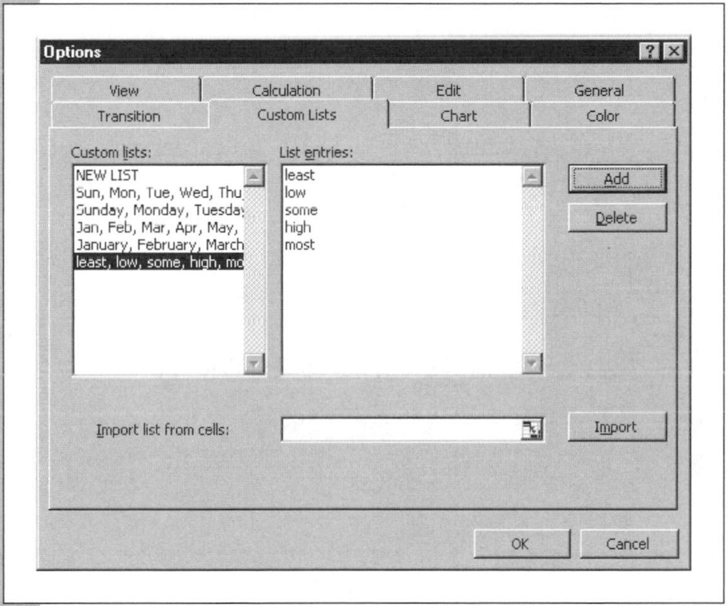

10   Click **OK**.

Now create a pivot table of Interest values to see whether Excel automatically applies the sort order you just created.

To create the pivot table:

1 Click **Data > PivotTable and PivotChart Report** from the Excel menu.

2 Click the **Another PivotTable or PivotChart** option button and click the **Next** button twice.

3 Click the **Layout** button and drag the **Interest** button to the Row area and then again to the Data area.

4 Click **OK**.

5 Verify that the **New Worksheet** option button is selected and click **Finish** to start creating the pivot table.

The resulting pivot table appears, as in Figure 7-29.

Figure 7-29 Custom sort order applied to the Interest table

Count of Interest	
Interest	Total
least	22
low	44
some	131
high	110
most	71
	14
Grand Total	392

Excel automatically sorts the Interest categories in the pivot table in the proper order—least, low, some, high, and most—rather than alphabetically.

You've completed your work with categorical data with Excel.

To save your workbook:

1 Save your workbook on your Student disk.

2 Click **File > Exit** to exit Excel.

Exercises

1. Use Excel to calculate the following p-values for the χ^2-distribution.
   a. $\chi^2 = 4$ with 4 degrees of freedom
   b. $\chi^2 = 4$ with 1 degree of freedom

    c. $\chi^2 = 10$ with 6 degrees of freedom

    d. $\chi^2 = 10$ with 3 degrees of freedom

2. Use Excel to calculate the following critical values for the χ^2-distribution.

    a. $\alpha = 0.10$, degrees of freedom = 4

    b. $\alpha = 0.05$, degrees of freedom = 4

    c. $\alpha = 0.05$, degrees of freedom = 9

    d. $\alpha = 0.01$, degrees of freedom = 9

3. *True or false (and why):* "The Pearson chi-square test allows us to measure the degree of association between one categorical variable and another."

4. Reopen the original Survey workbook and continue your exploration of the data.

    a. Create a pivot table listing the departments responding to the survey, removing all missing categories. Print versions of the table showing counts and percentages.

    b. Create a pie chart and bar chart (use the Column chart type) of the total counts reported in the Department pivot table. Include the counts on the bar chart and the percentages on the pie chart. Print the resulting charts. Which chart gives a clearer indication of the comparative sizes of the groups and why?

    c. Examine whether there is relationship between courses having a calculus prerequisite and the type of computer used in the course by constructing a pivot table and doing the appropriate analysis (remove any blank categories from the table). State your null and alternative hypotheses. Did you encounter any problems with sparseness in the table? How did you deal with the problems? Summarize your results and conclusions. Does a calculus prerequisite affect the type of computer used?

    d. Is the type of computer used related to class size? Create a table of total enrollment (Enroll_B) versus whether the course used a PC or a non-PC. Pool class sizes from 201 to 500 students in one group. Analyze the table and summarize your conclusions. On what test statistics are you basing your conclusions and why? Is class size a relevant factor, or are there other, competing factors (*Hint*: Are more advanced courses more likely to be smaller in class size?)

    e. Print all of the relevant tables and charts. Save your workbook as E7SURVEY.XLS.

5. The JRCOL.XLS workbook contains information on hiring practices at a junior college.

    a. Create a customized list of teaching ranks sorted in the following order: instructor, assistant professor, associate professor, full professor.

    b. Create a pivot table with Rank Hired as the row variable and Sex as the column variable.

c. Explore the question of whether there is a relationship between teaching rank and gender. What are your hypotheses? Generate the table statistics for this pivot table. Which statistics are appropriate to use with the table? Is there any difficulty with the data in the table? How would you correct these problems?

d. Group the associate professor and full professor groups together and redo your analysis.

e. Group the three professor ranks into one and redo your analysis, relating gender to the instructor/professor split.

f. Write a report summarizing your results, displaying the relevant tables and statistics. How do your three tables differ with respect to your conclusions? Discuss some of the problems one could encounter when trying to eliminate sparse data. Which of the three tables best describes the data, in your opinion, and would you conclude that there is a relationship between teacher rank and gender? What pieces of information is this analysis missing?

g. Create another pivot table with Degree as the page field, Rank Hired as the row field, and Sex as the column field. (Because you are obtaining counts, you can use either Sex or Rank Hired as the data field.)

h. Using the drop-down arrows on the Page field button, display the table of persons hired with a Master's degree.

i. Generate table statistics for this group. Is the rank when hired independent of gender for candidates with Master's degrees? Redo the analysis if necessary to remove sparse cells.

j. Write a report summarizing your conclusions, including any relevant tables and statistics. Save your workbook as E7JRCOL.XLS.

6. The COLD.XLS workbook contains data from a 1961 French study of 279 skiers during two 5–7 day periods. One group of skiers received a placebo (an ineffective saline solution), and another group received 1 gram of ascorbic acid per day. The study was designed to measure the incidence of the common cold in the two groups.
   a. State the null and alternative hypotheses of this study.
   b. Open the workbook and analyze the cold data.
   c. Summarize your results. Save the workbook as E7COLD.XLS.

7. The data in MARRIAGE.XLS contains information on the heights of newly married couples. The study is designed to test whether people tend to choose marriage partners similar in height to themselves.
   a. State the null and alternative hypotheses of this study.
   b. Open the workbook and analyze the data in the table. What test statistics are appropriate to use with this data? Do you accept or reject the null hypothesis?
   c. Summarize your results and print the relevant statistics supporting your conclusion. Save the workbook as E7MARRIAGE.XLS.

8. Open the GENDEROP.XLS workbook. The workbook contains polling data on how males and females respond to various social and political issues. Each worksheet in the workbook contains a table showing the responses to a particular question.

   a. On each worksheet, calculate the table statistics for the Opinion Poll table.
   b. For which question is the gender of the respondent associated with the outcome?
   c. Summarize your results and speculate on the types of issues that men and women might agree or disagree on.
   d. Save your workbook as E7GENDEROP.XLS.

9. Open the RACEOP.XLS workbook. This workbook contains additional polling data on how different races respond to social and political questions.

   a. On each worksheet, calculate the table statistics for the Opinion Poll table. Resolve any problem with sparse data by combining the Black and Other categories.
   b. For which question is the gender of the respondent associated with the outcome?
   c. Summarize your results and speculate on the types of questions that blacks and members of other races might agree or disagree on.
   d. Save your workbook as E7RACEOP.XLS.

10. Open the CARS.XLS workbook containing information about various models of cars manufactured from 1970 to 1982. You're exploring whether there is a difference between the number of cylinders in each car, based on the country that produced the car.

    a. State your null and alternative hypotheses.
    b. Create a pivot table with Origin in the row area and Cylinders in the column area.
    c. Calculate tables statistics for the pivot table. Which statistics are most relevant to your analysis?
    d. Are there any problems with the resulting test statistics? Correct the problem and redo the analysis.
    e. Summarize your conclusions. Save your workbook as E7CARS.XLS.

11. Open the HOMEDATA.XLS workbook (see Chapter 4 for a discussion of this workbook.)

    a. Analyze the data to determine whether there is evidence that houses in the NE sector are more likely to have offers pending than houses outside the NE sector.
    b. Summarize your conclusions and save your workbook as E7HOME-DATA.XLS.

Statistical Methods

Chapter 8

Regression and Correlation

Objectives

In this chapter you will learn to:

- Fit a regression line and interpret the coefficients

- Understand regression statistics

- Use residuals to check the validity of the assumptions needed for statistical inference

- Obtain and interpret correlations and their statistical significance

- Understand the relationship between correlation and simple regression

- Obtain a correlation matrix and apply hypothesis tests

- Obtain and interpret a scatterplot matrix

T his chapter examines the relationship between two variables using linear regression and correlation. Linear regression estimates a linear equation that describes the relationship, whereas correlation measures the strength of that linear relationship.

Simple Linear Regression

When you plot two variables against each other in a scatterplot, the values usually don't fall exactly in a perfectly straight line. When you perform a linear regression analysis, you attempt to find the line that best estimates the relationship between two variables (the y, or dependent, variable, and the x, or independent, variable). The line you find is called the **fitted regression line**, and the equation that specifies the line is called the **regression equation**.

The Regression Equation

If the data in a scatterplot fall approximately in a straight line, you can use linear regression to find an equation for the regression line drawn over the data. Usually, you will not be able to fit the data perfectly, so some points will lie above and some below the fitted regression line.

The regression line that Excel fits will have an equation of the form $y = a + bx$. Here y is the **dependent variable**, the one you are trying to predict, and x is the **independent**, or **predictor**, **variable**, the one that is doing the predicting. Finally, a and b are called **coefficients**. Figure 8-1 shows a line with $a = 10$ and $b = 2$. The short vertical line segments represent the errors, also called **residuals**, which are the gaps between the line and the points. The residuals are the differences between the observed dependent values and the predicted values. Because a is where the line intercepts the vertical axis, a is sometimes called the **intercept** or **constant term** in the model. Because b tells how steep the line is, b is called the **slope**. It gives the ratio between the vertical change and the horizontal change along the line. Here y increases from 10 to 30 when x increases from 0 to 10, so the slope is

$$b = \frac{\text{Vertical change}}{\text{Horizontal change}} = \frac{30 - 10}{10 - 0} = 2$$

**Figure 8-1
A fitted
regression
line**

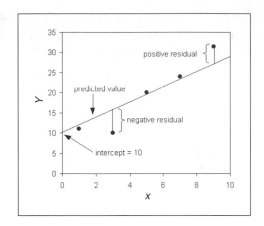

Suppose that x is years on the job and y is salary. Then the y-intercept ($x = 0$) is the salary for a person with zero years' experience, the starting salary. The slope is the change in salary per year of service. A person with a salary above the line would have a positive residual, and a person with a salary below the line would have a negative residual.

If the line trends downward so that y decreases when x increases, then the slope is negative. For example, if x is age and y is price for used cars, then the slope gives the drop in price per year of age. In this example, the intercept is the price when new, and the residuals represent the difference between the actual price and the predicted price. All other things being equal, if the straight line is the correct model, a positive residual means a car costs more than it should, and a negative residual means a car costs less than it should (that is, it's a bargain).

Fitting the Regression Line

When fitting a line to data, you assume that the data follow the **linear model**:

$$y = \alpha + \beta x + \varepsilon$$

where α is the "true" intercept, β is the "true" slope, and ε is an error term. When you fit the line, you'll try to estimate α and β, but you can never know them exactly. The estimates of α and β, we'll label "a" and "b." The predicted values of y using these estimates, we'll label $\hat{y}$, so that

$$\hat{y} = a + bx$$

To get estimates for α and β, we use values of a and b that result in a minimum value for the sum of squared residuals. In other words, if y_i is an observed value of y, we want values of a and b such that

$$\text{Sum of Squared Residuals} = \sum_{i=1}^{n}(y_i - \hat{y}_i)^2$$

is as small as possible. This procedure is called the **least squares** method. The values a and b that result in the smallest possible sum for the squared residuals can be calculated from the following formulas:

$$b = \frac{\sum_{i=1}^{n}(x_i - \overline{x})(y_i - \overline{y})}{\sum_{i=1}^{n}(x_i - \overline{x})^2}$$

$$a = \overline{y} - b\overline{x}$$

These are called the **least-squares estimates**. For example, say our data set contains the following values:

Table 8-1 Data for least-squares estimates

x	y
1	3
2	4
1	3
3	4
2	5

The sample averages for x and y are 1.8 and 3.4, and the estimates for a and b are

$$b = \frac{\sum_{i=1}^{n}(x_i - \overline{x})(y_i - \overline{y})}{\sum_{i=1}^{n}(x_i - \overline{x})^2}$$

$$= \frac{(1 - 1.8)(3 - 3.4) + (2 - 1.8)(4 - 3.4) + \cdots + (2 - 1.8)(5 - 3.4)}{(1 - 1.8)^2 + (2 - 1.8)^2 + \cdots + (2 - 1.8)^2}$$

$$= 0.5$$

✓ $a = \overline{y} - b\overline{x}$

$$= 3.4 - 0.5 \times 1.8$$

$$= 2.5$$

Thus the least-squares estimate of the regression equation is $y = 2.5 + 0.5x$.

Regression Functions in Excel

Excel contains several functions to help you calculate the least-squares estimates. Two of these are shown in Table 8-2.

Table 8-2 Calculating least-squares estimates

Function	Description
INTERCEPT(y, x)	Calculates the least squares estimate, a, for known values y and x.
SLOPE(y, x)	Calculates the least squares estimate, b, for known values y and x.

For example, if the y-values are in the cell range A2:A11, and the x-values lie in the range B2:B11, then the function INTERCEPT(A2:A11, B2:B11) will display the value of a, and the function SLOPE(A2:A11, B2:B11) will display the value of b.

EXCEL TIPS

- You can also calculate linear regression values using the LINEST and LOGEST functions, but this is a more advanced topic. Both of these functions use arrays. You can learn more about these functions and about array functions in general by using Excel's online Help.

Performing a Regression Analysis

The workbook BCANCER.XLS contains data from a 1965 study analyzing the relationship between mean annual temperature and the mortality rate for a certain type of breast cancer in women. The subjects came from 16 different regions in Great Britain, Norway, and Sweden. Table 8-3 presents the data.

Table 8-3 Data for BCANCER.XLS workbook

Range Name	Range	Description
Region	A2:A17	A number indicating the region where the data has been collected
Temperature	B2:B17	The mean annual temperature of the region
Mortality	C2:C17	Mortality index for neoplasms of the female breast for the region

You've been asked to determine whether there is evidence of a linear relationship between the mean annual temperature in the region and the mortality index. Is the mortality index different for women who live in regions with different temperatures?

To open the BCancer.xls workbook:

1 Open the **BCancer** workbook from the folder containing your student files.

2 Click **File** > **Save As** and save the workbook as **BCancer2**. The workbook appears as shown in Figure 8-2.

**Figure 8-2
The BCancer
workbook**

annual mean
temperature of the
region in degrees
Fahrenheit

breast cancer
mortality index

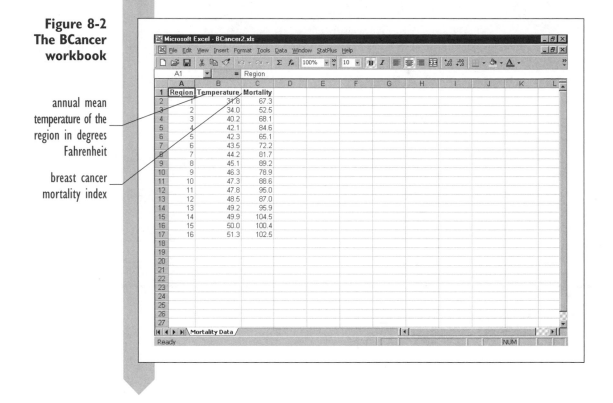

Plotting Regression Data

Before you calculate any regression statistics, you should always plot your data. A scatterplot can quickly point out obvious problems in assuming that a linear model fits your data (perhaps the scatterplot will show that the data values do *not* fall along a straight line). Scatterplots in Excel also allow you to superimpose the regression line on the plot along with the regression equation. From this information, you can get a pretty good idea whether a straight line fits your data or not.

To create a scatterplot of the mortality data:

1 Select the range, **B1:C17**.

2 Click **Insert** and **Chart**.

3 Click **XY(Scatter)** from the list of chart types and click **Next** twice.

4 Click the **Titles** tab and enter **Mortality vs. Temp** in the Chart title box, **Temperature** in the Value (X) axis box, and **Mortality Index** in the Value (Y) axis box.

5 Click the **Gridlines** tab and remove the major gridlines from the plot.

6 Click the **Legend** tab and remove the legend from the plot and click **Next**.

7 Click the **As new sheet** option button and send the output to a chart sheet named **Mortality Scatterplot**. Click the **Finish** button.

8 In the scatterplot that's created, resize the horizontal scale so that the lower boundary is **30**, and then resize the vertical scale so that the lower boundary is **50**. Figure 8-3 shows the final version of the scatterplot.

Figure 8-3 Scatterplot of the mortality index vs. mean annual temperature

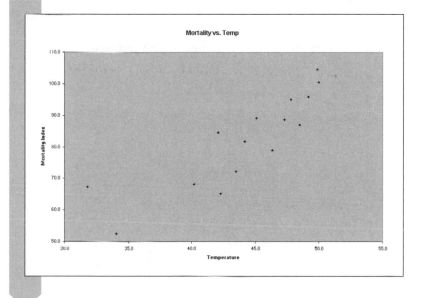

Now you'll add a regression line to the data.

To add a regression line:

1 Right-click any of the data points in the graph and click **Add Trendline** from the menu. See Figure 8-4.

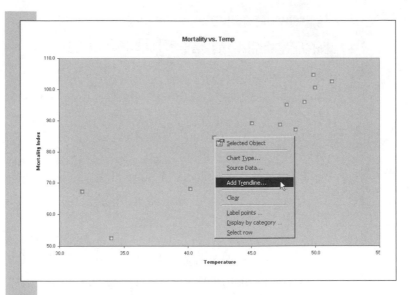

**Figure 8-4
Accessing
the Add
Trendline
command**

2 Click the **Type** tab and verify that the Linear Trend/Regression option is selected. See Figure 8-5.

**Figure 8-5
Choosing
the linear
regression
line**

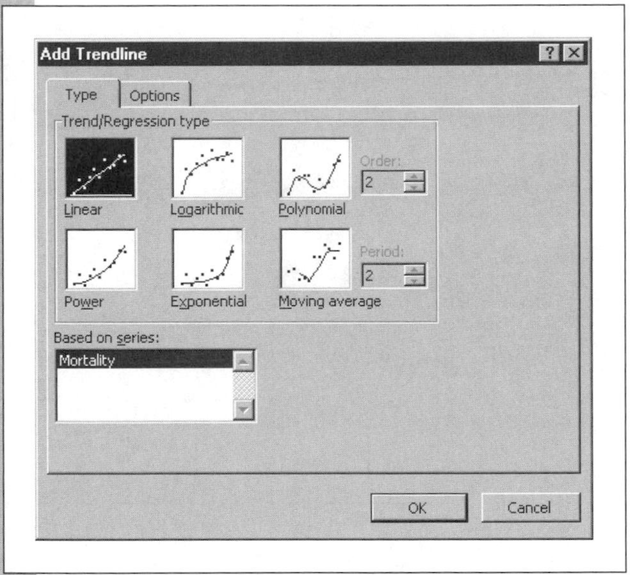

3 Click the **Options** tab and click the **Display equation on chart** checkbox and then click the **Display R-squared value on chart** checkbox. See Figure 8-6.

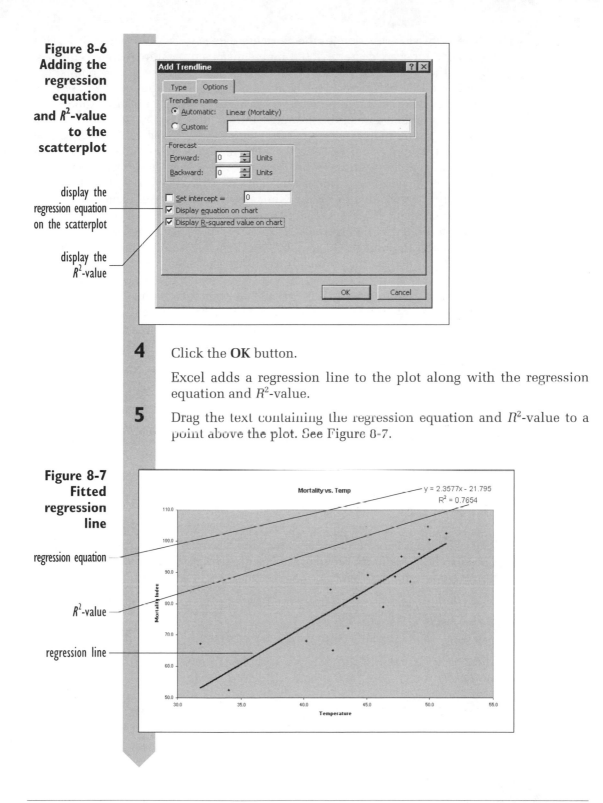

**Figure 8-6
Adding the
regression
equation
and R^2-value
to the
scatterplot**

display the
regression equation
on the scatterplot

display the
R^2-value

4 Click the **OK** button.

Excel adds a regression line to the plot along with the regression equation and R^2-value.

5 Drag the text containing the regression equation and R^2-value to a point above the plot. See Figure 8-7.

**Figure 8-7
Fitted
regression
line**

regression equation

R^2-value

regression line

The regression equation for the mortality data is $y = -21.795 + 2.3577x$. This means that for every degree that the annual mean temperature increases in these regions, the breast cancer mortality index increases by about 2.3577 points.

How would you interpret the constant term in this equation (−21.795)? At first glance, this is the y-intercept, and it means that if the mean annual temperature is 0, the value of the mortality index would be −21.795. Clearly this is absurd; the mortality index can't drop below zero. In fact, any mean annual temperature of less than 9.24 degrees Fahrenheit will result in a negative estimate of the mortality index. This does not mean that the linear equation is useless, but it means you should be cautious in making any predictions for temperature values that lie outside the range of the observed data.

The R^2-value is 0.7654. What does this mean? The R^2-value, also known as the **coefficient of determination**, measures the percentage of variation in the values of the dependent variable (in this case, the mortality index) that can be explained by the change in the independent variable (temperature). R^2-values vary from 0 to 1. A value of 0.7654 means that 76.54% of the variation in the mortality index can be explained by the change in annual mean temperature. The remaining 23.46% of the variation is presumed to be due to random variability.

EXCEL TIPS

- You can use the "Add Trendline" command to add other types of least-squares curves, including logarithmic, polynomial, exponential, and power curves. For example, instead of fitting a straight line, you can fit a second-degree curve to your data.

Calculating Regression Statistics

The regression equation in the scatterplot is useful information, but it does not tell you whether the regression is statistically significant. At this point, you have two hypotheses to choose from:

H_0: There is no linear relationship between the mortality index and the mean annual temperature.

H_a: There is a linear relationship.

and the linear relationship we're testing is expressed in terms of the regression equation.

In order to analyze our regression, we need to use the Analysis ToolPak, an add-in that comes with Excel and provides tools for analyzing regression. If you do not have the Analysis ToolPak loaded on your system, you should install it now.

To install the Analysis ToolPak:

1 Click **Tools > Add-Ins**.

2 Click the **Analysis ToolPak** checkbox from the Add-Ins list box and click **OK**.

Note: Excel may prompt you to insert your installation disk. If so, insert your installation disk and follow the onscreen directions.

Now create a table of regression statistics.

To create a table of regression statistics:

1 Click the **Mortality Data** worksheet tab.

2 Click **Tools > Data Analysis** from the Excel menu.

3 Scroll down the Analysis Tools list box, click **Regression**, and then click the **OK** button.

4 Enter the cell range **C1:C17** in the Input Y Range box.

5 Enter the cell range **B1:B17** in the Input X Range box.

6 Because the first cell in these ranges contains a text label, click the **Labels** checkbox.

7 Click the **New Worksheet Ply** option button and type **Regression Statistics** in the accompanying text box.

8 Click all four of the Residuals checkboxes.

Your Regression dialog box should appear as shown in Figure 8-8.

**Figure 8-8
The Analysis
ToolPak's
Regression
command**

selected range
contains column
labels

send output to
the "Regression
Statistics"
worksheet

display residuals

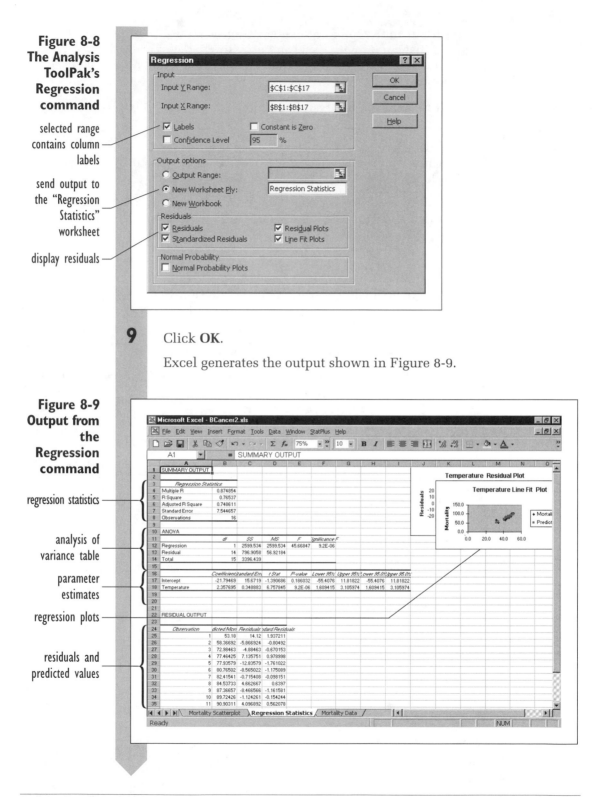

9   Click **OK**.

Excel generates the output shown in Figure 8-9.

**Figure 8-9
Output from
the
Regression
command**

regression statistics

analysis of
variance table

parameter
estimates

regression plots

residuals and
predicted values

The output is divided into six areas: regression statistics, analysis of variance (ANOVA), parameter estimates, residual output, probability output (not shown in Figure 8-9), and plots. Let's examine these areas more closely. The Regression command doesn't format the output for us, so you may want to do that yourself on your worksheet.

Interpreting Regression Statistics

Figure 8-10
Regression
statistics

3	Regression Statistics	
4	Multiple R	0.875
5	R Square	0.765
6	Adjusted R Square	0.749
7	Standard Error	7.5447
8	Observations	16

You've seen some of the regression statistics shown in Figure 8-10 before. The R^2-value of 0.765 you've seen in the scatterplot. The multiple R-value is equal to the square root of the R^2 value. The multiple R is equal to the absolute value of the correlation between the dependent variable and the predictor variable. You'll learn about correlation later in this chapter. The adjusted R^2 is used when performing a regression with several predictor variables. This statistic will be covered more in depth in Chapter 9. Finally, the standard error measures the size of a typical deviation of an observed value (x, y) from the regression line. Think of the standard error as a way of averaging the size of the deviations from the regression line. The typical deviation of an observed point from the regression line in this example is about 7.5447. The observations value is the size of the sample used in the regression. In this case, the regression is based on the values from 16 regions.

Interpreting the Analysis of Variance Table

Figure 8-11 shows the ANOVA table output from the Analysis ToolPak Regression command.

Figure 8-11
Analysis of
variance
(ANOVA)
table

11		df	SS	MS	F	Significance F
12	Regression	1	2599.53	2599.53	45.7	0.0000092
13	Residual	14	796.91	56.92		
14	Total	15	3396.44			

The ANOVA table analyzes the variability of the mortality index. The variability is divided into two parts: the first is the variability due to the regression line and the second is due to random variability.

The values in the *df* column of the table indicate the number of degrees of freedom for each part. The total degrees of freedom are equal to the number of observations minus 1. In this case the total degrees of freedom are 15. Of those 15 degrees of freedom, 1 degree of freedom is attributed to the regression, and the remaining 14 degrees of freedom are attributed to random variability.

The *SS* column gives you the sums of squares. The total sum of squares is the sum of the squared deviations of the mortality index from the overall mean. This total is also divided into two parts. The first part, labeled in the table as the regression sum of squares, is the sum of squared deviations between the regression line and the overall mean. The second part, labeled the residual sum of squares, is equal to the sum of the squared deviations of the mortality index from the regression line. Recall that this is the value that we want to make as small as possible in the regression equation. In this example, the total sum of squares is 3396.44, of which 2599.53 is attributed to the regression and 796.91 is attributed to the residual sum of squares.

What percentage of the total sum of squares can be attributed to the regression? In this case, it is $2599.53/3396.44 = 0.7654$, or 76.54%. This is equal to the R^2 value, which, as you learned earlier, measures the percentage of variability explained by the regression. Note also that the total sum of squares (3396.44) divided by the total degrees of freedom (15) equals 226.43, which is the variance of the mortality index. The square root of this value is the standard deviation of the mortality index.

The *MS* (mean square) column displays the sum of squares divided by the degrees of freedom. Note that the mean square for the residual is equal to the square of the standard error in cell B7 ($7.5447^2 = 56.9218$). Thus you can use the mean square for the residual to derive the standard error.

The next column displays the ratio of the mean square for the regression to the mean square error of the residuals. This value is called the **F-ratio**. A large *F*-ratio indicates that the regression may be statistically significant. In this example, the ratio is 45.7. The *p*-value is displayed in the next column and equals 0.0000092. Thus the regression is statistically significant from this table. You'll learn more about analysis of variance and interpreting ANOVA tables in an upcoming chapter.

Parameter Estimates and Statistics

The file output table created by the Analysis ToolPak Regression command displays the estimates of the regression parameters along with statistics measuring their significance. See Figure 8-12.

Figure 8-12
Parameter
estimates
and
statistics

16		Coefficients	Standard Error	t Stat	P-value	Lower 95%	Upper 95%
17	Intercept	-21.79	15.672	-1.39	0.19	-55.41	11.82
18	Temperature	2.36	0.349	6.76	0.0000092	1.61	3.11

As you've already seen, the constant coefficient, or intercept, equals about −21.79, and the slope based on the temperature variable is about 2.36. The standard errors for these values are shown in the Standard Error column and are 15.672 and 0.349, respectively. The ratio of the parameter estimates to their standard errors follows a t-distribution with $n - 2$ or 14, degrees of freedom. The ratios for each parameter are shown in the t Stat column, and the corresponding p-values are shown in the P-value column. In this example, the p-value for the intercept term is 0.186, and the p-value for the slope term (labeled "temperature") is 9.2×10^{-6}, or 0.0000092 (note that this is the same p-value that appeared in the ANOVA table.)

The final part of this table displays the 95% confidence interval for each of the terms. In this case, the 95% confidence interval for the intercept term is about (−55.41, 11.82), and the 95% confidence interval for the slope is (1.61, 3.11).

Note: The confidence intervals appear twice. The first pair, a 95% interval, always appears. The second pair always appears, but with the confidence level you specify in the Regression dialog box. In this case, you used the default 95% value, so that interval appears in both pairs.

What have you learned from the regression statistics? First of all, you would decide to reject the null hypothesis and accept the alternative hypothesis that a linear relationship exists between the mortality index and temperature. Based on the confidence interval for the slope parameter, you can report with 95% confidence that for each degree increase in the mean annual temperature, the mortality index for the region increases between 1.61 to 3.11 points.

Residuals and Predicted Values

The last part of the output from the Analysis ToolPak's Regression command consists of the residuals and the predicted values. See Figure 8-13 (the values have been reformatted to make them easier to view).

Figure 8-13
Residuals
and
predicted
values

	Observation	Predicted Mortality	Residuals	Standard Residuals
24				
25	1	53.180	14.120	1.937
26	2	58.367	-5.867	-0.805
27	3	72.985	-4.885	-0.670
28	4	77.464	7.136	0.979
29	5	77.936	-12.836	-1.761
30	6	80.765	-8.565	-1.175
31	7	82.415	-0.715	-0.098
32	8	84.537	4.663	0.640
33	9	87.367	-8.467	-1.162
34	10	89.724	-1.124	-0.154
35	11	90.903	4.097	0.562
36	12	92.553	-5.553	-0.762
37	13	94.204	1.696	0.233
38	14	95.854	8.646	1.186
39	15	96.090	4.310	0.591
40	16	99.155	3.345	0.459

As you've learned, the residuals are the differences between the observed values and the regression line (the predicted values). Also included in the output are the standardized residuals. From the values shown in Figure 8-13, you see that there is one residual that seems larger than the others, it is found in the first observation and has a standardized residual value of 1.937. You'll want to keep an eye on this observation as you continue to explore this regression model. As you'll see shortly, the residuals play an important role in determining the appropriateness of the regression model.

Checking the Regression Model

As in any statistical procedure, when you perform regression on a set of data, you are making some important assumptions. There are four:

1. The straight-line model is correct.
2. The error term, ε, is normally distributed with mean 0.
3. The errors have constant variance.
4. The errors are independent of each other.

Whenever you use regression to fit a line to data, you should consider these assumptions. Fortunately, regression is somewhat robust, so the assumptions do not need to be perfectly satisfied.

One point that cannot be emphasized too strongly is that *a significant regression is not proof that these assumptions haven't been violated*. To verify that your data does not violate these assumptions is to go through a series of tests, called **diagnostics**.

Testing the Straight-line Assumption

To test whether or not the straight-line model is correct, you should first create a scatterplot of the data to inspect visually whether the data departs from this assumption in any way. Figure 8-14 shows a classic problem that you may see in your data.

Figure 8-14
A curved
relationship

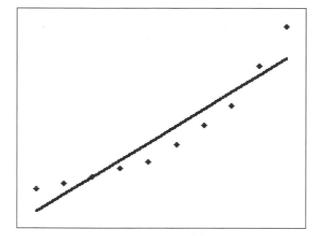

Another sharper way of seeing whether the data follows a straight line is to fit the regression line and then plot the residuals of the regression against the values of the predictor variable. A U-shaped (or upside-down U) pattern to the plot, as shown in Figure 8-15, is a good indication that the data actually follows a curved relationship and that the straight-line assumption is wrong.

Figure 8-15
Residuals
showing a
curved
relationship

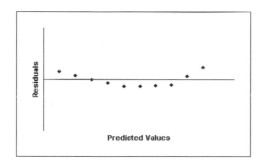

Let's apply this diagnostic to the mortality index data. The Regression command creates this plot for you, but it can be difficult to read because of its size. We'll move the plot to a chart sheet and reformat the axes for easier viewing.

To create a plot of the residuals versus the predictor variable:

1   Scroll to cell J1 in the Regression Statistics worksheet and click the **Temperature Residual Plot**.

2   Click **Chart > Location**.

3   Click the **As new sheet** option button and then type **Residuals vs. Temperature** and click **OK**.

4   Rescale the horizontal axes of the temperature variable, so that the lower boundary is **30**. The revised plot appears in Figure 8-16.

**Figure 8-16
Residuals vs.
temperature**

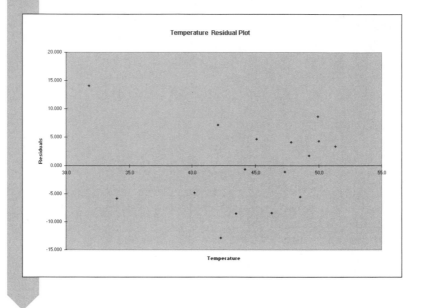

The plot shows that most of the positive residuals tend be located at the lower and higher temperatures and most of the negative residuals are concentrated in the middle temperatures. This may indicate a curve in the data. The large first observation is influential here. Without it, there would be less indication of a curve.

Testing for Normal Distribution of the Residuals

The next diagnostic is a normal plot of the residuals.

To create a Normal Probability Plot of the residuals:

1   Return to the Regression Statistics worksheet.

2   Click **StatPlus > Single Variable Charts > Normal P-Plots**.

3   Click the **Data Values** button.

4   In the Input Options dialog box, click the **Use Range References** option button and then select the range **C24:C40**. Verify that the Range Include a Row of Column Labels checkbox is selected and click **OK**.

5   Click the **Output** button, verify that the As a New Chart sheet option button is selected, and type **Residual Normal Plot** in the accompanying text box. Click the **OK** button.

6   Click the **OK** button to start generating the Normal Probability plot. See Figure 8-17.

**Figure 8-17
Normal
probability
plot of
residuals**

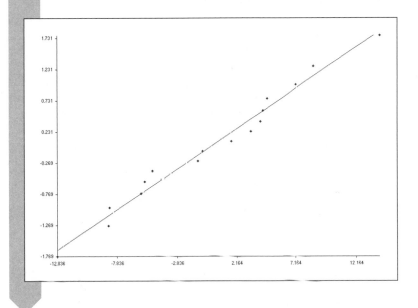

Recall that if the residuals follow a normal distribution, they should fall evenly along the superimposed line on the normal probability plot. Although the points in Figure 8-17 do not fall perfectly on the line, the departure is not strong enough to invalidate our assumption of normality.

Testing for Constant Variance in the Residuals

The next assumption you should always investigate is the assumption of constant variance in the residuals. A commonly used plot to help verify this assumption is the plot of the residuals versus the predicted values. This plot will also highlight any problems with the straight-line assumption.

**Figure 8-18
Residuals
showing
nonconstant
variance**

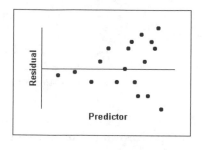

If the constant variance assumption is violated, you may see a plot like the one shown in Figure 8-18. In this example, the variance of the residuals is larger for larger predicted values. It's not uncommon for variability to increase as the value of the response variable increases. If that happens, you might remove this problem by using the log of the response variable and performing the regression on the transformed values.

With one predictor variable in the regression equation, the scatterplot of the residuals versus the predicted values is identical to the scatterplot of the residuals versus the predictor variable (shown earlier in Figure 8-16). The scatterplot indicates that there *may* be a decrease in the variability of the residuals as the predicted values increase. Once again, though, this interpretation is influenced by the presence of the possible outlier in the first observation. Without this observation, there might be no reason to doubt the assumption of constant variance.

Testing for the Independence of Residuals

The final regression assumption is that the residuals are independent of each other. This assumption is of concern only in situations where there is a defined order for the observations. For example, if we do a regression of a predictor variable versus time, the observations will follow a sequential order. The assumption of independence can be violated if the value of one observation influences the value of the next observation. For example, a large value might be followed by a small value, or large and small values could be clustered together (see Figure 8-19). In these cases, the residuals do not show independence, because you can predict what the sign of the next value will be on the basis of the current value.

Figure 8-19
Residuals vs.
predicted
values

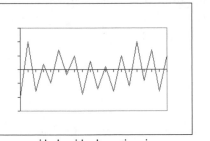

residuals with alternating signs

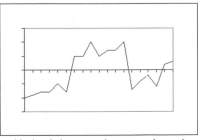

residuals of the same sign grouped together

In examining residuals, we can examine the sign of the values (either positive or negative) and determine how many values with the same sign are clustered together. These groups of similarly signed values are called runs. For example, consider a data set of 10 residuals containing 5 positive values and 5 negative values. The values could follow an order with only two runs, such as

$$+ + + + + - - - - -$$

In this case, we would suspect that the residuals were not independent, because the positives and negatives are clustered together in the sequence. On the other hand, we might have the opposite problem, where there could be as many as ten runs, such as

$$+ - + - + - + - + -$$

Here, we suspect the residuals are not independent, because the residuals are constantly switching sign. Finally, we might have something in-between, such as

$$+ + - - + + - - +$$

which has five runs. If the number of runs is very large or very small, we would suspect that the residuals are not independent. How large (or how small) does this value have to be? Using probability theory, statisticians have calculated the p-values for a **runs test**, associated with the number of runs observed for different sample sizes. If we let n be the sample size, n_+ be the number of positive values, and n_- be the number of negative values, the expected number of runs, μ, is

$$\mu = \frac{2n_+n_-}{n} + 1$$

and the standard deviation, σ, is

$$\sigma = \sqrt{\frac{2n_+n_-}{n(n-1)}\left(\frac{2n_+n_-}{n} - 1\right)}$$

If r is the observed number of runs, then the value

$$z = \frac{(r - \mu + {}^1/_2)}{\sigma}$$

approximately follows a standard normal distribution for large sample sizes (where n_+ and n_- are both > 10). For example, if $n = 10$, $n_+ = 5$, and $n_- = 5$, then $\mu = 6$ and $\sigma = \sqrt{20/9} = 1.49$. If 5 runs have been observed, $z = -0.335$ and the p-value is 0.368. This is very close to the exact p-value of 0.357, so we would not find this an extremely unusual number of runs. On the other hand, if we observe only 3 runs, then $z = -2.012$ and the p-value is .022 (the exact value is 0.04).

Another statistic used to test the assumption of independence is the **Durbin-Watson** test statistic. In this test, we calculate the value

$$DW = \frac{\displaystyle\sum_{i=2}^{n}(e_i - e_{i-1})^2}{\displaystyle\sum_{i=1}^{n}e_i^2}$$

where e_i is the ith residual in the data set. The value of DW is then compared to a table of Durbin-Watson values to see whether there is evidence of a lack of independence in the residuals. Generally, a value of DW approximately equal to 0 or 4 suggests that the residuals are not independent. A value of DW near 2 suggests independence. Values in between may be inconclusive.

Because the mortality index data is *not* sequential, you shouldn't apply the runs test or the Durbin-Watson test. Remember, these statistics are most useful when the residuals have a definite sequential order.

After performing the diagnostics on the residuals, you conclude that there is no hard evidence to suggest that the regression assumptions have been violated. On the other hand, there is a problematic large residual in the first observation to consider. You should probably redo the analysis without the first observation to see what effect (if any) this has on your model. You'll have a chance to do that in the first exercise at the end of the chapter.

STATPLUS TIPS

- Excel does not include a function to perform the runs test, but you can use the StatPlus command "StatPlus > Time Series > Runs Test" on your time-ordered residuals to perform this analysis.

- Use the functions RUNS(*range*, *center*) and RUNSP(*range*, *center*) to calculate the number of runs in a data set and the corresponding p-value for a set of data in the cell range, *range*, around the central line, *center* (*StatPlus required*).

- Use the function DW(*range*) to calculate the Durbin-Watson test statistic for the values in the cell range, *range* (*StatPlus required*).

Correlation

The value of the slope in our regression equation is a product of the scale in which we measure our data. If, for example, we had chosen to express the temperature values in degrees Centigrade, we would naturally have a different value for the slope (though, of course, the statistical significance of the regression would not change). Sometimes, it's an advantage to express the strength of the relationship between one variable and another in a "dimensionless" number, one that does not depend on scale. One such value is the correlation. The **correlation** expresses the strength of the relationship on a scale ranging from −1 to 1.

A positive correlation indicates a strong positive relationship, in which an increase in one variable implies an increase in the value of the second variable. This might occur in the relationship between height and weight. A negative correlation indicates that an increase in the first variable signals a decrease in the second variable. An increase in price for an object could be negatively correlated with sales. See Figure 8-20. A correlation of zero does *not* imply there is no relationship between the two variables. One can construct a nonlinear relationship that produces a correlation of zero.

Figure 8-20
Correlations

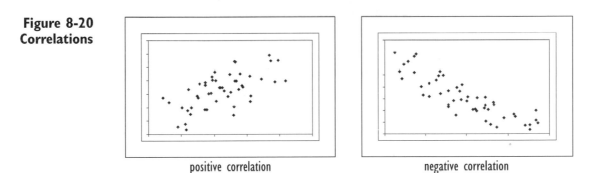

positive correlation   negative correlation

The most often used measure of correlation is the **Pearson Correlation coefficient**, which is usually identified with the letter r. The formula for r is

$$r = \frac{\sum_{i=1}^{n}(x_i - \overline{x})(y_i - \overline{y})}{\sqrt{\sum_{i=1}^{n}(x_i - \overline{x})^2}\sqrt{\sum_{i=1}^{n}(y_i - \overline{y})^2}}$$

For example, the correlation of the data in Table 8-1 is

$$r = \frac{(1 - 1.8)(3 - 3.4) + (2 - 1.8)(4 - 3.4) + \cdots + (2 - 1.8)(5 - 3.4)}{\sqrt{(1 - 1.8)^2 + \cdots + (2 - 1.8)^2} \times \sqrt{(3 - 3.4)^2 + \cdots + (5 - 3.4)^2}}$$

$$= \frac{1.4}{\sqrt{2.8} \times \sqrt{1.2}}$$

$$= 0.763$$

which indicates a high positive correlation.

Correlation and Slope

Notice that the numerator in the equation for r is exactly the same as the numerator for the slope b in the regression equation shown earlier. This is important because it means that the slope = 0 when r = 0 and that the sign of the slope is the same as the sign of the correlation. The slope can be any real number, but the correlation must always be between −1 and +1. A correlation of +1 means that all of the data points fall perfectly on a line of positive slope. In such a case, all of the residuals would be 0, and the line would pass right through the points; it would have a perfect fit.

In terms of hypothesis testing, the following statements are equivalent:

H_0: There is no linear relationship between the predictor variable and the dependent variable.

H_0: There is no population correlation between the two variables.

In other words, the correlation is zero if the slope is zero, and vice versa. When you do a statistical test for correlation, the assumptions are the same as the assumptions for linear regression.

Correlation and Causality

Correlation indicates the relationship between two variables without assuming that a change in one causes a change in the other. For example, if you learn of a correlation between the number of extracurricular activities and grade-point average for high school students, does this imply that if you raise a student's GPA, he or she will participate in more after-school activities? Or that if you ask the student to get more involved in extracurricular activities, his or her grades will improve as a result? Or is it more likely that if this correlation is true, the type of people who are good students also tend to be the type of people who join after-school groups? You should therefore be careful never to confuse correlation with "cause and effect," or causality.

Spearman's Rank Correlation Coefficient s

Pearson's correlation coefficient is not without problems. It can be susceptible to the influence of outliers in the data set, and it assumes that a straight-line relationship exists between the two variables. In the presence of outliers or a curved relationship, Pearson's r may not detect a significant correlation. In those cases, you may be better off using a nonparametric measure of correlation, **Spearman's rank correlation**, which is usually shown with the symbol s. As with the nonparametric tests in Chapter 7, you replace observed values with their ranks and calculate the value of s on the ranks. Spearman's rank correlation, like many other nonparametric tests, is less susceptible to the influence of outliers and is better than Pearson's correlation for nonlinear relationships. The downside to the Spearman correlation is that it is not as powerful as the Pearson correlation in detecting significant correlations in situations where the parametric assumptions are satisfied.

Correlation Functions in Excel

To calculate correlation values in Excel, you can use some of the functions shown in Table 8-4. Note that Excel does not include functions to calculate Spearman's rank correlation or the p-values for the two types of correlation measures.

Table 8-4 Calculating correlation values

Function	Description
CORREL(x, y)	Calculates Pearson's correlation, r, for the values in x and y.
CORRELP(x, y)	Calculates the p-value of Pearson's correlation for the values in x and y. *StatPlus required.*
SPEARMAN(x, y)	Calculates Spearman's rank correlation, s, for the values in x and y. *StatPlus required.*
SPEARMANP(x, y)	Calculates the p-value of Spearman's rank correlation, s, for the values in x and y. *StatPlus required.*

Let's use these functions to calculate the correlation between the mortality index and the mean annual temperature for the breast cancer data.

To calculate the correlations and p-values:

1 Return to the Mortality Data worksheet.

2 Enter the labels **Pearson's r**, **p-value**, **Spearman's s**, and **p-value** in the cell range **A19:A22**. Enlarge the width of column A to fit the size of the new labels.

3 Click cell **B19**, type **=CORREL(temperature, mortality)**, and press **Enter**.

4 In cell B20, type **=CORRELP(temperature, mortality)** and press **Enter**.

5 In cell B21, type **=SPEARMAN(temperature, mortality)** and press **Enter**.

6 In cell B22, type **=SPEARMANP(temperature, mortality)** and press **Enter**.

The correlation values are shown in Figure 8-21.

**Figure 8-21
Correlations
and *p*-values**

19	Pearson's r	0.874854404
20	p-value	9.20185E-06
21	Spearman's s	0.902941176
22	p-value	1.67773E-06

The values in Figure 8-22 indicate a strong positive correlation between the mortality index and the mean annual temperature. The *p*-values for both measures are also very significant, indicating that this correlation is statistically different from zero. Note that the *p*-value for Pearson's *r* is equal to the *p*-value for the linear regression shown earlier in Figure 8-11. One more important point: the value of *r*, 0.875, is equal to the square root of the R^2 statistic, computed earlier in Figure 8-10. It will always be the case that R^2 is equal to the square of Pearson's correlation coefficient between two variables.

You can close the BCancer2 workbook now. You've completed your analysis of the data, but you'll return to it in the chapter exercises.

Creating a Correlation Matrix

When you have several variables to study, it's useful to calculate the correlations between the variables. In this way, you can get a quick picture of the relationships between the variables, determining which variables are highly correlated and which are not. One way of doing this is to create a **correlation matrix**, in which the correlations (and associated *p*-values) are laid out in a square grid.

To illustrate the use of a correlation matrix, consider the CALC.XLS workbook. This file contains data collected to see how performance in a freshman calculus class is related to various predictors (Edge and Friedberg, 1984). Here are the variables in the CALC workbook.

Table 8-5 CALC.XLS workbook variables

Range Name	Range	Description
Calc_HS	A2:A81	Indicates whether calculus was taken in high school (0 = no; 1 = yes)
ACT_Math	B2:B81	The student's score on the ACT mathematics exam
Alg_Place	C2:C81	The student's score on the algebra placement exam given in the first week of classes
Alg2_Grade	D2:D81	The student's grade point in second-year high school algebra
HS_Rank	E2:E81	The student's rank in high school
Gender	F2:F81	The student's gender
Gender_Code	G2:G81	The student's gender code (0 = female; 1 = male)
Calc	H2:H81	The student's grade in calculus

To open the Calc.xls workbook:

1 Open the **Calc** workbook from the folder containing your student files.

2 Click **File > Save As** and save the workbook as **Calc2**. The workbook appears as shown in Figure 8-22.

Figure 8-22 Calc workbook

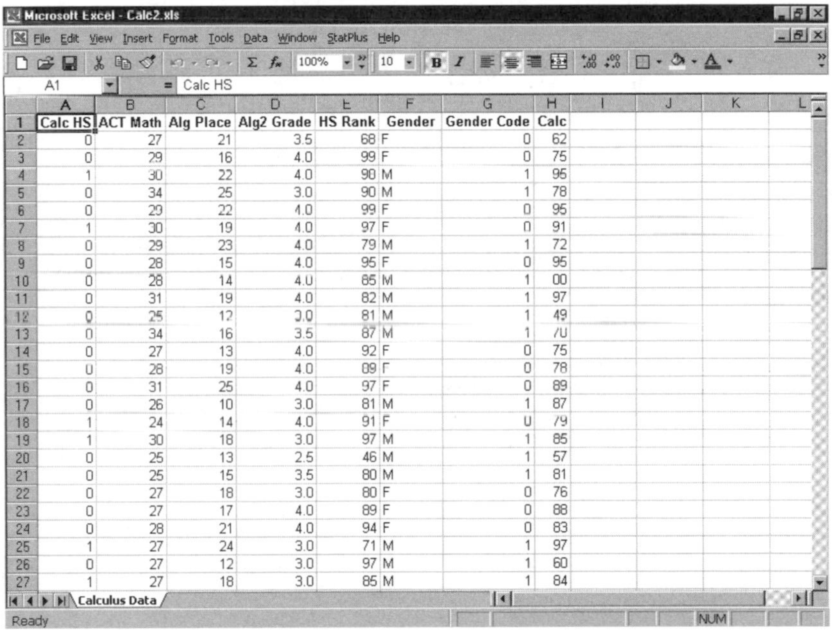

	A	B	C	D	E	F	G	H
1	Calc HS	ACT Math	Alg Place	Alg2 Grade	HS Rank	Gender	Gender Code	Calc
2	0	27	21	3.5	68	F	0	62
3	0	29	16	4.0	99	F	0	75
4	1	30	22	4.0	90	M	1	95
5	0	34	25	3.0	90	M	1	78
6	0	29	22	4.0	99	F	0	95
7	1	30	19	4.0	97	F	0	91
8	0	29	23	4.0	79	M	1	72
9	0	28	15	4.0	95	F	0	95
10	0	28	14	4.0	85	M	1	00
11	0	31	19	4.0	82	M	1	97
12	0	25	12	3.0	81	M	1	49
13	0	34	16	3.5	87	M	1	70
14	0	27	13	4.0	92	F	0	75
15	0	28	19	4.0	89	F	0	78
16	0	31	25	4.0	97	F	0	89
17	0	26	10	3.0	81	M	1	87
18	1	24	14	4.0	91	F	0	79
19	1	30	18	3.0	97	M	1	85
20	0	25	13	2.5	46	M	1	57
21	0	25	15	3.5	80	M	1	81
22	0	27	18	3.0	80	F	0	76
23	0	27	17	4.0	89	F	0	88
24	0	28	21	4.0	94	F	0	83
25	1	27	24	3.0	71	M	1	97
26	0	27	12	3.0	97	M	1	60
27	1	27	18	3.0	85	M	1	84

Now let's create a matrix of the correlations for all of the variables in the workbook.

To create a correlation matrix of the numeric variables:

1 Click **StatPlus > Multivariate > Correlation Matrix**.

2 Click the **Data Values** button and select all of the variables in the workbook *except* the Gender variable.

3 Click the **Output** button and send the output to a new worksheet named **Corr Matrix**. Click the **OK** button. Figure 8-23 shows the completed dialog box.

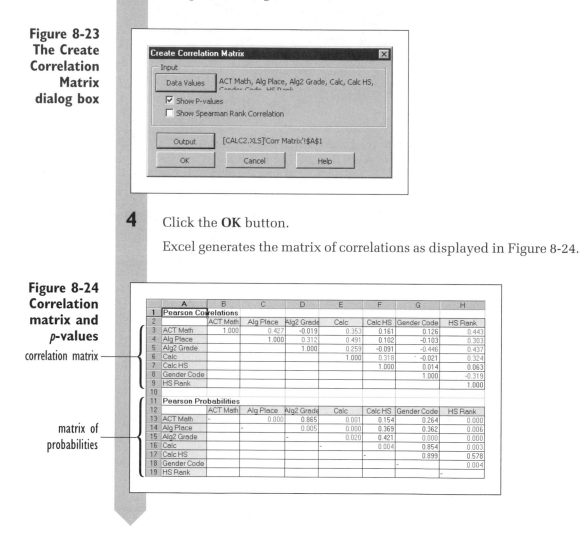

**Figure 8-23
The Create
Correlation
Matrix
dialog box**

4 Click the **OK** button.

Excel generates the matrix of correlations as displayed in Figure 8-24.

**Figure 8-24
Correlation
matrix and
p-values**

correlation matrix —

matrix of
probabilities

	A	B	C	D	E	F	G	H
1	Pearson Correlations							
2		ACT Math	Alg Place	Alg2 Grade	Calc	Calc HS	Gender Code	HS Rank
3	ACT Math	1.000	0.427	-0.019	0.353	0.161	0.126	0.443
4	Alg Place		1.000	0.312	0.491	0.102	-0.103	0.303
5	Alg2 Grade			1.000	0.259	-0.091	-0.446	0.437
6	Calc				1.000	0.318	-0.021	0.324
7	Calc HS					1.000	0.014	0.063
8	Gender Code						1.000	-0.319
9	HS Rank							1.000
10								
11	Pearson Probabilities							
12		ACT Math	Alg Place	Alg2 Grade	Calc	Calc HS	Gender Code	HS Rank
13	ACT Math	-	0.000	0.865	0.001	0.154	0.264	0.000
14	Alg Place		-	0.005	0.000	0.369	0.362	0.006
15	Alg2 Grade			-	0.020	0.421	0.000	0.000
16	Calc				-	0.004	0.854	0.003
17	Calc HS					-	0.899	0.578
18	Gender Code						-	0.004
19	HS Rank							-

Figure 8-24 shows two matrices. The first, in cells A1:H9, is the correlation matrix, which shows the Pearson correlations. The second, the matrix of probabilities in cells A11:H19, gives the corresponding p-values. P-values less than 0.05 are highlighted in red.

The most interesting numbers here are the correlations with the calculus score, because the object of the study was to predict this score. The highest correlation appears in cell E4 (0.491), with Alg Place, the algebra placement test score. The other correlations, ACT Math, HS Rank, and Calc HS, are not impressive predictors when you consider that the squared correlation gives R^2, the percentage of variance explained by the variable as a regression predictor.

For example, the correlation between Calc and HS Rank is 0.324 (cell H6); the square of this is 0.105, so regression on HS Rank would account for only 10.5% of the variation in the calculus score. Another way of saying this is that using HS Rank as a predictor improves by 10.5% the sum of squared errors, as compared with using just the mean calculus score as a predictor. Note that the p-value for this correlation is 0.003 (cell H16), which is less than 0.05, so the correlation is significant at the 5% significance level.

Just because the taking of high school calculus and the subsequent college calculus score have a significant correlation, you cannot conclude that taking calculus in high school *causes* a better grade in college. The stronger math students tend to take calculus in high school, and these students also do well in college. Only if a fair assignment of students to classes could be guaranteed (so that the students in high school calculus would be no better or worse than others) could the correlation be interpreted in terms of causation.

Correlation with a Two-Valued Variable

You might reasonably wonder about using Calc HS here. After all, it assumes only the two values 0 and 1. Does the correlation between Calc and Calc HS make sense? The positive correlation of 0.324 indicates that if the student has taken calculus in high school, the student is more likely to have a high calculus grade.

Another categorical variable in this correlation matrix is Gender Code, which has a significant negative correlation with the Alg2 Grade ($r = -0.446$, p-value = 0.000) and HS Rank ($r = -0.319$, p-value = 0.004). Recall that in the gender code, 0 = female and 1 = male. A negative correlation here means that females tended to have higher grades in second-year algebra and were ranked higher in high school.

Adjusting Multiple p-Values with Bonferroni

The second matrix in Figure 8-24 gives the p-values for the correlations. Except for Gender, all of the correlations with Calc are significant at the 5% level because all the p-values are less than 0.05.

Some statisticians believe that the *p*-values should be adjusted for the number of tests, because conducting several hypothesis tests raises above 5% the probability of rejecting at least one true null hypothesis. The **Bonferroni** approach to this problem is to multiply the *p*-value in each test by the total number of tests conducted. With this approach, the probability of rejecting one or more of the true hypotheses is less than 5%.

Let's apply this approach to correlations of Calc with the other variables. Because there are six correlations, the Bonferroni approach would have us multiply each *p*-value by 6 (equivalent to decreasing the *p*-value required for statistical significance to $0.05/6 = 0.0083$). Alg2 Grade has a *p*-value of 0.020, and because $6 \times (0.020) = 0.120$, the correlation is no longer significant from this point of view. Instead of focusing on the individual correlation tests, the Bonferroni approach rolls all of the tests into one big package, with 0.05 referring to the whole package.

Bonferroni makes it much harder to achieve significance, and many researchers are reluctant to use it because it is so conservative. In any case, this is a controversial area, and professional statisticians argue about it.

EXCEL TIPS

- To create a correlation matrix with Excel, you can use the Data Analysis ToolPak. Click "Tools > Data Analysis" from the menu and then "Correlation" from the Data Analysis dialog box. Complete the Correlation dialog box to create the matrix.

Creating a Scatterplot Matrix

The Pearson correlation measures the extent of the linear relationship between two variables. To see whether the relationship between the variables is really linear, you should create a scatterplot of the two variables. In this case, that would mean creating 15 different scatterplots, a time-consuming task! To speed up the process, you can create a scatterplot matrix. In a **scatterplot matrix**, or **SPLOM**, you can create a matrix containing the scatterplots between the variables. By viewing the matrix, you can tell at a glance the nature of the relationships between the variables.

To create a scatterplot matrix:

1 Click **StatPlus > Multi-variable Charts > Scatterplot Matrix**.

2 Click the **Data Values** button and select the range names **ACT Math**, **Alg Place**, **Alg2 Grade**, **Calc**, and **HS Rank**.

3 Click the **Output** button, and send the output to the worksheet **SPLOM**. Click the **OK** button twice.

Excel generates the scatterplot matrix shown in Figure 8-25.

**Figure 8-25
Scatterplot
matrix**

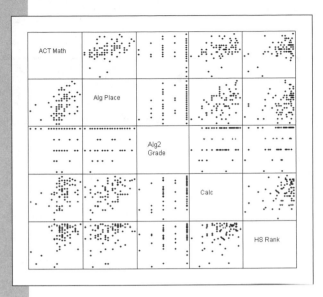

Depending on the number of variables you are plotting, SPLOMs can be difficult to view on the screen. If you can't see the entire SPLOM on your screen, consider reducing the value in the Zoom Control box. You can also reduce the SPLOM by selecting it and dragging one of the resizing handles to make it smaller.

How should you interpret the SPLOM? Each of the five variables is plotted against the other four variables, with the four plots displayed in a row. For example, ACT Math is plotted as the y-variable against the other four variables in the first row of the SPLOM. The first plot in the first row is ACT Math vs. Alg Place, and so on. On the other hand, the first plot in the first column displays Alg Place as the y-variable and is plotted against the x-variable, ACT Math. The scales of the plot are not shown in order to save space. If you find a plot of interest, you can recreate it using Excel's Chart Wizard to show more details and information.

Carefully consider the plots in the bottom row, which show Calc against the other variables. Each plot shows a roughly linear upward trend. It would be reasonable to conclude here that correlation and linear regression are appropriate when predicting Calc from ACT Math, Alg2 Grade, Alg Place, and HS Rank.

Recall from Figure 8-24 that Alg Place had the highest correlation with Calc. How is that evident here? A good predictor has good accuracy, which means that the range of y is small for each x. Of the four plots in the fourth row, the plot of Calc against Alg Place has the narrowest range of *y*-values for each *x*. However, Alg Place is the best of a weak lot. None of the plots shows that really accurate prediction is possible. None of these plots shows a relationship anywhere near as strong as the relationship between mortality index and temperature that you worked with earlier in the chapter.

Save your work and close the Calc2 workbook.

Exercises

1. *True or false (and why)*: If the slope of a regression line is large, the correlation between the variables will also be large.

2. *True or false (and why)*: If the correlation between two variables is near 1, the slope will be a large positive number.

3. *True or false (and why)*: If the *p*-value of the Pearson's correlation coefficient is low, the *p*-value of the slope parameter of the regression equation will also be low.

4. *True or false (and why)*: A correlation of zero means that the two variables are unrelated.

5. *True or false (and why)*: The runs test is one of the diagnostic tests you should always apply to the residuals in your regression analysis.

6. In a time-ordered study, you have 25 residuals from the regression model. There are 10 negative residuals and 15 positive ones. There are a total of 10 runs. Is this an unusual number of runs? What is the level of statistical significance?

7. Using the following ANOVA table for the regression of variable *y* on variable *x*, answer the questions below.

Table 8-6 Regression of variable *y* on *x*

ANOVA	df	SS	MS	F	Significance F
Regression	1	129.6	129.6	4.91	0.057
Residual	8	210.9	26.4		
Total	9	340.5			

    a. How many observations are in the data set?
    b. What is the variance of *y*?
    c. What is the value of R^2?
    d. What percentage of the variability in *y* is explained by the regression?

e. What is the absolute value of the correlation of x and y?

f. What is the p-value of the correlation of x and y?

g. What is the standard error (the typical deviation of an observed point from the regression line)?

8. Reopen the BCancer workbook discussed in this chapter. As shown in the chapter, there may be an outlier in the data set. Perform the following analysis:

a. Remove the observation for the first region from the data set.

b. Create a scatterplot of mortality vs. temperature and add a linear trend line to the plot, showing both the R^2 value and the regression equation.

c. Calculate the regression statistics for the new data set.

d. Create scatterplots of the residuals of the regression equation vs. temperature and the predicted values. Also create a normal probability plot of the residuals.

e. Calculate the Pearson and Spearman correlation coefficients, including the p-values.

f. Report your results, including a description of the diagnostic tests you performed. How does this regression compare to the regression you performed earlier that included the possible outlier? How do the diagnostic plots compare?

g. Save your workbook as E8BCANCER.XLS.

9. The WHEAT.XLS workbook contains nutritional information on 10 wheat products. Open the workbook and perform the following analysis:

a. Plot calories vs. serving oz, adding a regression line, equation, and R^2 value to the plot.

b. Compute Pearson's correlation and p-value on the Nutrition Info worksheet.

c. Calculate the statistics for the regression equation.

d. Create diagnostic plots of residuals vs. serving oz, and the normal probability plot of the residuals. Do the regression assumptions seem to be satisfied?

e. In the plot of residuals versus predicted values, label each point with the food type (pretzel, bagel, bread, etc). Where do the residuals for the breads appear?

f. Breads are often low in calories because of high moisture content. One way of removing the moisture content from the equation is to create a new variable that sums up the total of the nutrient weights. With this in mind, create a new variable, total, which is the sum of the weights of carbohydrates, proteins, and fats.

g. Redo your regression equation, regressing the calories variable on the new variable, total. How does the diagnostic plot of residuals vs. predicted values compare to the earlier plot? How do the R^2-values compare? Where are the residuals for the bread values located?

h. Report your results. Save the workbook as E8WHEAT.XLS.

10. Reopen the WHEAT.XLS workbook. Perform the following analysis:
    a. Create scatterplot and correlation matrices (Pearson correlation only) for the variables serving ounces, calories, protein, carbohydrate, and fat.
    b. Why is fat so weakly related to the other variables? Given that fat is supposed to be very important in calories, why is the correlation so weak here?
    c. Would the relationship between the fat and calories variables be stronger if we used foods that cover a wider range of fat-content values?

11. The MUSTANG.XLS workbook displays prices and ages of used Mustangs from the San Francisco *Chronicle*, November 25, 1990. Open the workbook and perform the following analysis:
    a. Compute the Pearson and Spearman correlations (and the *p*-values) between price and age.
    b. Plot price against age. Does this scatterplot cause you any concern about the validity of the correlations you calculated?
    c. How do the correlation values change if you concentrate only on cars that are less than 10 years in age?
    d. Excluding the old classic cars (older than 10 years), perform a regression of price against age and find the drop in price per year of age.
    e. Do you see any problems in the diagnostic plots?
    f. Report your results and save your workbook as E8MUSTNG.XLS.

12. Reopen the CALC.XLS workbook that you used in this chapter.
    a. Regress Calc on Alg Place and obtain a 95% confidence interval for the slope.
    b. Interpret the slope in terms of the increase in final grade when the placement score increases by one point.
    c. Do the residuals give you any cause for concern about the validity of the model?
    d. Report your results and save your workbook as E8CALC.XLS.

13. The BOOTH.XLS workbook gives total assets and net income for 45 of the largest American banks in 1973. Open the workbook and perform the following analysis.
    a. Plot net income against total assets and notice that the points tend to bunch up toward the lower left, with just a few big banks dominating the upper part of the graph. Add a linear trend line to the plot.
    b. Regress net income against total assets and plot the standard residuals against the predictor values. (The standard residuals appear with the regression output when you select the Standardized Residuals check box in the Regression dialog box.)

c. Given that the residuals tend to be bigger for the big banks, you should be concerned about the assumption of constant variance. Try taking logs of both variables. Now repeat the plot of one against the other, repeat the regression, and again look at the plot of the residuals against the predicted values. Does the transformation help the relationship? Is there now less reason to be concerned about the assumptions? Notice that some banks have strongly positive residuals, indicating good performance, and some banks have strongly negative residuals, indicating below-par performance. Indeed, bank 20, Franklin National Bank, has the second most negative residual and failed the following year. Booth (1985) suggests that regression is a good way to locate problem banks before it is too late.

d. Report your results and save your workbook as E8BOOTH.XLS.

14. Open the ALUM.XLS workbook, which contains mass and volume measurements on eight chunks of aluminum from a high school chemistry class.

   a. Plot mass against volume, and notice the outlier.

   b. After excluding the outlier, regress mass on volume, without the constant term (select the Constant is Zero check box in the Regression dialog box), because the mass should be 0 when the volume is 0. The slope of the regression line is an estimate of the density (not a statistical word here but a measure of how dense the metal is) of aluminum.

   c. Give a 95% confidence interval for the true density. Does your interval include the accepted true value, which is 2.699?

   d. Save your workbook as E8ALUM.XLS.

15. The STATE.XLS workbook contains data from the 1986 Metropolitan Area Data Book of the U.S. Census Bureau on the 1980 death rates per 100,000 people, for each of the 50 states. There are two variables you should consider: Cardio, the cardiovascular death rate, and Pulmon, the pulmonary (lung) death rate.

   a. Compute the Pearson correlation and the Spearman rank correlation between them. How does the Spearman rank correlation differ from the Pearson correlation?

   b. Create the corresponding scatterplot. Label each point on the scatterplot with the name of the state. Which states are outliers on the lower left of the plot?

   c. Copy the data to a new worksheet, removing the most extreme outlier. Redo the correlations.

   d. Copy the data to a new worksheet again without the next most extreme point in the lower left-hand corner of the plot, and redo the correlations.

e. How are the size and significance of the correlations influenced by these deletions? Make a case for the deletions on the basis of the plot and some geography. Why should Alaska be low on both death rates? Does the original correlation give an exaggerated notion of the relationship between the two variables? Does the nonparametric correlation coefficient solve the problem? Explain. Would you say that a correlation without a plot can be deceiving?

f. Save your workbook as E8STATE.XLS.

16. Open the FIDELITY.XLS workbook, which contains figures from 1989, 1990, and 1991 for 33 Fidelity sector funds. The source is the Morningside Mutual Fund Sourcebook 1992, Equity Mutual Funds. The name of the fund is given in the Sector column. The TOTL90 column is the percentage total return during the year 1990, and TOTL91 is the percentage total return for the year 1991. NAV90 is the percentage increase in net asset value during 1990, and similarly, NAV91 is the percentage change in net asset value during 1991. INC90 is the percentage net income for 1990, and similarly, INC91 is the percentage net income for 1991. CAPRET90 is the percentage capital gain for 1990, and CAPRET91 is the percentage capital gain for 1991.

a. What is the correlation between the percentage capital gains for 1990 and 1991?

b. What is the correlation between the percentage net income for 1990 and 1991?

c. Create a scatterplot for the two correlations in parts a and b. Label each point on the scatterplot with labels from the Sector column.

d. You should get a stronger correlation for income than for capital gain. How do you explain this?

e. Calculate the correlation between NAV90 and NAV91 and then generate the scatterplot, labeling the points with the sector names. The Biotechnology Fund stands out in the plot. It was the only fund that performed well in both years.

f. See what happens to the Pearson correlation if this fund is excluded (copy the data to a second worksheet and delete that row in the new worksheet).

g. Repeat the analysis with the Spearman rank correlation.

h. If the correlation is this weak, what does it suggest about using performance in one year as a guide to performance in the following year?

i. Save your workbook as E8FIDEL.XLS.

17. The file DRAFT.XLS contains information on the 1970 military draft lottery. Draft numbers were determined by placing all 366 possible birth dates in a rotating drum and selecting them one by one. The first birth date drawn received a draft number of 1 and men born on that date were drafted first, the second birth date entered received a draft number of 2, and so forth. Is there any relationship between the draft number and the birth date?

a. Using the values in the Draft Numbers worksheet, calculate the Pearson correlation coefficient and p-value between the Day_of_the_Year and the Draft number. Is there a significant correlation between the two? Using the value of the correlation, would you expect higher draft numbers to be assigned to people born earlier in the year or later?

b. Create a scatterplot of Number versus Day_of_the_Year. Is there an obvious relationship between the two in the scatterplot?

c. Add a trend line to your scatterplot and include both the regression equation and the R^2-value. How much of the variation in draft number is explained by the Day_of_the_Year variable?

d. Calculate the average draft number for each month and then calculate the correlation between the month number and the average draft number. How do the values of the correlation in this analysis compare to the correlation you performed earlier?

e. Create a scatterplot of average draft number vs. month number. Add a trend line and include the regression equation and R^2-value. How much of the variability in the average draft number per month is explained by the month?

f. Summarize your conclusions. Which analysis (looking at daily values or looking at monthly averages) better describes any problem with the draft lottery? Save your workbook as E8DRAFT.XLS.

18. The Emerald health-care providers claim that the cost of their health plan has risen more slowly than overall health costs. You decide to investigate Emerald's claim. You have data stored in the EMERALD.XLS file on Emerald costs over the past seven years, along with the consumer price index (CPI) for all urban consumers and the medical component of the CPI. Open this workbook and save it as EMERALD8.XLS on your Student disk.

a. Using the Analysis ToolPak's Regression command, calculate the regression equation for each of the three price indexes against the year variable. What are the values for the three slopes? Express the slope in terms of the increase in the index per year. How does Emerald's change in cost compare to the other two indexes?

b. Look at the 95% confidence intervals for the three slopes. Do the confidence intervals overlap? Does there appear to be a significant difference in the rate of increase under the Emerald plan as compared to the other two indexes?

c. Summarize your conclusions. Do you see evidence to substantiate Emerald's claim?

d. Save your workbook as E8EMERALD.XLS.

19. The TEACHER.XLS workbook contains data on the relationship between teachers' salaries and the spending on public schools per pupil in 1985. Open the workbook and perform the following analysis.

a. Create a scatterplot of spending per pupil vs. teacher salary. Add a trend line containing the R^2-value and regression to the plot.

b. Compute the regression statistics for the data, and then create the diagnostic plots discussed in this chapter. Is there any evidence of a problem in the diagnostic plots?

c. Copy the spending per pupil vs. teacher salary scatterplot to a new chart sheet and then break down the points in the plot on the basis of the values of the area variable. For each of the three series in the chart, add a linear trend line and compute the R^2 value and regression equation. How do the least-squares lines compare among the three regions? What do you think accounts for any difference in the trend lines?

d. Redo the regression statistics, performing three regressions, one for each of the three areas in the data set. Compare the regression equations. What are the 95% confidence intervals for the slope parameters in the three areas?

e. Summarize and report your conclusions. Save your workbook as E8TEACHER.XLS.

Chapter 9

Multiple Regression

Objectives

In this chapter you will learn to:

- Use the F distribution

- Fit a multiple regression equation and interpret the results

- Use plots to help understand a regression relationship

- Validate a regression using residual diagnostics

Regression Models with Multiple Parameters

In Chapter 8, you used simple linear regression to predict a dependent variable (y) from a single independent variable (x, a predictor variable). In multiple regression, you predict a dependent variable from several independent variables. For three predictors, x_1, x_2, and x_3, the multiple regression model takes the form where the coefficients $\beta_0, \beta_1, \beta_2$, and β_3 are unknown parameter values that you can estimate and ε is random error, which follows a normal distribution with mean 0 and variance σ^2. Note that the predictors can also be functions of variables. The following are also examples of models whose parameters you can estimate with multiple regression:

$$\text{Polynomial: } y = \beta_0 + \beta_1 x + \beta_2 x^2 + \beta_2 x^3 + \varepsilon$$

$$\text{Trigonometric: } y = \beta_0 + \beta_1 \sin(x) + \beta_2 \cos(x) + \varepsilon$$

$$\text{Logarithmic: } y = \beta_0 + \beta_1 \log(x_1) + \beta_2 \log(x_2) + \varepsilon$$

Note that all of these equations are examples of linear models, even though they use various trigonometric and logarithmic functions. The "linear" in *linear model* refers to the error term, ε, and the parameter's β_i. The equations are linear in those terms. For example, one could create new variables, $l = \sin(x)$ and $k = \cos(x)$, and then the second model is the linear equation $y = \beta_0 + \beta_1 l + \beta_2 k + \varepsilon$.

After computing estimated values for the β coefficients, you can plug them into the equation to get predicted values for y. The estimated regression model is expressed as

$$y = b_0 + b_1 x_1 + b_2 x_2 + b_3 x_3$$

where the b_i's are the estimated parameter values, and the residuals correspond to the error term, ε.

CONCEPT TUTORIALS:
The *F*-distribution

The *F-distribution* is basic to regression and analysis of variance as studied in this chapter and the next. An example of the *F*-distribution is shown in the Distributions workbook.

To view the *F*-distribution:

1 Open the file **Distributions**, located in the Explore folder of your Student files. Enable the macros in the workbook.

2 Click **F** from the Table of Contents column. Review the material and scroll to the bottom of the worksheet. See Figure 9-1.

Figure 9-1
The
***F*-distribution**
worksheet

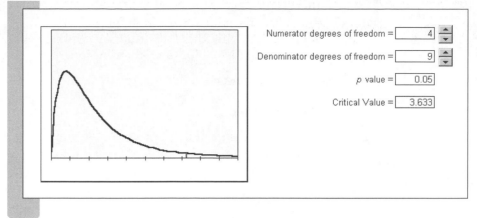

The *F-distribution* has two degrees-of-freedom parameters: the numerator and denominator degrees of freedom. The distribution is usually referred to as *F(m,n)*—that is, the *F-distribution* with *m* numerator degrees of freedom and *n* denominator degrees of freedom. The Distribution workbook opens with an *F*(4,9) distribution.

Like the χ^2-distribution, the *F-distribution* is skewed. To help you better understand the shape of the *F-distribution*, the worksheet lets you vary the degrees of freedom of the numerator and the denominator by clicking the degrees-of-freedom scroll arrows. Experiment with the worksheet to view how the distribution of the *F* changes as you increase the degrees of freedom.

To increase the degrees of freedom in the numerator and denominator:

1 Click the up spin arrow to increase the numerator degrees of freedom to **10**.

2 Click the up spin arrow to increase the denominator degrees of freedom to **15**. Then watch how the distribution changes.

In this book, hypothesis tests based on the *F-distribution* always use the area under the upper tail of the distribution to determine the *p*-value.

To change the *p*-value:

1 Click the **Critical Value** box, type 0.10, and then press **Enter**. This gives you the location of the critical value for the *F*-test at the 10% significance level.

Notice that the critical value shifts to the left, telling you that 10% of the values of the *F*-distribution lie to the right of this point.

Continue working with the *F*-distribution worksheet, trying different parameter values to get a feel for the *F*-distribution.

To close the worksheet:

1 Click **File > Close**.

2 Click **No**, because you don't need to save your work in the Distributions workbook.

Using Regression for Prediction

One of the goals of regression is prediction. For example, you could use regression to predict what grade a student will get in a college calculus course. (This is the dependent variable, the one being predicted.) The predictors (the independent variables) might be ACT or SAT math score, high school rank, and a placement test score from the first week of class. Students with low predictions might be asked to take a lower-level class.

However, suppose the dependent variable is the price of a four-unit apartment building, and the independent variables are the square footage, the age of the building, the total current rent, and a measure of the condition of the building. Here you might use the predictions to find a building that is undervalued, with a price that is much less than its prediction. This analysis was actually carried out by some students, who found that there was a bargain building available. The owner needed to sell quickly as a result of "cash flow problems."

You can use multiple regression to see how several variables combine to predict the dependent variable. How much of the variability in the dependent variable is accounted for by the predictors? Do the combined independent variables do better or worse than you might expect, based on their individual correlations with the dependent variable? You might be interested in the individual coefficients and in whether they seem to matter in the prediction equation. Could you eliminate some of the predictors without losing much prediction ability?

When you use regression in this way, the individual coefficients are important. Rosner and Woods (1988) compiled statistics from baseball box scores, and they regressed runs on singles, doubles, triples, home runs, and walks (walks are combined with hit-by-pitched-ball). Their estimated prediction equation is

$$\text{Runs} = -2.49 + 0.47 \text{ singles} + 0.76 \text{ doubles} + 1.14 \text{ triples} + 1.54 \text{ home runs} + 0.39 \text{ walks}$$

Notice that walks have a coefficient of 0.39, and singles have a coefficient of 0.47, so a walk has more than 80% of the weight of a single. This is in contrast to the popular slugging percentage used to measure the offensive production of players, which gives weight 0 to walks, 1 to singles, 2 to doubles, 3 to triples, and 4 to home runs. The Rosner–Woods equation gives relatively more weight to singles, and the weight for doubles is less than twice as much as the weight for singles. Similar comparisons are true for triples and home runs. Do baseball general managers use equations like the Rosner–Woods equation to evaluate ball players? If not, why not?

You can also use regression to see whether a particular group is being discriminated against. A company might ask whether women are paid less than men with comparable jobs. You can include a term in a regression to account for the effect of gender. Alternatively, you can fit a regression model for just men, apply the model to women, and see whether women have salaries that are less than would be predicted for men with comparable positions. It is now common for such arguments to be offered as evidence in court, and many statisticians have experience in legal proceedings.

Regression Example: Predicting Grades

For a detailed example of a multiple regression, consider the CALC.XLS data from Chapter 8, which were collected to see how performance in freshman calculus is related to various predictors (Edge and Friedberg, 1984).

To open the CALC workbook:

1 Open **CALC.XLS.** (Be sure you select the drive and directory containing your Student files.)

2 Click **File > Save As**, select the folder containing your Student files, and then save your workbook on your Student disk as **CALC3.XLS**.

In Chapter 8, it appeared from the correlation matrix and scatterplot matrix that the algebra placement test is the best individual predictor of the first-semester calculus score (although it is not very successful). Multiple regression gives a measure of how good the predictors are when used together. The model is

$$\text{Calculus score} = \beta_0 + \beta_1(\text{Calc HS}) + \beta_2(\text{ACT Math}) + \beta_3(\text{Alg Place}) +$$
$$\beta_4(\text{Alg2 Grade}) + \beta_5(\text{HS Rank}) + \beta_6(\text{Gender Code}) + \varepsilon$$

You can use the Analysis ToolPak Regression command to perform a multiple regression on the data, but the predictor variables must occupy a contiguous range. You will be using columns A, B, C, D, E, and G as your predictor variables, so you need to move column G, Gender Code, next to columns A:E.

To move column G next to columns A:E:

1 Click the **G** column header to select the entire column.

2 Right-click the selection to open the shortcut menu; then click **Cut**.

3 Click the F column header.

4 Right-click the selection to open the shortcut menu; then click **Insert Cut Cells.** You can now identify the contiguous range of columns A:F as your predictor variables.

To perform a multiple regression on the calculus score based on the predictor variables Calc HS, ACT Math, Alg Place, Alg2 Grade, HS Rank, and Gender Code, use the Regression command found in the Analysis ToolPak provided with Excel.

To perform the multiple regression:

1 Click **Tools > Data Analysis**, click **Regression** in the Analysis Tools list box, and then click **OK**.

2 Type **H1:H81** in the Input Y Range text box, press **Tab**, and then type **A1:F81** in the Input X Range text box.

3 Click the **Labels** checkbox and the **Confidence Level** checkbox to select them, and then verify that the **Confidence Level** box contains 95.

4 Click the **New Worksheet Ply** option button, click the corresponding text box, and then type **Mult Reg**.

5 Click the **Residuals**, **Standardized Residuals**, **Residual Plots**, and **Line Fit Plots** checkboxes to select them. Your Regression dialog box should look like Figure 9-2.

Figure 9-2
The completed Regression dialog box

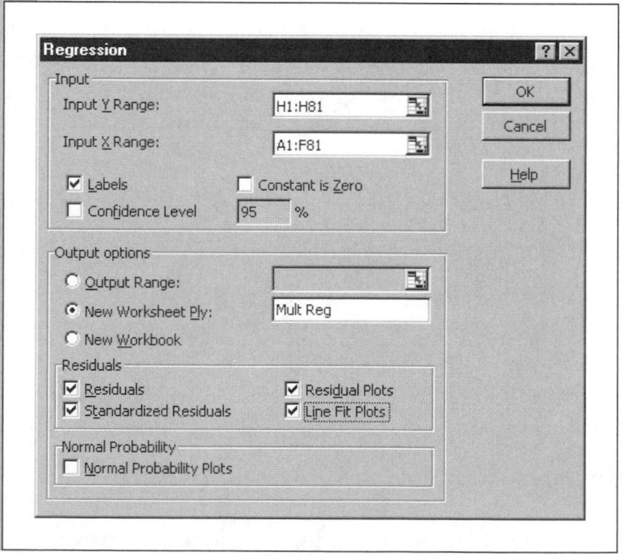

6 Click **OK**.

Excel creates a new sheet, "Mult Reg," which contains the summary output and the residual plots.

Interpreting the Regression Output

To interpret the output, look first at the analysis of variance (ANOVA) table found in cells A10:F14. Figure 9-3 shows this range with the columns widened to display the labels and the values reformatted. The analysis of variance table shows you whether the fitted regression model is significant.

The analysis of variance table helps you choose between two hypotheses:

H_0: The coefficients of all six predictor variables = 0

H_a: At least one of the six coefficients ≠ 0

**Figure 9-3
ANOVA table
from
multiple
regression,
cells
A10:F14**

10	ANOVA					
11		*df*	*SS*	*MS*	*F*	*Significance F*
12	Regression	6	3840.164	640.027	7.197	4.70E-06
13	Residual	73	6492.036	88.932		
14	Total	79	10332.200			

There are many different parts to an ANOVA table, some of which you might have considered already in the optional section of the previous chapter. At this point, you should just concentrate on the *F*-ratio and its *p*-value, which tell you whether the regression is significant. This ratio is large when the predictor variables explain much of the variability of the response variable, and hence it has a small *p*-value as measured by the *F*-distribution. A small value for this ratio indicates that much of the variability in *y* is due to random error (as estimated by the residuals of the model) and is not due to the regression. The next chapter, on analysis of variance, contains a more detailed description of the ANOVA table.

The *F*-ratio, 7.197, is located in cell E12. Under the null hypothesis, you assume that there is no relationship between the six predictors and the calculus score. If the null hypothesis is true, the *F*-ratio in the ANOVA table follows the *F*-distribution, with 6 numerator degrees of freedom and 73 denominator degrees of freedom. You can test the null hypothesis by seeing whether this observed *F*-ratio is much larger than you would expect in the *F*-distribution. If you want to get a visual picture of this hypothesis test, use the *F*-distribution worksheet from the Distributions workbook and display the $F(6,73)$ distribution.

The Significance F column gives a p-value of 4.69×10^{-6} (cell F12), representing the probability that an F-ratio with 6 degrees of freedom in the numerator and 73 in the denominator has a value 7.197 or more. This is much less than 0.05, so the regression is significant at the 5% level. You could also say that you reject the null hypothesis at the 5% level and accept the alternative that at least one of the coefficients in the regression is not zero. If the F-ratio were not significant, there would not be much interest in looking at the rest of the output.

Multiple Correlation

The regression statistics appear in the range A3:B8, shown in Figure 9-4 (formatted to show column labels and the values to three decimal places).

**Figure 9-4
Regression
statistics in
cells A3:B8**

3	*Regression Statistics*	
4	Multiple R	0.610
5	R Square	0.372
6	Adjusted R Square	0.320
7	Standard Error	9.430
8	Observations	80

The R Square value in cell B5 (0.372) is the coefficient of determination, R^2, discussed in the previous chapter. This value indicates that 37% of the variance in calculus scores can be attributed to the regression. In other words, 37% of the variability in the final calculus score is due to differences among students (as quantified by the values of the predictor variables), and the rest is due to random fluctuation. Although this value might seem low, it is an unfortunate fact that decisions are often made on the basis of weak predictor variables, including decisions about college admissions and scholarships, freshman eligibility in sports, and placement in college classes.

The Multiple R (0.610) in cell B4 is just the square root of the R^2; this is also known as the **multiple correlation**. It is the correlation among the response variable, the calculus score, and the linear combination of the predictor variables as expressed by the regression. If there were only one predictor, this would be the absolute value of the correlation between the predictor and the dependent variable. The Adjusted R Square value in cell B6 (0.320) attempts to adjust the R^2 for the number of predictors. You look at the adjusted R^2 because the unadjusted R^2 value either increases or stays the same when you add predictors to the model. If you add enough predictors to the model, you can reach some very high R^2 values, but not much is to be gained by analyzing a data set with 200 observations if the regression model has 200 predictors, even if the R^2 value is 100%. Adjusting the R^2 compensates for this effect and helps you determine whether adding additional predictors is worthwhile.

The standard error value, 9.430 (cell B7), is the estimated value of σ, the standard deviation of the error term ε—in other words, the standard deviation of the calculus score once you compensate for differences in the predictor variables. You can also think of the standard error as the typical error for prediction of the 80 calculus scores. Because a span of 10 points corresponds to a difference of one letter grade (A vs. B, B vs. C, and so on), the typical error of prediction is about one letter grade.

Coefficients and the Prediction Equation

At this point you know the model is statistically significant and accounts for about 37% of the variability in calculus scores. What is the regression equation itself and which predictor variables are most important?

You can read the estimated regression model from cells A16:I23, shown in Figure 9-5, where the first column contains labels for the predictor variables.

**Figure 9-5
Parameter
estimates
and *p*-values**

16		Coefficients	Standard Error	t Stat	P-value	Lower 95%	Upper 95%
17	Intercept	27.943	12.438	2.247	0.028	3.155	52.732
18	Calc HS	7.192	2.488	2.891	0.005	2.233	12.151
19	ACT Math	0.352	0.430	0.817	0.417	-0.506	1.209
20	Alg Place	0.827	0.268	3.092	0.003	0.294	1.360
21	Alg2 Grade	3.683	2.441	1.509	0.136	-1.182	8.548
22	HS Rank	0.111	0.116	0.953	0.344	-0.121	0.342
23	Gender Code	2.627	2.469	1.064	0.291	-2.294	7.548

The Coefficients column (B16:B23) gives the estimated coefficients for the model. The corresponding prediction equation is

$$\text{Calc} = 27.943 + 7.192(\text{Calc HS}) + 0.352(\text{ACT Math}) + 0.827\,(\text{Alg Place})$$
$$+ \; 3.683\,(\text{Alg2 Grade}) \; + \; 0.111(\text{HS Rank}) + 2.627\,(\text{Gender Code})$$

The coefficient for each variable estimates how much the calculus score will change if the variable is increased by 1 and the other variables are held constant. For example, the coefficient 0.352 of ACT Math indicates that the calculus score should increase by 0.352 point if the ACT math score increases by 1 point and all other variables are held constant.

Some variables, such as Calc HS, have a value of either 0 or 1—in this case, to indicate the absence or presence of calculus in high school. The coefficient 7.192 is the estimated effect on the calculus score of taking high school calculus, other things being equal. Because 10 points correspond to one letter grade, the coefficient 7.192 for Calc HS is almost one letter grade.

Using the coefficients of this regression equation, you can forecast what a particular student's calculus score might be, given background information on the student. For example, consider a male student who did not take calculus in high school, scored 30 on his ACT Math exam, scored 23 on his algebra placement test, had a 4.0 grade in second-year high school algebra, and was ranked in the 90th percentile in his high school graduation class. You would predict that his calculus score would be

$$\text{Calc} = 27.943 + 7.192(0) + 0.352(30) + 0.827(23) + 3.683(4.0) +$$

$$0.111(90) + 2.627(1) = 74.87 \text{ or about } 75 \text{ points}$$

Notice the Gender Code coefficient, 2.627, which shows the effect of gender if the other variables are held constant. Because the males are coded 1 and the females are coded 0, if the regression model is true, a male student will score 2.627 points higher than a female student, even when the backgrounds of both students are equivalent (equivalent in terms of the predictor variables in the model).

Whether you can trust that conclusion depends partly on whether the coefficient for Gender Code is significant. For that you have to determine the precision with which the value of the coefficient has been determined. You can do this by examining the estimated standard deviations of the coefficients, displayed in the Standard Error column.

t-Tests for the Coefficients

The t Stat column shows the ratio between the coefficient and the standard error. If the population coefficient is 0, then this has the t-distribution with degrees of freedom $n - p - 1 = 80 - 6 - 1 = 73$. Here n is the number of cases (80) and p is the number of predictors (6). The next column, P-value, is the corresponding p-value—the probability of a t-value this large or larger in absolute value. For example, the t-value for Alg Place is 3.092, so the probability of a t this large or larger in absolute value is about 0.003. The coefficient is significant at the 5% level because this is less than 0.05. In terms of hypothesis testing, you would reject the null hypothesis that the coefficient is 0 at the 5% level and accept the alternative hypothesis. This is a two-tailed test—it rejects the null hypothesis for either large positive or large negative values of t—so your alternative hypothesis is that the coefficient is not zero. Notice that only the coefficients for Alg Place and Calc HS are significant. This suggests that you not devote a lot of effort to interpreting the others. In particular, it would not be appropriate to assume from the regression that male students perform better than equally qualified female students.

The range F17:G23 indicates the 95% confidence intervals for each of the coefficients. You are 95% confident that having calculus in high school is associated with an increase in the calculus score of at least 2.233 points and not more than 12.151 points in this particular regression equation.

Is it strange that the ACT math score is nowhere near significant here, even though this test is supposed to be a strong indication of mathematics achievement? Looking back at the correlation matrix in Chapter 8, you can see that it has correlation 0.353 with Calc, which is highly significant ($p = 0.001$). Why is it not significant here? The answer involves other variables that contain some of the same information. In using the t-distribution to test the significance of the ACT Math term, you are testing whether you can get away with deleting this term. If the other predictors can take up the slack and provide most of its information, then the test says that this term is not significant and therefore is not needed in the model. If each of the predictors can be predicted from the others, any single predictor can be eliminated without losing much.

You might think that you could just drop from the model all the terms that are not significant. However, it is important to bear in mind that the individual tests are correlated, so each of them changes when you drop one of the terms. If you drop the least significant term, others might then become significant. A frequently used strategy for reducing the number of predictors involves the following steps:

NOT VALID

1. Eliminate the least significant predictor if it is not significant.
2. Refit the model.
3. Repeat Steps 1 and 2 until all predictors are significant.

In the exercises, you'll get a chance to rerun this model and eliminate all non-significant variables. For now, examine the model and see whether any assumptions have been violated.

Testing Regression Assumptions

There are a number of useful ways to look at the results produced by multiple linear regression. This section reviews the four common plots that can help you assess the success of the regression.

1. Plotting dependent variables against the predicted values shows how well the regression fits the data.
2. Plotting residuals against the predicted values magnifies the vertical spread of the data so you can assess whether the regression assumptions are justified. A curved pattern to the residuals indicates that the model does not fit the data. If the vertical spread is wider on one side of the plot, it suggests that the variance is not constant.
3. Plotting residuals against individual predictor variables can sometimes reveal problems that are not clear from a plot of the residuals vs. the predicted values.
4. Creating a normal plot of the residuals helps you assess whether the regression assumption of normality is justified.

Observed vs. Predicted Values

How successful is the regression? To see how well the regression fits the data, plot the actual Calc values against the predicted values stored in B29:B109. (You can scroll down to view the residual output.) To plot the observed calculus scores vs. the predicted scores, you must first place the data on the same worksheet.

To copy the observed scores:

1 Select the range B29:B109 and click the **Copy** button 🖺 on the Standard toolbar.

2 Click the **Calculus Data** sheet tab.

3 Select the range **H1:H81**; then click **Insert > Copied Cells** to paste the predicted values into column H.

4 Click the **Shift Cells Right** option button to move the observed calculus scores into column I; then click **OK**.

The predicted calculus scores appear in column H, as shown in Figure 9-6 (formatted to show the column labels).

predicted scores   observed scores

Figure 9-6 Predicted and observed calculus scores

	A	B	C	D	E	F	G	H	I	J	K
1	Calc HS	ACT Math	Alg Place	Alg2 Grade	HS Rank	Gender Code	Gender	*Predicted Calc*	Calc		
2	0	27	21	3.5	68	0	F	75.21286866	62		
3	0	29	16	4.0	99	0	F	77.05046265	75		
4	1	30	22	4.0	98	1	M	92.07306E4	95		
5	0	34	25	3.0	90	1	M	84.20016977	78		
6	0	29	22	4.0	99	0	F	82.01272956	95		
7	1	30	19	4.0	97	0	F	86.85422E9	91		
8	0	29	23	4.0	79	1	M	83.25509849	72		
9	0	28	15	4.0	95	0	F	75.42955084	95		
10	0	28	14	4.0	85	1	M	76.12372456	88		
11	0	31	19	4.0	82	1	M	80.98170744	97		
12	0	25	12	3.0	81	1	M	69.28970058	49		
13	0	34	16	3.5	87	1	M	78.266524E3	70		
14	0	27	13	4.0	92	0	F	73.09218395	75		
15	0	28	19	4.0	89	0	F	78.07419223	78		
16	0	31	25	4.0	97	0	F	84.97570343	89		
17	0	26	10	3.0	81	1	M	67.98712125	87		
18	1	24	14	4.0	91	0	F	79.946413C6	79		
19	1	30	18	3.0	97	1	M	84.97124865	85		
20	0	25	13	2.5	46	1	M	64.404589E5	57		
21	0	25	15	3.5	80	1	M	73.5017656	81		
22	0	27	18	3.0	80	0	F	72.2172842	76		
23	0	27	17	4.0	89	0	F	76.06859375	88		
24	0	28	21	4.0	94	0	F	80.281228E8	83		
25	1	27	24	3.0	71	1	M	86.003661E2	97		
26	0	27	12	3.0	97	1	M	71.76215125	60		
27	1	27	18	3.0	85	1	M	82.589647C8	84		

Mult Reg \ Calculus Data

Ready   NUM

Now create a scatterplot of the data in the range H1:I81.

To create the scatterplot of the observed scores vs. the predicted scores:

1 Select the range **H1:I81** and click **Insert > Chart**.

2 Click **XY (Scatter)** as the chart type; then click **Next** twice.

3 Click the **Title** tab, and type **Calculus Scores** in the Chart title box, **Predicted** in the Value(X) axis box, and **Observed** in the Value(Y) axis box.

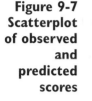

4 Remove gridlines and the legend from the plot and click **Next**.

5 Click the **As new sheet** option button and type **Observed vs. Predicted** in the accompanying text box. Click **Finish**.

6 Rescale the *x*-axis and *y*-axis in the plot so that the ranges go from 40 to 100 rather than from 0 to 100 or 0 to 120.

The final form of the scatterplot should look like Figure 9-7.

Figure 9-7 Scatterplot of observed and predicted scores

How good is the prediction shown here? Is there a narrow range of observed values for a given predicted value? This plot is a slight improvement on the plot of Calc vs. Alg Place from the scatterplot matrix in Chapter 8. Figure 9-7 should be better because Alg Place and five other predictors are being used here.

Does it appear that the range of values is narrower for large values of predicted calculus score? If the error variance were lower for students with high predicted values, it would be a violation of the fourth regression assumption, which requires a constant error variance. Consider the students predicted to have a grade of 80 in calculus. These students have actual grades of around 65 to around 95, a wide range. Notice that the variation is lower for students predicted to have a grade of 90. Their actual scores are all in the 80s and 90s. There is a barrier at the top—no score can be above 100—and this limits the possible range. In general, when a barrier limits the range of the dependent variable, it can cause nonconstant error variance. This issue is considered further in the next section.

Plotting Residuals vs. Predicted Values

The plot of the residuals vs. the predicted values shows another view of the variation in Figure 9-7 because the residuals are the differences between the actual calculus scores and the predicted values.

To make the plot:

1 Click the **Mult Reg** sheet tab to return to the regression output.

2 Select the range **B29:C109**, and click **Insert > Chart**.

3 Following the Chart Wizard directions as you did for the previous plot, create a scatterplot with no gridlines or legends. Enter a chart title of **Residual Plot**, and label the x-axis **Predicted** and the y-axis **Residuals**. Save the scatterplot to a chart sheet named **Residuals vs. Predicted**.

4 Change the scale of the x-axis from 0–100 to 60–100. Your chart sheet should look like Figure 9-8.

Figure 9-8 Scatterplot of residuals and predicted scores

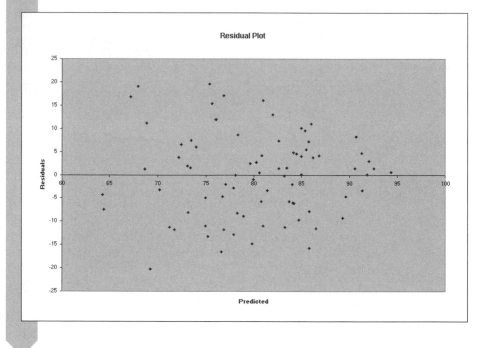

This plot is useful for verifying the regression assumptions. For example, the first assumption requires that the form of the model be correct. A violation of this assumption might be seen in a curved pattern. No curve is apparent here.

If the assumption of constant variance is not satisfied, then it should be apparent in Figure 9-8. Look for a trend in the vertical spread of the data.

For example, the data may widen out as the predicted value increase. There appears to be a definite trend toward a narrower spread on the right, and it is cause for concern about the validity of the regression—although regression does have some robustness with respect to the assumption of constant variance.

For data that range from 0 to 100 (such as percentages), the arcsine–square-root transformation sometimes helps fix problems with nonconstant variance. The transformation involves creating a new column of transformed calculus scores where

$$\text{Transformed calc score} = \sin^{-1}\left(\sqrt{\text{calculus score}/100}\right)$$

Using Excel, you would enter the formula

$$= \text{ASIN}(\text{SQRT}(x/100))$$

where x is the value or cell reference of a value you want to transform.

If you were to apply this transformation here and use the transformed calculus score in the regression in place of the untransformed score, you would find that it helps to make the variance more constant, but the regression results are about the same. Calc HS and Alg Place are still the only significant coefficients, and the R^2 value is almost the same as before. Of course, it is much harder to interpret the coefficients after transformation. Who would understand if you said that each point in the algebra placement score is worth 0.012 point in the arcsine of the square root of the calculus score divided by 100? From this perspective, the transformed regression is useful mainly to validate the original regression. If it is valid and it gives essentially the same results as the original regression, then the original results are valid.

Plotting Residuals vs. Predictor Variables

It is also useful to look at the plot of the residuals against each of the predictor variables because a curve might show up on only one of those plots or there might be an indication of nonconstant variance. Such plots are created automatically with the Analysis ToolPak Add-Ins.

To view one of these plots:

1   Click the **Mult Reg** sheet tab to return to the regression output.

The plots generated by the add-in start in cell J1 and extend to Z32. Two types of plots are generated: scatterplots of the regression residuals vs. each of the regression variables, and the observed and predicted values of the response variable (calculus score) against each of the regression variables. See Figure 9-9. (You might have to scroll up and right to see the charts.)

**Figure 9-9
Scatterplots
created
with the
Regression
command**

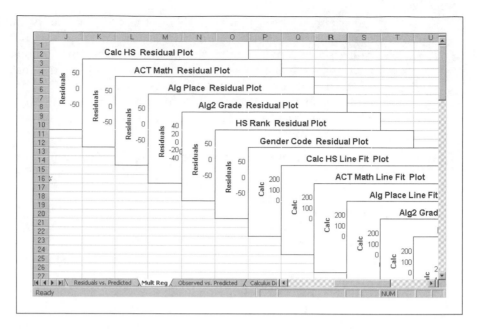

The plots are shown in a cascading format, in which the plot title is often the only visible element of a chart. When you click a chart title, the chart goes to the front of the stack. The charts are small and hard to read, however. You can better view each chart by placing it on a chart sheet of its own. Try doing this with the plot of the residuals vs. Alg Place.

To view the chart:

1 Click the chart **Alg Place Residual Plot** (located in the range L5:Q14).

2 Click **Chart > Location**.

3 Click the **As new sheet** option button and type **Alg Place Residual Plot** in the accompanying text box. Click **OK**.

4 The scatterplot is moved to a chart sheet shown in Figure 9-10.

Figure 9-10
Alg Place
Residual
Plot

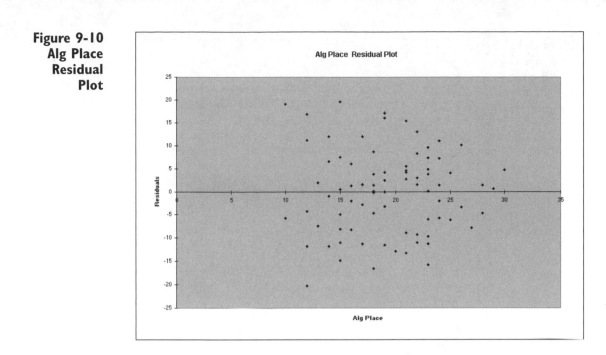

Alg Place Residual Plot

Does the spread of the residual values appear constant for differing values of the algebra placement score? It appears that the spread of the residuals is wider for lower values of Alg Place. This might indicate that you have to transform the data, perhaps using the arcsine transformation just discussed.

Normal Errors and the Normal Plot

What about the assumption of normal errors? Usually, if there is a problem with non-normal errors, extreme values show up in the plot of residuals vs. predicted values. In this example there are no residual values beyond 25 in absolute value, as shown in Figure 9-8.

How large should the residuals be if the errors are normal? You can decide whether these values are reasonable with a normal probability plot.

To make a normal plot of the residuals:

1 Return to the Mult Reg worksheet.

2 Click **StatPlus > Single Variable Charts > Normal P-plots**.

3 Click the **Data Values** button, click the **Use Range References** option button, and select the range **C29:C109**. Click **OK**.

4 Click the **Output** button and specify the new chart sheet **Residual P-plot** as the output destination. Click **OK**.

5 Click **OK** to start generating the plot. See Figure 9-11.

**Figure 9-11
Normal
P-plot
of the
residuals**

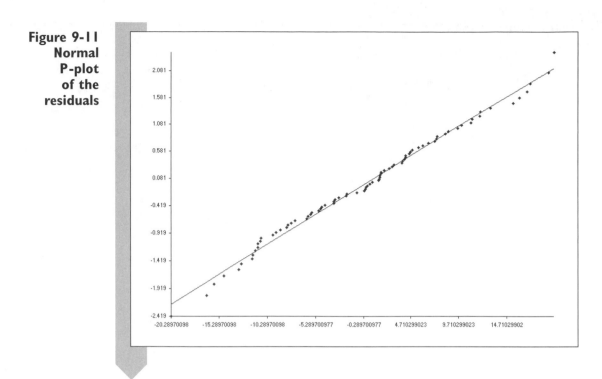

The plot is quite well behaved. It is fairly straight, and there are no extreme values (either in the upper right or lower left corners) at either end. It appears there is no problem with the normality assumption.

Summary of Calc Analysis

What main conclusions can you make about the calculus data, now that you have done a regression, examined the regression residual file, and plotted some of the data? With an R^2 of 0.37 and an adjusted R^2 of 0.320, the regression accounts for only about one-third of the variance of the calculus score. This is disappointing, considering all the weight that college scholarships, admissions, placement, and athletics place on the predictors. Only the algebra placement score and whether or not calculus was taken in high school have significant coefficients in the regression. There is a slight problem with the assumption of a constant variance, but that does not affect these conclusions.

To close the workbook and save the changes you've made to CALC3.XLS:

1 Click **File > Close**.

2 Click **Yes** when prompted to save changes.

Regression Example: Sex Discrimination

In this next example, you'll use regression analysis to determine whether a particular group is being discriminated against. For example, some of the female faculty at a junior college felt underpaid, and they sought statistical help in proving their case. The college collected data for the variables that influence salary for 37 females and 44 males. The data are stored in the file DISCRIM.XLS. (These data were also discussed in Chapter 4 exercises in the workbook JRCOL.XLS. DISCRIM.XLS orders the data differently so that you can use the Analysis ToolPak Add-Ins more easily.)

To open the file:

1 Open **DISCRIM.XLS** from the folder or disk containing your Student data files.

2 Click **File > Save As**, and save the workbook as **DISCRIM2.XLS**.

Table 9-1 shows the variables in the workbook.

Table 9-1 The Discrim workbook

Range Name	Range	Description
Gender	A2:A82	Gender of faculty member (F = female, M = male)
MS_Hired	B2:B82	1 for Master's degree when hired; 0 for no Master's degree when hired
Degree	C2:C82	Current degree: 1 for Bachelor's, 2 for Master's, 3 for Master's plus 30 hours, and 4 for Ph.D.
Age_Hired	D2:D82	Age when hired
Years	E2:E82	Number of years the faculty member has been employed at the college
Salary	F2:F82	Current salary of faculty member

In this example, you will use salary as the dependent variable, using four other quantitative variables as predictors. One way to see whether female faculty have been treated unfairly is to do the regression using just the male data and then apply the regression to the female data. For each female faculty member, this predicts what a male faculty member would make with the same years, age when hired, degree, and Master's degree status. The residuals are interesting because they are the difference between what each woman makes and her predicted salary if she were a man. This assumes that all of the relevant predictors are being used, but it would be the college's responsibility to point out all the variables that influence salary in an important way. When there is a union contract, which is the case here, it should be clear which factors influence salary.

Regression on Male Faculty

To do the regression on just the male faculty and then look at the residuals for the females, use Excel's AutoFilter capability and copy the male rows to a new worksheet.

To create a worksheet of salary information for male faculty only:

1   Click **Data > Filter > AutoFilter**.

2   Click the **Gender** drop-down arrow; then click **M**.

3   Select the range **A1:F82**; then click the **Copy** button on the Standard toolbar.

4   Right-click the **Salary Data** sheet tab to open the shortcut menu; then click **Insert**.

5   Click **Worksheet** in the General sheet of the Insert dialog box; then click **OK**.

6   Click the **Paste** button on the Standard toolbar.

7   Double-click the **Sheet1** sheet tab, and type **Male Data**. The salary data for male faculty now occupy the range A1:F45 on the Male Data worksheet. Now you are ready to analyze this subset of data.

Using a SPLOM to See Relationships

To get a sense of the relationships among the variables, it is a good idea to compute a correlation matrix and plot the corresponding scatterplot matrix.

To create the SPLOM:

1   Click **StatPlus > Multi-variable Charts > Scatterplot Matrix**.

2   Click the **Data Values** button, click the **Use Range References** option button, and select the range **B1:F45**. Click **OK**.

3   Click the **Output** button, click the **New Worksheet** option button, and type **Male SPLOM** in the accompanying text box. Click **OK**.

4   Click **OK** to start generating the scatterplot matrix. See Figure 9-12 for the completed SPLOM.

Figure 9-12 SPLOM of variables for male faculty salary data

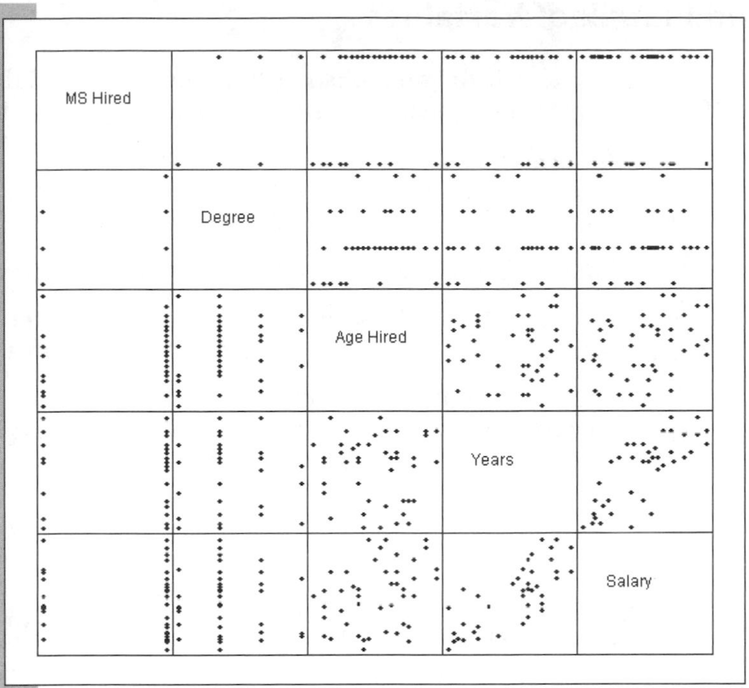

Years employed is a good predictor because the range of salary is fairly narrow for each value of years employed (although the relationship is not perfectly linear). Age at which the employee was hired is not a very good predictor because there is a wide range of salary values for each value of age hired. There is not a significant relationship between the two predictors years employed and age hired. What about the other two predictors? Looking at the plots of salary against degree and MS hired makes it clear that neither of them is closely related to salary. The people with higher degrees do not seem to be making higher salaries. Those with a Master's degree when hired do not seem to be making much more, either. Therefore, the correlations of degree and MS hired with salary should be low.

You might have some misgivings about using degree as a predictor. After all, it is only an ordinal variable. There is a natural order to the four levels, but it is arbitrary to assign the values 1, 2, 3, and 4. This says that the spacing from Bachelor's to Master's (1 to 2) is the same as the spacing from Master's plus 30 hours to Ph.D. (3 to 4). You could instead assign the values 1, 2, 3, and 5, which would mean greater space from Master's plus 30 hours to the Ph.D. In spite of this arbitrary assignment, ordinal variables are frequently used as regression predictors. Usually, it does not make a significant difference whether the numbers are 1, 2, 3, and 4 or 1, 2, 3, and 5. In the present situation, you can see from Figure 9-12 that salaries are about the same in all four degree categories, which implies that the correlation of salary and degree is close to 0. This is true no matter what spacing is used.

Correlation Matrix of Variables

The SPLOM shows the relationships between salary and the other variables. To quantify this relationship, create a correlation matrix of the variables.

To form the correlation matrix:

1 Click the **Male Data** sheet tab.

2 Click **StatPlus > Multivariate > Correlation Matrix**.

3 Click the **Data Values** button, click the **Use Range References** option button, select the range, **B1:F45**, and click **OK**.

4 Click the **Output** button, click the New Sheet option button, and type **Male Corr Matrix** in the New Sheet text box; then click **OK** twice. The resulting correlation matrix appears on its own sheet, as shown in Figure 9-13.

Figure 9-13
Correlation matrix for the male data

Pearson Correlations

	MS Hired	Degree	Age Hired	Years	Salary
MS Hired	1.000	0.520	0.219	-0.099	0.009
Degree		1.000	0.215	-0.103	-0.072
Age Hired			1.000	-0.064	0.325
Years				1.000	0.765
Salary					1.000

Pearson Probabilities

	MS Hired	Degree	Age Hired	Years	Salary
MS Hired	-	0.000	0.153	0.525	0.952
Degree		-	0.161	0.505	0.643
Age Hired			-	0.681	0.032
Years				-	0.000
Salary					-

You might wonder why the variable age hired is used instead of employee age. The problem with using the employee age is one of collinearity. **Collinearity** means that one or more of the predictor variables are highly correlated with each other. In this case, the age of the employee is highly correlated with the number of years employed because there is some overlap between the two. (People who have been employed more years are likely to be older.) This means that the information those two variables provide is somewhat redundant. However, you can tell from Figure 9-13 that the relationship between years employed and age when hired is negligible because the p-value is 0.681 (cell E13). Using the variable age hired instead of age gives the advantage of having two nearly uncorrelated predictors in the model. When predictors are only weakly correlated, it is much easier to interpret the results of a multiple regression.

The correlations for salary show a strong relationship to the number of years employed and some relationship to age when hired, but there is little relationship to a person's degree. This is in agreement with the SPLOM in Figure 9-12.

Multiple Regression

What happens when you throw all four predictors into the regression pot?

To specify the model for the regression:

1 Click the **Male Data** sheet tab.

2 Click **Tools > Data Analysis**, click **Regression** in the Analysis Tools list box, and then click **OK**. The Regression dialog box might contain the options you selected for the previous regression.

3 Type **F1:F45** in the Input Y Range text box, press **Tab**, and then type **B1:E45** in the Input X Range text box.

4 Verify that the **Labels** checkbox is selected and that the **Confidence Level** checkbox is selected and contains a value of 95.

5 Click the **New Worksheet Ply** option button and type **Male Reg** in the corresponding text box (replace the current contents if necessary).

6 Verify that the **Residuals, Standardized Residuals, Residual Plots,** and **Line Fit Plots** checkboxes are selected.

7 Click **OK**.
The first portion of the summary output is shown in Figure 9-14, with columns resized and values reformatted.

**Figure 9-14
Regression
output for
male data**

	A	B	C	D	E	F	G
1	SUMMARY OUTPUT						
2							
3	*Regression Statistics*						
4	Multiple R	0.856					
5	R Square	0.732					
6	Adjusted R Square	0.705					
7	Standard Error	3168.434					
8	Observations	44					
9							
10	ANOVA						
11		*df*	*SS*	*MS*	*F*	*Significance F*	
12	Regression	4	1,071,001,258.77	267,750,314.69	26.671	1.05610E-10	
13	Residual	39	391,519,916.86	10,038,972.23			
14	Total	43	1,462,521,175.64				
15							
16		*Coefficients*	*Standard Error*	*t Stat*	*P-value*	*Lower 95%*	*Upper 95%*
17	Intercept	12,900.67	3,178.17	4.059	0.000	6,472.22	19,329.11
18	MS Hired	744.48	1,304.01	0.571	0.571	-1,893.12	3,382.09
19	Degree	-783.53	746.09	-1.050	0.300	-2,292.63	725.57
20	Age Hired	373.74	83.20	4.492	0.000	205.45	542.02
21	Years	606.18	64.53	9.393	0.000	475.65	736.70

Interpreting the Regression Output

The R^2 of 0.732 shows that the regression explains 73.2% of the variance in salary. However, when this is adjusted for the number of predictors (four), the adjusted R^2 is about 0.705 = 70.5%. The standard error is 3,168.434, so salaries vary roughly plus or minus $3,000 from their predictions. The overall F-ratio is about 26.67, with a p-value in cell F12 of 1.063×10^{-10}, which rules out the hypothesis that all four population coefficients are 0. Looking at the coefficient values and their standard errors, you see that the coefficients for the variables Degree and MS hired have values that are not much more than 1 times their standard errors. Their t-statistics are much less than 2, and their p-values are much more than 0.05; therefore, they are not significant at the 5% level. On the other hand, Years employed and Age Hired do have coefficients that are much larger than their standard errors, with t-values of 9.39 and 4.49, respectively. The corresponding p-values are significant at the 0.1% level.

The coefficient estimate of 606 for years employed indicates that each year on the job is worth $606 in annual salary if the other predictors are held fixed. Correspondingly, because the coefficient for age hired is about $374, all other factors being equal, an employee who was hired at an age 1 year older than another employee will be paid an additional $374.

Residual Analysis of Discrimination Data

Now check the assumptions under which you performed the regression.

To create a plot of residuals vs. predicted salary values:

1   Select the range **B27:C71** and click **Insert > Chart**.

2   Following the Chart Wizard directions, create a scatterplot with no gridlines or legends. Enter a chart title of **Residual Plot**, and label the x-axis **Predicted** and the y-axis **Residuals**. Save the scatterplot to a chart sheet named **Residuals vs. Predicted**.

3   Change the scale of the x-axis from 0–45,000 to 20,000–45,000. Your chart sheet should look like Figure 9-15.

**Figure 9-15
Residuals vs.
predicted
values for
the male
salary data**

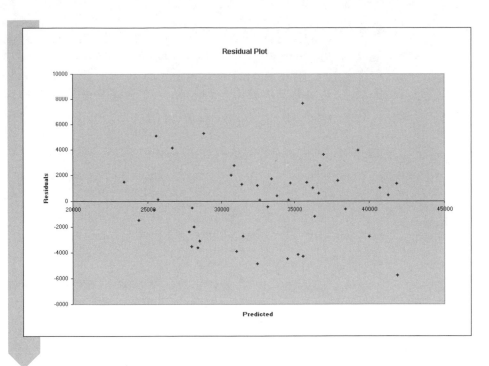

There does not appear to be a problem with nonconstant variance. At least, there is not a big change in the vertical spread of the residuals as you move from left to right. However, there are two points that look questionable. The one at the top has a residual value near 8,000 (indicating that this individual is paid $8,000 more than predicted from the regression equation), and at the bottom of the plot an individual is paid about $6,000 less than predicted from the regression.

Except for these two, the points have a somewhat curved pattern—high on the ends and low in the middle—of the kind that is sometimes helped by a log transformation. As it turns out, the log transformation would straighten out the plot, but the regression results would not change much. For example, if log(salary) is used in place of salary, the R^2-value changes only from 0.732 to 0.733. When the results are unaffected by a transformation, it is best not to bother because it is much easier to interpret the untransformed regression.

Normal Plot of Residuals

What about the normality assumption? Are the residuals reasonably in accord with what is expected for normal data?

To create a normal probability plot of the residuals:

1 Click the **Male Reg** sheet tab.

2 Click **StatPlus > Single Variable Charts > Normal P-plots**.

3 Click the **Data Values** button, click the **Use Range References** option button, and select the range **C27:C71**. Click **OK**.

4 Click the **Output** button and specify the new chart sheet **Male Residual P-plot** as the output destination. Click **OK**.

5 Click **OK** to start generating the plot. See Figure 9-16.

Figure 9-16 Normal P-plot of residuals for the male faculty data

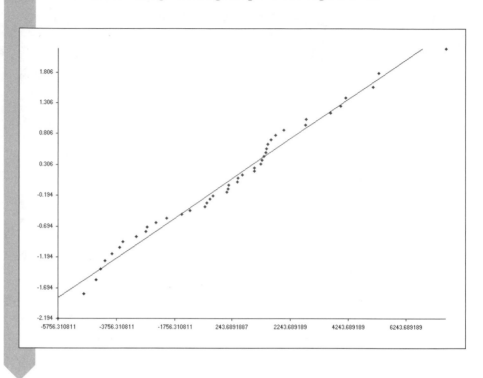

The plot is reasonably straight, although there is a point at the upper right that is a little farther to the right than expected. This point belongs to the employee whose salary is $8,000 more than predicted, but it does not appear to be too extreme. You can conclude that the residuals seem consistent with the normality assumption.

Are Female Faculty Underpaid?

Being satisfied with the validity of the regression on males, let's go ahead and apply it to the females to see whether they are underpaid. The idea is to look at the differences between what female faculty members were paid and what we would predict they would be paid on the basis of the regression model for male faculty. Your ultimate goal is to choose between two hypotheses:

H_0: The salaries of females are equal to the salaries predicted from the population model for males.

H_a: The salaries of females are lower than the salaries predicted from the population model for males.

To obtain statistics on the salaries for females relative to males, you must create new columns of predicted values and residuals.

To create new columns of predicted values and residuals:

1 Return to the Salary Data worksheet.

2 Click **F** from the Gender drop-down list arrow.
Verify that the range A1:F38 is selected.

3 Copy the selection and paste it to a new worksheet named **Female Data**.

4 In the Female Data worksheet, click cell **G1**, type **Pred Sal**, press **Tab**, type **Resid**, and then press **Enter**.

5 Select the range **G2:H38**.

6 In cell G2, type
=12900.67+744.4821*B2−783.529*C2+373.7354*D2+606.1759*E2
(the regression equation for males), and press **Tab**.

7 Type **=F2−G2** in cell H2, and press **Enter**.

8 Click **Edit > Fill > Down**. The data should appear as in Figure 9-17.

Figure 9-17 Predicted salaries and residuals for female faculty

	A	B	C	D	E	F	G	H	I	J	K	L
1	Gender	MS Hired	Degree	Age Hired	Years	Salary	Pred Sal	Resid				
2	F	1	2	35	0	26,209	25158.83	1,050				
3	F	0	1	37	0	23,253	25945.35	-2,692				
4	F	0	1	31	1	26,399	24309.11	2,090				
5	F	0	1	25	1	19,876	22066.7	-2,191				
6	F	1	2	32	1	21,619	24643.8	-3,025				
7	F	1	2	41	2	23,602	28613.6	-5,012				
8	F	0	1	39	2	23,602	27905.17	-4,303				
9	F	1	2	37	2	22,447	27118.66	-4,672				
10	F	0	1	37	2	21,864	27157.7	-5,294				
11	F	1	2	39	2	23,602	27866.13	-4,264				
12	F	1	2	27	3	23,413	23987.48	-574				
13	F	0	1	38	3	19,313	28137.61	-8,825				
14	F	1	2	38	3	21,455	28098.57	-6,644				
15	F	1	2	42	4	25,072	30199.68	-5,128				
16	F	0	1	40	4	22,981	29491.26	-6,510				
17	F	1	2	24	4	21,669	23472.45	-1,803				
18	F	1	2	33	5	24,740	27442.24	-2,702				
19	F	1	2	30	5	23,602	26321.04	-2,719				
20	F	1	3	40	6	24,772	29881.04	-5,109				
21	F	1	2	39	6	25,784	30290.83	-4,507				
22	F	1	2	36	6	26,120	29169.62	-3,050				
23	F	1	2	27	6	23,449	25806.01	-2,357				
24	F	0	1	38	8	25,110	31168.49	-6,058				
25	F	1	2	25	11	29,598	28089.41	1,509				
26	F	0	2	36	12	33,675	32062.2	1,613				
27	F	0	2	21	13	27,129	27062.34	67				

Residuals vs. Predicted / Male Reg / Male Data / **Female Data** / Salary D.

Ready   NUM

To see whether females are paid about the same salary that would be predicted if they were males, create a scatterplot of residuals vs. predicted salary.

To create the scatterplot:

1 Select the range **G1:H38**, and click **Insert > Chart**.

2 Follow the Chart Wizard directions and create a scatterplot showing points only, with no lines. Give a chart title of **Female Residual Plot**, and label the x-axis **Predicted** and the y-axis **Residuals**. Save the chart to a chart sheet named **Female Residual Plot**.

3 Change the scale of the x-axis from 0–45,000 to 20,000–40,000. Your chart sheet should look like Figure 9-18.

Figure 9-18
Scatterplot of residuals for females' salaries vs. predicted values

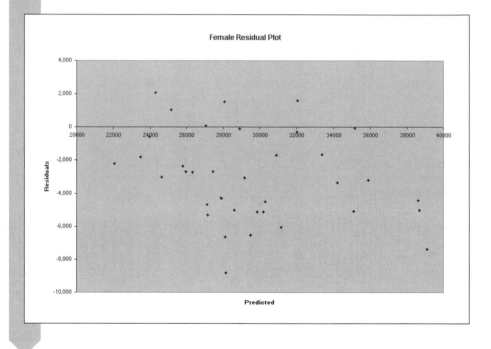

Out of 37 female faculty, only 5 have salaries greater than what would be predicted if they were males, whereas 32 have salaries less than predicted. Calculate the descriptive statistics for the residuals to determine the average discrepancy in salary.

To calculate descriptive statistics for female faculty's salaries:

1 Click the **Female Data** worksheet tab.

2 Click **StatPlus > Univariate Statistics**.

3 Click the **All summary statistics** and **All variability statistics** checkboxes.

4 Click the **Input** button, click the **Use Range References** option button, and select the range **H1:H38**. Click **OK**.

5 Click the **Output** button, and select the new worksheet **Female Residual Stats** as the output destination. Click **OK**.

6 Click **OK** to generate the table of descriptive statistics shown in Figure 9-19.

Figure 9-19
Descriptive statistics of the residuals for female faculty

	A	B
1		**Univariate Statistics**
2		Resid
3	Count	37
4	Sum	-113,355
5	Average	-3,063.64
6	Median	-3,050
7	Mode	#N/A
8	Trimmed Mean (0.2)	-3,087.60
9	Minimum	-8,825
10	Maximum	2,090
11	Range	10,914
12	Standard Deviation	2,662.033
13	Variance	7,086,418.085
14	Standard Error	437.635
15	Skewness	0.130
16	Kurtosis	-0.503

On the basis of the descriptive statistics, you can conclude that the female faculty are paid, on average, $3,063.64 less than what would be expected for equally qualified male faculty members (as quantified by the predictor variables). The largest discrepancy is for a female faculty member who is paid $8,825 less than expected (cell B9). Of those with salaries greater than predicted, there is a female faculty member who is paid $2,090 more than expected (cell B10).

To understand the salary deficit better, you can plot residuals against the relevant predictor variables. Start by plotting the female salary residuals vs. age when hired. (You could plot residuals vs. years, but you would see no particular trend.)

To plot the residuals against Age Hired:

1 Click the **Female Data** sheet tab to return to the data worksheet.

2 Select the range **D1:D38** and **H1:H38** (use the **CTRL** key to select noncontiguous ranges); then click **Insert > Chart**.

3 Follow the Chart Wizard directions and create a scatterplot showing points only, with no lines, gridlines, or legends. Give a chart title of **Residuals vs. Age Hired**, and label the x-axis **Age Hired** and the y-axis **Residuals**. Save the chart in a new chart sheet named **Female Resid vs. Age Hired**.

4 Change the scale of the x-axis from 0–60 to 20–50. Your chart should look like Figure 9-20.

Figure 9-20 Scatterplot of residuals vs. age hired, for female faculty

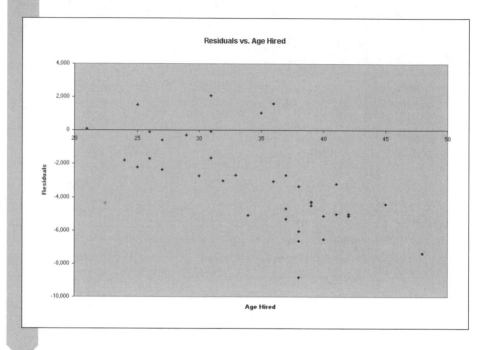

There seems to be a downward trend to the scatterplot, indicating that the greater discrepancies in salaries occur for older female faculty. Add a linear regression line to the plot, regressing residuals vs. age when hired.

To add a linear regression line to the plot:

1 Right-click the data series (any one of the data points in Figure 9-20), and click **Insert Trendline** in the shortcut menu.

2 Click the **Type** tab, verify that the Linear Trend/Regression Type option is selected, and then click **OK**. Your plot should now look like Figure 9-21.

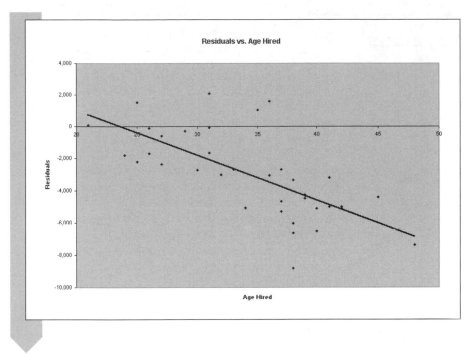

**Figure 9-21
Scatterplot
with trend
line added**

This plot shows a salary deficiency that depends very much on the age at which a female was hired. Those who were hired under the age of 25 have residuals that average around 0, or a little below. Those who were hired over the age of 40 are underpaid by more than $5,000, on average. The most underpaid female has a deficit of nearly $9,000.

Drawing Conclusions

Why should age make a difference in the discrepancies? One possibility is that women are more likely than men to take time off from their careers to raise their children. If this is the case, an older male faculty member would have more job experience and thus be paid more. However, this might not be true of all women, yet all of the females who were hired over the age of 36 were underpaid.

To summarize, the female faculty are underpaid an average of about $3,000. However, there is a big difference depending on how old they were when hired. Those who were hired after the age of 40 have an average deficit of more than $5,000. It should be noted that when the case was eventually settled out of court, each woman received the same compensation, regardless of age.

To save and close the DISCRIM2 workbook:

1   Click the **Save** button to save your file.

2   Click **File > Exit**.

Exercises

1. The PCSURV.XLS data set has survey results on 35 models of personal computers, taken from *PC Magazine*, February 9, 1993. The magazine sent 17,000 questionnaires to randomly chosen subscribers, and the data are based on 8,176 responses. There are five columns: Company, the name of the company; Reliability, the overall reliability; Repairs, satisfaction with repair experience; Support, satisfaction with technical support; and Buy Again, future likelihood of buying from the vendor.
   a. Open the workbook and create a correlation matrix and scatterplot matrix for the four numeric variables.
   b. Which vendors have the highest Buy Again scores? The lowest Buy Again scores?
   c. Regress Buy Again on the other three numeric variables. Plot the residuals to check the assumptions. How successful are the other variables at predicting the willingness of customers to buy again? Does it appear that the regression assumptions are satisfied?
   d. Determine the sign of the correlation between the Buy Again variable and the Repairs variable. What does this imply about the nature of the relationship? Are high Buy Again values associated with high marks for repair work? Now look at the sign of the Repair variable in the regression equation. What does this imply about the relationship between the Repair variable and the Buy Again variable? How does this compare with what you earlier concluded from the correlation of the two variables? How would you explain this?
   e. Summarize your conclusions and save your workbook as E9PCSURV.XLS.

2. Open the workbook WHEAT.XLS.
   a. Generate the correlation matrix for the variables Calories, Carbo, Protein, and Fat. Also create the corresponding scatterplot matrix.
   b. Regress Calories on the other three variables and obtain the residual output. How successful is the regression? It is known that carbohydrates have 4 calories per gram, protein has 4 calories per gram, and fats have 9 calories per gram. How do the coefficients compare with the known values?
   c. Explain why the coefficient for fat is the least accurate, in terms of its standard error and in comparison with the known value of 9. (*Hint*: Examine the data and notice that the fat content is specified with the least precision.)
   d. Plot the residuals against the predicted values. Is there an outlier? Label the points of the scatterplot to see which case is most extreme. Do the calories add up correctly for this case? That is, when you multiply the carbohydrate content by 4, the protein content by 4, and the fat content by 9, does it add up to more calories than are stated on the package? Notice also that another case has the same values of Carbo, Protein, and

Fat, but the Calories value is 10 higher. How do you explain this? Would a company understate the calorie content?

   e. Write your conclusions and then save your workbook as E9WHEAT.XLS.

3. The WHEATDAN.XLS data set is a slight modification of WHEAT.XLS, with an additional case, the Apple Danish from McDonald's. It is included because it has a substantial fat content, in contrast to the foods in WHEAT.XLS. Because none of the foods in WHEAT.XLS have much fat, it is impossible to see from WHEAT.XLS how much fat contributes to the calories in the foods.

   a. Repeat the regression of Exercise 2 for WHEATDAN.XLS and see whether the coefficient for fat is now estimated more accurately. Use both the known value of 9 for comparison and the standard error of the regression that is printed in the output.

   b. Save your workbook as E9WHEATD.XLS.

4. BASE26.XLS includes data from major league baseball games played Tuesday, July 29, 1992, as reported the next day in the Bloomington, Illinois, *Pantagraph*. For each of the 26 major league teams, the data include Runs, Singles, Doubles, Triples, Home Runs, and Walk HBP. This last variable combines walks and hit-by-pitched-ball.

   a. Regress Runs on the other variables and compare with the results obtained by Rosner and Woods (1988), as quoted in the beginning of this chapter. Can the differences be explained in terms of the standard errors of the coefficients?

   b. Do the Rosner–Woods coefficients make more sense in terms of which should be largest and which should be smallest?

   c. How would you expect your answers to change if you obtained data for several more days?

   d. Save your workbook as E9BASE26.XLS.

5. The HONDACIV.XLS data include Price, Age, and Miles for used Honda Civics advertised in the San Francisco *Chronicle*, November 25, 1990. Notice that there is a problem with missing data because Miles were not included in many of the advertisements.

   a. Because of problems that can occur with missing data, copy only the rows with non-missing values for mileage to a new worksheet. In this new worksheet, find correlations and plots to relate Price to the predictors Age and Miles. (You will have to copy age to the column that is adjacent to the Miles column.)

   b. Regress Price on Age and Miles, and examine the residuals. Is there any evidence from the residuals that the regression assumptions have been violated?

   c. Notice that one car is much older than the others, and its high residual suggests it might be an outlier. Do a new regression without this observation to see how much difference it makes. Which regression is better? Interpret your regression results in terms of the effect of one more year of Age on the price.

d. What about Miles? Why does the number of Miles not seem to matter? Notice that when you look at the data, the Miles values tend to be low relative to the age of the car. Would you advertise the number of miles on your car if it were high relative to the car's age? If people advertise only low mileage, how would this affect the regression?

e. Write your report and save your workbook as E9HONDACIV.XLS.

6. Reopen the HONDACIVIC.XLS workbook.

a. Repeat the last regression model from Exercise 5b using log price instead of price.

b. Does this improve the multiple correlation? Does the old car no longer have a high residual, so there is no longer any need to do a new regression without this car? Is it still true that the Miles variable is not significant in the regression?

c. When Log Price is used as the dependent variable, the regression can be interpreted in terms of percentage drop in Price per year of age, instead of a fixed drop per year of Age when Price is used as the dependent variable. Does it make more sense to have the price drop by 16.5% each year or to have the price drop by $721 per year? In particular, would an old car lose as much value per year as it did when it was young?

d. Save your workbook as E92HONDAC.XLS.

7. Open the workbook CARS.XLS. The data file is based on material collected by Donoho and Ramos (1982). The workbook contains observations from 392 car models on the following eight variables: MPG (miles per gallon), Cylinders (number of cylinders), Engine Disp (engine displacement in cubic inches), Horsepower, Weight (vehicle weight in pounds), Accelerate (time to accelerate from 0 to 60 mph in seconds), Year (model year), and Origin (origin of car: American, European, or Japanese).

a. Create a correlation matrix (excluding Spearman's rank correlation) and a scatterplot matrix of the seven quantitative variables.

b. Regress MPG on Cylinders, Engine Disp, Horsepower, Weight, Accelerate, and Year.

c. Note that the regression coefficients for Engine Disp and Horsepower are nonsignificant. Compare this to the p-values for these variables in the correlation matrix. What accounts for the lack of significance? (*Hint*: Look at the correlations between Engine Disp, Horsepower, and Weight.)

d. Create a scatterplot of the regression residuals vs. the predicted values. Judging by the scatterplot, do the assumptions of the regression appear to be violated? Why or why not?

e. Create a new variable, Log MPG, that displays the logarithm of the miles per gallon. Redo your regression model with this new dependent variable in place of MPG. How does the residual vs. predicted value plot compare to the earlier one?

f. Save your workbook as E9CARS.XLS.

8. Reopen the CARS.XLS workbook.

a. Recreate the Log MPG variable described in the previous exercise and then regress Log MPG on the numeric variables. Try to reduce the number of variables in the model using the following algorithm:

  i) Perform the regression.

  ii) If any coefficients in the regression are nonsignificant, redo the regression with the least significant variable removed.

  iii) Continue until all coefficients remaining are significant.

  To do this, you may have to move the columns around because the Regression command requires that all predictor variables lie in adjacent columns.

b. How does the R^2-value for this reduced model compare to the full model that you started the process with?

c. Report your final model and save the workbook as E92CARS.XLS.

9. Reopen the CARS.XLS workbook.

a. In the workbook CARS.XLS, regress the variable Log MPG that you created in Exercise 8 on Cylinders, Engine Disp, Horsepower, Weight, Accelerate, and Year for *only* the American cars. (You will have to copy the data to a new worksheet using the AutoFilter function.)

b. Analyze the residuals of the model. Do they follow the assumptions reasonably well?

c. In the Car Data worksheet, add a new column containing the predicted Log MPG values for all car models using the regression equation you created for only the American cars. Create another new column containing the residuals.

d. Plot the predicted values against the residuals for all of the cars, and then break down the scatterplot into categories based on origin. Rescale the x-axis so that it ranges from 1 to 1.6.

e. Calculate descriptive statistics (include the summary, variability, and 95% confidence *t*-confidence intervals) for the residuals column, broken down by country of origin.

f. Summarize your conclusions, answering the question whether Japanese and European cars appear to have a different MPG after correction for the other factors.

g. Save your workbook as E93CARS.XLS.

10. Open the workbook TEMP.XLS. The file contains average January temperatures for 56 cities in the United States, along with the cities' latitude and longitude.

a. Create a chart sheet containing a scatterplot of latitude vs. longitude. Modify the scales for the horizontal and vertical axes to go from 60 to 120 degrees in longitude and from 20 to 50 degrees in latitude. Reverse the orientation of the x-axis so that it starts from 120 degrees on the left and goes down to 60 degrees on the right. Add labels to the points, showing the temperature for each city.

b. Construct a regression model that relates average temperature to latitude and longitude.

c. Examine the results of the regression. Are both predictor variables statistically significant at the 5% level? What is the R^2-value? How much of the variability in temperature is explained by longitude and latitude?

d. Format the regression values generated by the Analysis ToolPak to display residual values as integers. Copy the map chart from part b to a new chart sheet, and delete the temperature labels. Now label the points using the residual values.

e. Interpret your findings. Where are the negative values clustered? Where do you usually find positive residuals?

f. Summarize your findings, discussing where the linear model fails and why. Save the workbook as E9TEMP.XLS.

11. Open the HOMEDAT.XLS workbook, which contains information on home prices in Albuquerque, New Mexico.

a. Regress the price of the houses in the sample on three predictor variables: Square Feet, Age, and number of features.

b. Examine the plot of residuals vs. predicted values. Is there any violation of the regression assumptions evident in this plot?

c. Redo the regression analysis, this time regressing the Log Price on the three predictor variables. How does the plot of residuals vs. predicted values appear in this model? Did the logarithm correct the problem you noted earlier?

d. There is an outlier in the plot. Identify the point and describe what this tells us about the price of the house if the model is correct.

e. Save the workbook as E9HOMEDAT.XLS.

12. Open the UNEMP.XLS workbook. This workbook contains the United States unemployment rate, Federal Reserve Board index of industrial production, and year of the decade for 1950–1959. Unemployment is the dependent variable; Industrial Production and Year of the Decade are the predictor variables.

a. Create a chart sheet showing the scatterplot of Unemployment versus FRB_Index. Add a linear trendline to the chart. Does unemployment appear to rise along with production?

b. Using the Analysis ToolPak, run a simple linear regression of Unemployment versus FRB_Index. What is the regression equation? What is the R^2-value? Does the regression explain much of the variability in unemployment values during the 1950s?

c. Rerun the regression, adding Years to the regression equation. How does the R^2-value change with the addition of the Years factor? What is the regression equation?

d. Compare the parameter value for FRB_Index in the first equation with that in the second. How are they different? Does your interpretation of the effect of production on unemployment change from one regression to the other?

e. Calculate the correlation between FRB_Index and Year. How signficant is the correlation?

f. Save your workbook as E9UNEMP.XLS.

Chapter 10

Analysis of Variance

Objectives

In this chapter you will learn to:

- Compare several groups graphically

- Compare the means of several groups using analysis of variance

- Correct for multiple comparisons using the Bonferroni test

- Find which pairs of means differ significantly

- Compare analysis of variance to regression analysis

- Perform a two-way analysis of variance

- Create and interpret an interaction plot

- Check the validity of assumptions

One-Way Analysis of Variance

Earlier we used the *t*-test to compare two treatment groups, such as two groups taught by two different methods. What if there are four treatment groups? We might have 40 subjects split into four groups, with each group receiving a different treatment. The treatments might be four different drugs, four different diets, or four different advertising videos. **Analysis of variance**, or **ANOVA**, provides a test to determine whether to accept or reject the hypothesis that all of the group means are equal.

The model we'll use for analysis of variance, called a means model, is

$$y = \mu_i + \varepsilon$$

Here, μ_i is the mean of the *i*th group, and ε is a random error following a normal distribution with mean 0 and variance σ^2. If there are P groups, the null and alternative hypotheses for the means model are

$$H_0: \mu_1 = \mu_2 = \cdots = \mu_P$$

$$H_a: \text{Not all of the } \mu_i \text{ are equal}$$

Note that the assumptions for the means model are similar to those used for regression analysis:

- The errors are normally distributed.
- The errors are independent.
- The errors have constant variance σ^2.

The similarity to regression is no accident. As you will see later in this chapter, analysis of variance can be thought of as a special case of regression.

To verify analysis of variance assumptions, it is helpful to make a plot that shows the distribution of observations in each of the treatment groups. If the plot shows large differences in the spread among the treatment groups, there might be a problem of nonconstant variance. If the plot shows outliers, there might be a problem with the normality assumption. Independence could also be a problem if time is important in the data collection, in which case consecutive observations might be correlated. However, there are usually no problems with the independence assumption in the analysis of variance.

Analysis of Variance Example: Comparing Hotel Prices

Some professional associations are reluctant to hold meetings in New York City because of high hotel prices and taxes. The American Statistical Association, for example, has not met in New York since 1973. Are hotels in New York City more expensive than hotels in other major cities?

To answer this question, let's look at hotel prices in four major cities: Los Angeles, New York City, San Francisco, and Washington, D.C. For each city, a random sample of eight hotels was taken from the 1992 *Mobil Travel Guide to Major Cities* and stored in the workbook HOTEL.XLS. The workbook contains the following variables, shown in Table 10-1:

Table 10-1 The Hotel workbook

Range Name	Range	Description
City	A2:A33	City of each hotel
Hotel	B2:B33	Name of hotel
Stars	C2:C33	*Mobil Travel Guide* rating (1992), on a scale from 1 to 5
Price	D2:D33	Price of a single room

To open the Hotel.xls workbook:

1 Open the **Hotel** workbook from the folder containing your student files.

2 Click **File > Save As** and save the workbook as **Hotel2**. The workbook appears as shown in Figure 10-1.

Figure 10-1 The Hotel2 workbook

Los Angeles

San Francisco

Washington, D.C.

New York

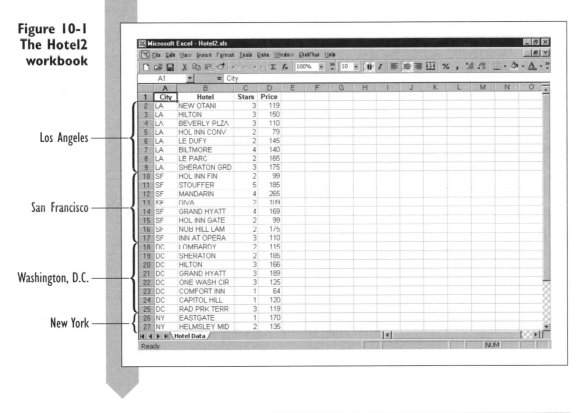

We have to decide between two hypotheses:

H_0: The mean hotel price is the same for each city.

H_a: The mean hotel prices are not the same.

Graphing the Data to Verify ANOVA Assumptions

It is best to begin with a graph that shows the distribution of hotel prices in each of the four cities. To do this, you can use the multiple histograms command available in StatPlus.

To create the graphs:

1 Click **StatPlus > Multi-variable Charts > Multiple Histograms**.

2 Because the workbook is laid out, with the variable values in one column and the categories in another, verify that the **Use a column of category levels** option button is selected.

3 Click the **Data Values** button, and select **Price** from the list of range names. Click **OK**.

4 Click the **Categories** button, and select **City** from the range names list. Click **OK**.

5 Click the **Display Normal Curve** checkbox, and verify that the **Frequency** option button is selected.

6 Click the **Output** button, and send the output to a new worksheet named **Histograms**. Click **OK**.

Your completed dialog box should appear as shown in Figure 10-2.

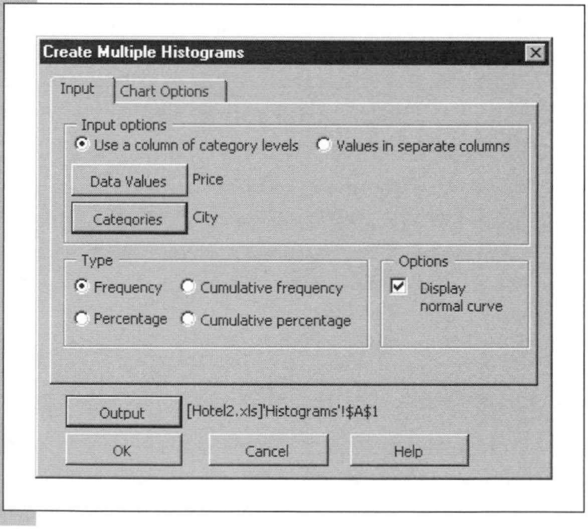

**Figure 10-2
Completed
Create
Multiple
Histograms
dialog box**

7 Click **OK**. StatPlus generates the histograms shown in Figure 10-3.

**Figure 10-3
Multiple
histograms
of hotel
prices in
each city**

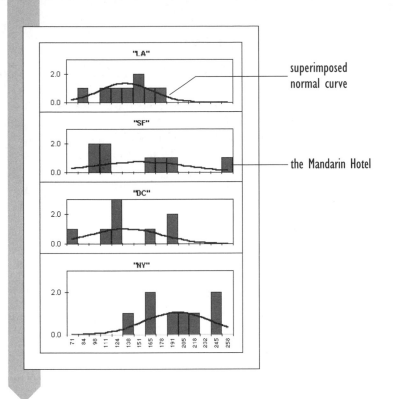

What do these histograms tell you about the analysis of variance assumptions? One of the assumptions states that the variance is the same in each group. If one city has prices that are all bunched together and another has a very wide spread of prices, unequal variances could be a problem. The plot shows similar spreads in three of the groups, whereas a fourth, San Francisco, appears to cover a slightly wider range of values. This appears to be due to one hotel that is much more expensive than the others in that city. If you examine that value, you'll discover that it represents the Mandarin, costing more than $250 per night (the highest in the whole sample). The other hotels in San Francisco all cost less than $200. Of all the 32 hotels, this one is the most influential in the sense that its removal would have the greatest effect on its group mean and variance. Excluding this hotel would change its city mean and variance more than excluding any of the other hotels. It would be a good idea to do the analysis both with and without this hotel. If exclusion makes little difference in the results, there is not much to worry about. However, if exclusion makes a big difference, our results rest on shaky ground. After all, the hotels were chosen randomly; the Mandarin was included only by chance.

What about the assumption of normal data? The analysis of variance is robust to the normality assumption, so only major departures from normality would cause trouble. In any case, eight observations in each group may be too few to determine whether or not the normality assumption is violated.

Computing the Analysis of Variance

From the histograms, it appears that New York has the highest mean hotel price. Still, there is a lot of overlap between the New York prices and the others. The Mandarin in San Francisco is more expensive than any of the New York City hotels. Do you think that New York City is significantly more expensive than the other cities? We'll soon find out by performing an analysis of variance. To do so, we'll have to use the Analysis of Variance command available from the Analysis ToolPak, the statistical add-in supplied with Excel. The Analysis ToolPak requires that the group values be placed in separate columns. In this workbook, groups are identified by a category variable, so you'll have to "unstack" the price values based on the city variable, creating four separate price columns.

To unstack the Price column:

1 Click **StatPlus > Manipulate Columns > Unstack Column**.

2 Click the **Data Values** button, and select **Price** from the range names list. Click **OK**.

3 Click the **Categories** button, select **City** from the range name list, and click **OK**.

4 Click the **Output** button and send the unstacked values to a new worksheet named **Price Data**. Click **OK**.

5 Figure 10-4 shows the completed Unstack Column dialog box. Click **OK**.

**Figure 10-4
The Unstack
Column
dialog box**

data values to be placed in separate columns

columns will be based on different values of the city variable

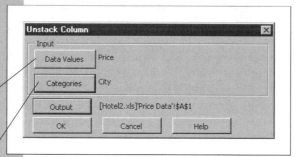

The unstacked data are shown in Figure 10-5.

Figure 10-5
The
unstacked
data

city abbreviations ——

hotel prices
for each city ——

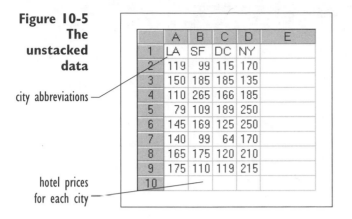

	A	B	C	D	E
1	LA	SF	DC	NY	
2	119	99	115	170	
3	150	185	185	135	
4	110	265	166	185	
5	79	109	189	250	
6	145	169	125	250	
7	140	99	64	170	
8	165	175	120	210	
9	175	110	119	215	
10					

STATPLUS TIPS _____

- You can use the "StatPlus > Manipulate Columns > Stack Columns" command to stack a series of columns. The resulting data set will contain two columns: a column of data values and a column containing the category labels.

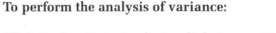

Now you can perform the analysis of variance on the price data.

To perform the analysis of variance:

1 Click **Tools > Data Analysis**, click **Anova: Single Factor** in the Analysis Tools list box, and then click **OK**.

2 Type **A1:D9** in the Input Range text box, and verify that the **Grouped By Columns** option button is selected.

3 Click the **Labels in First Row** checkbox to select it.

4 Click the **New Worksheet Ply** option button and type **Price ANOVA** in the corresponding text box. Your dialog box should look like Figure 10-6.

Figure 10-6
The Anova:
Single Factor
dialog box

Figure 10-6 The Anova: Single Factor dialog box

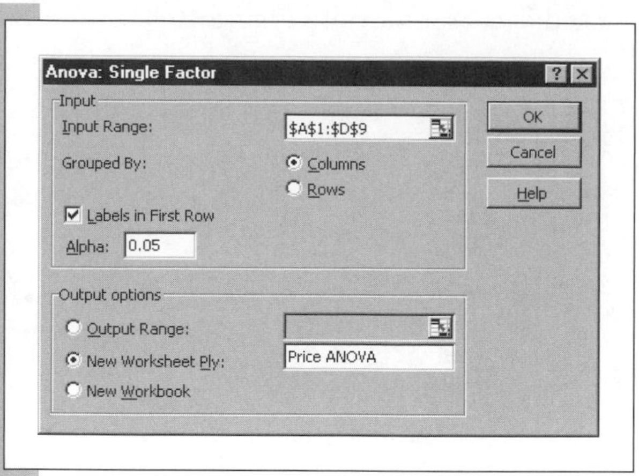

5 Click **OK**.

Figure 10-7 shows the resulting analysis of variance output, with some minor formatting.

Figure 10-7
Analysis of
variance
output

Figure 10-7 Analysis of variance output

	A	B	C	D	E	F	G
1	Anova: Single Factor						
2							
3	SUMMARY						
4	Groups	Count	Sum	Average	Variance		
5	LA	8	1083	135.375	980.839		
6	SF	8	1211	151.375	3414.839		
7	DC	8	1083	135.375	1771.125		
8	NY	8	1585	198.125	1649.554		
9							
10							
11	ANOVA						
12	Source of Variation	SS	df	MS	F	P-value	F crit
13	Between Groups	21,145.38	3	7048.46	3.61	0.025	2.947
14	Within Groups	54,714.50	28	1954.09			
15							
16	Total	75,859.88	31				

Interpreting the Analysis of Variance Table

In performing an analysis of variance, you determine what part of the variance you should attribute to randomness and what part you can attribute to other factors. Analysis of variance does this by splitting the total sum of squares (the sum of squared deviations from the mean) into two parts: a part attributed to differences between the groups and a part due to random error or random chance. To see how this is done, we'll first recall that the formula for the total sum of squares is

$$\text{Total SS} = \sum_{i=1}^{n}(y_i - \overline{y})^2$$

Here, the total number of observations is n, and the average of *all* observations is $\overline{y}$. The value for the hotel data is 75,859.88 and is shown in cell B16. The sample average (not shown) is 155.0625.

Let's express the total SS in a different way. We'll break the calculation down by the various groups. Assume that there are a total of P groups and that the size of each group is n_i (groups need not be equal in size, so n_i would indicate the sample size of the ith group), and we calculate the total sum of squares for each group separately. This can be written as

$$\text{Total SS} = \sum_{i=1}^{P} \sum_{j=1}^{n_i}(y_{ij} - \overline{y})^2$$

Here, y_{ij} identifies the jth observation from the ith group (for example, y_{23} would mean the third observation from the second group). Notice that we haven't changed the value; all we've done is specify the order in which we'll calculate the total sum of squares. We'll calculate the sum of squares in the first group, and then in the second and so forth, adding up all of the sums of squares in each group to arrive at the total sum of squares.

Next we'll calculate the sample average for *each* group, labeling it as $\overline{y}_i$, which is the sample average of the ith group. For example, in the hotel data, the values (shown in cells D5:D8) are

LA	135.375
SF	151.375
DC	135.375
NY	198.125

Using the group averages, we can calculate the total sum of squares within each group. This is equal to the sum of the squared deviations, where the deviation is from each observation to its group average. We'll call this value the **error sum of squares**, or **SSE**, and it can be expressed as

$$\text{SSE} = \sum_{i=1}^{P} \sum_{j=1}^{n_i}(y_{ij} - \overline{y}_i)^2$$

Another term for this value is the **within-groups sum of squares** because it is the sum of squares within each group. The value for SSE in the hotel data is 54,714.50 (shown in cell B14).

The final piece of the analysis of variance is to calculate the sum of squares between each of the group averages and the overall average. This value, called the **between-groups sum of squares** and otherwise known as the **treatment sum of squares**, or **SST**, is

$$\text{SST} = \sum_{i=1}^{P} n_i(\overline{y}_i - \overline{y})^2$$

Note that we take each squared difference and multiply it by the number of observations in the group. In this hotel data, each group has eight observations, so the value of n_i is always eight. The between-groups sum of squares for the hotel data is equal to 21,145.38 (cell B13).

But note that the total sum of squares is equal to the within-groups sum of squares plus the between-groups sum of squares, because 75,859.88 = 21,145.38 + 54,714.50. In general terms,

$$\text{Total SS} = \text{SSE} + \text{SST}$$

Let's try to relate this to the price of staying at hotels in various cities. If the average prices in the various cities are very different, the between-groups sum of squares will be a large value. However, if the city averages are close in value, the between-groups sum of squares will be near zero. The argument goes the other way, too; a large value for the between-groups sum of squares could indicate that the city averages are very different, whereas a small value might show that they are not so different.

A large value for the between-groups sum of squares could also be due to a large number of groups, so you have to adjust for the number of groups in the data set. The degrees of freedom (df) column in the ANOVA table (cells C13:C16) tells you that. The df for the city factor (in this case the between-groups term) is the number of groups minus 1, or $4 - 1 = 3$ (cell C13). The degrees of freedom for the total sum of squares is the total number of observations minus 1, or $32 - 1 = 31$ (cell C16). The remaining degrees of freedom are assigned to the error term (the within-groups term) and are equal to $31 - 3 = 28$ (cell C14).

The Mean Square (MS) column (cells D13:D14) shows the sum of squares divided by the degrees of freedom; you can think of the entries in this column as variances. The first value, 7048.458 (cell D13), measures the variance in hotel cost between the various cities; the second value, 1954.089 (cell D14), measures the variance of the cost within cities. The within-groups mean square also estimates the value of σ^2—the variance of the error term ε shown in the means model earlier in the chapter. If the variability in hotel prices between cities is large relative to the variability of hotel prices within the cities, then we might conclude that mean hotel price is not the same for each city.

To test this, we calculate the ratio of the two variances. Under the null hypothesis, this value should follow an F-distribution with n, m degrees of freedom, where n is the degrees of freedom for the between-groups variance and m is the degrees of freedom for the within-groups variance.

In the hotel data, the F-value is 3.61 (cell E13) and follows an $F(3,28)$ distribution. Excel calculates the p-value to be 0.025 (cell F13), which is less than 0.05. We reject the null hypothesis, accepting the alternative that there is a difference in the mean hotel price.

Although the output does not show it, you can use the values in the ANOVA table to derive some of the same statistics you used in regression analysis. For example, the ratio of the between-groups sum of squares to the total sum of squares equals R^2, the coefficient of determination discussed in some depth in Chapters 8 and 9. In this case $R^2 = 21,145.38/75,859.88 = 0.2787$. Thus about 27.9% of the variability in hotel price is explained by the city of origin.

Comparing Means

The ANOVA table has led you to reject the hypothesis that the mean single-room price is the same in all four cities and to accept the alternative that the four means are not all the same. Looking at the mean values, you might be tempted to conclude that the high price for New York City hotel rooms is the cause and leave it at that. This assumption would be unwarranted because you haven't tested for this specific hypothesis. Moreover, the price for a single room in San Francisco is higher than the price for a room in either Los Angeles or Washington, D.C. Are these differences also significant? To find out, you need to calculate the differences in mean value between all pairs of cities and then test the differences to discover their statistical significance.

Excel does not provide a function to test pairwise mean differences, but one has been provided for you with StatPlus.

To create a matrix of paired differences:

1 Click **StatPlus > Multivariate > Means Matrix**.

2 Click the **Data Values** button, and select **Price** from the list of range names. Click **OK**.

3 Click the **Categories** button, and select **City** from the range names list. Click **OK**.

4 Click the **Use Bonferroni Correction** checkbox.

5 Click the **Output** button, and direct the output to a new worksheet named **Means Matrix**. Click **OK**.

Figure 10-8 shows the completed dialog box.

Figure 10-8
The completed Create Matrix of Mean Differences dialog box

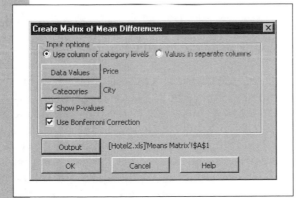

6 Click **OK**. Excel generates the output shown in Figure 10-9.

Figure 10-9
Pairwise
mean
differences

	A	B	C	D	E
1		Descriptive Statistics			
2		City = "LA"	City = "SF"	City = "DC"	City = "NY"
3		Price	Price	Price	Price
4	Count	8	8	8	8
5	Average	135.38	151.38	135.38	198.13
6	Standard Deviation	31.318	58.437	42.085	40.615
7	Sum of Squares	153,477.000	207,219.000	159,009.000	325,575.000
8					
9	Pairwise Mean Difference (row – column)				
10		"LA"	"SF"	"DC"	"NY"
11	"LA"	0.000	-16.000	0.000	-62.750
12	"SF"		0.000	16.000	-46.750
13	"DC"			0.000	-62.750
14	"NY"				0.000
15	MSE = 1954.08928571429				
16					
17	Pairwise Probabilities (Bonferroni Correction)				
18		"LA"	"SF"	"DC"	"NY"
19	"LA"	-	1.000	1.000	0.050
20	"SF"		-	1.000	0.261
21	"DC"			-	0.050
22	"NY"				-

You can tell from the Pairwise Mean Difference table that the mean cost for a single hotel room in Los Angeles is \$16 less than the mean cost in San Francisco. The largest difference is between Los Angeles or Washington, D.C., and New York City, with a single room in these cities costing about \$63 less than a single room in New York City hotels. Note that the output includes the mean squared error value from the ANOVA table, 1954.089, which is the estimate of the variance of hotel prices.

Using the Bonferroni Correction Factor

You also requested in the dialog box a table of p-values for these mean differences using the Bonferroni correction factor. Recall from Chapter 8 that the Bonferroni procedure is a conservative method for calculating the probabilities by multiplying the p-value by the total number of comparisons. Because the p-values are much higher than you would see if you compared the cities with t-tests, it is harder to get significant comparisons with the Bonferroni procedure. However, the Bonferroni procedure has the advantage of giving fewer false positives than t-tests would give.

With the Bonferroni procedure, the chances of finding at least one significant difference among the means is less than 5% if all of the four population means are the same. On the other hand, if you do six t-tests to compare the four cities at the 5% level, there is much more than a 5% chance of getting significance in at least one of the six tests if all four population means are the same. Other methods are available to help you adjust the p-value for multiple comparisons, including Tukey's and Scheffé's, but the Bonferroni method is the easiest to implement in Excel, which does not provide a correction procedure.

Note: Essentially, the difference between the Bonferroni procedure and a t-test is that for the Bonferroni procedure the 5% applies to all six comparisons together, but for t-tests the 5% applies to each of the six comparisons separately. In statistical language, the Bonferroni procedure is testing at the 5% level experimentwise, whereas the t-test is testing at the 5% level pairwise.

The pairwise comparison probabilities show that only the two biggest differences are significant (highlighted in red). Only the differences between NY and LA and between NY and DC have p-values at 0.05, so only these two differences are significant at the 5% level.

Although the NY vs. SF difference is not significant, it is interesting that the other two differences, NY vs. DC and NY vs. LA, are significant. Recall that the data collection was partly motivated by high prices in New York City.

When to Use Bonferroni

As the size of the means matrix increases, the number of comparisons increases as well. Consequently, the p-values for the pairwise differences are greatly inflated. As you can imagine, there might be a point where there are so many comparisons in the matrix that it is nearly impossible for any one of the comparisons to be statistically significant using the Bonferroni correction factor. Many statisticians are concerned about this problem and feel that although the Bonferroni correction factor does guard well against incorrectly finding significant differences, it is also too conservative and misses true differences in pairs of mean values.

In such situations, statisticians make a distinction between paired comparisons that are planned before the data are analyzed and those that occur only after we look at the data. For example, the planned comparisons here are the differences in hotel room price between New York City and the others. You should be careful with new comparisons that you come up with after you have collected the data. You should hold these comparisons to a much higher standard than the comparisons you've planned to make all along. This distinction is important in order to ward off the effects of data "snooping" (unplanned comparisons). Some statisticians recommend that you do the following when analyzing the paired means differences in your analysis of variance:

1. Conduct an F-test for equal means.
2. If the F-statistic is significant at the 5% level, make any planned comparisons you want without correcting the p-value. For data snooping, use a correction factor such as Bonferroni's on the p-value.
3. If the F-statistic for equal means is not significant, you can still consider any planned comparisons, but only with a correction factor to the p-value. Do not analyze any unplanned comparisons. (Milliken and Johnson 1984.)

It should be emphasized that although some statisticians embrace this approach, others question its validity.

Comparing Means with a Boxplot

Earlier you used multiple histograms to compare the distribution of hotel prices among the different cities. The boxplot is also very useful for this task because it shows the broad outline of the distributions and displays the medians for the four cities. Recall that if the data are very badly skewed, the mean might be strongly affected by outlying values. The median would not have this problem.

To create a boxplot of price versus city:

1 Click **StatPlus > Single Variable Charts > Boxplots**.

2 Click the **Data Values** button, and select **Price** from the list of range names. Click **OK**.

3 Click the **Categories** button, and select **City** from the range names list. Click **OK**.

4 Click the **Use Bonferroni Correction** checkbox.

5 Click the **Output** button, and direct the output to a new chart sheet named **Boxplots**. Click **OK**.

6 Click **OK**. The resulting boxplots are shown in Figure 10-10.

Figure 10-10 Boxplots of hotel prices for each city

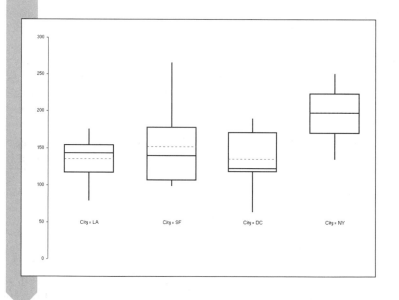

Compare the medians, indicated by the middle horizontal line (not dotted) in each box. The median for San Francisco is below the median for LA, even though you discovered from the pairwise mean difference matrix that the mean price in San Francisco is $16 above the mean in Los Angeles. The reason for the difference is the Mandarin Hotel, which is much more expensive than the other San Francisco hotels. This hotel has a big effect on the San Francisco mean price, but not on the median. The median is much more robust to the effect of outliers. On the other hand, although the cost of the Mandarin Hotel is much higher than the others, it is not shown as an outlier in the plot. For it to qualify as an outlier, its distance above the box would have to exceed the box's height by 50%.

This discussion might make you curious about what would happen if the Mandarin Hotel were excluded; you will have the opportunity to find out in the exercises.

One-Way Analysis of Variance and Regression

You can think of analysis of variance as a special form of regression. In the case of analysis of variance, the predictor variables are discrete rather than continuous. Still, you can express an analysis of variance in terms of regression and, in doing so, can get additional insights into the data. To do this you have to reformulate the model.

Earlier in this chapter you were introduced to the means model:

$$y = \mu_i + \varepsilon$$

for the ith treatment group. An equivalent way to express this relationship is with the **effects model**:

$$y = \mu + \alpha_i + \varepsilon$$

Here μ is a mean term, α_i is the effect from the ith treatment group, and ε is a normally distributed error term with mean 0 and variance σ^2.

Let's apply this equation to the hotel data. In this data set there are four groups representing the four cities, so you would expect the effects model to have a mean term μ and four effect terms, α_1, α_2, α_3, and α_4, representing the four cities. There is a problem, however: You have five parameters in your model, but you are estimating only four mean values. This is an example of an **overparametrized model**, where you have more parameters than response values. As a result, an infinite number of possible values for the parameters will "solve" the equation. To correct this problem, you have to reduce the number of parameters. Statistical packages generally do this in one of two ways: Either they constrain the values of the effect terms so that the sum of the terms is zero, or they define one of the effect terms to be zero (Milliken and Johnson, 1984). Let's apply this second approach to the hotel data and perform the analysis of variance using regression modeling.

Indicator Variables

To perform the analysis of variance using regression modeling, you can create indicator variables for the data. **Indicator variables** take on values of either 1 or 0, depending on whether the data belong to a certain treatment group or not. For example, you can create an indicator variable where the variable values are 1 if the observation comes from a hotel in San Francisco and 0 if the observation comes from a hotel not in San Francisco.

You'll use the indicator variables to represent the terms in the effects model.

To create indicator variables for the hotel data:

1. Click the **Hotel Data** worksheet tab (you might have to scroll to see it) to return to the worksheet containing the hotel data.
2. Click **StatPlus > Manipulate Columns > Create Indicator Columns**.
3. Click the **Categories** button, and select **City** from the list of range names. Click **OK**.
4. Click the **Output** button, click the **Cell** option button, and select cell **F1**. Click **OK**.
5. Click **OK**.

Excel generates the four new columns shown in Figure 10-11.

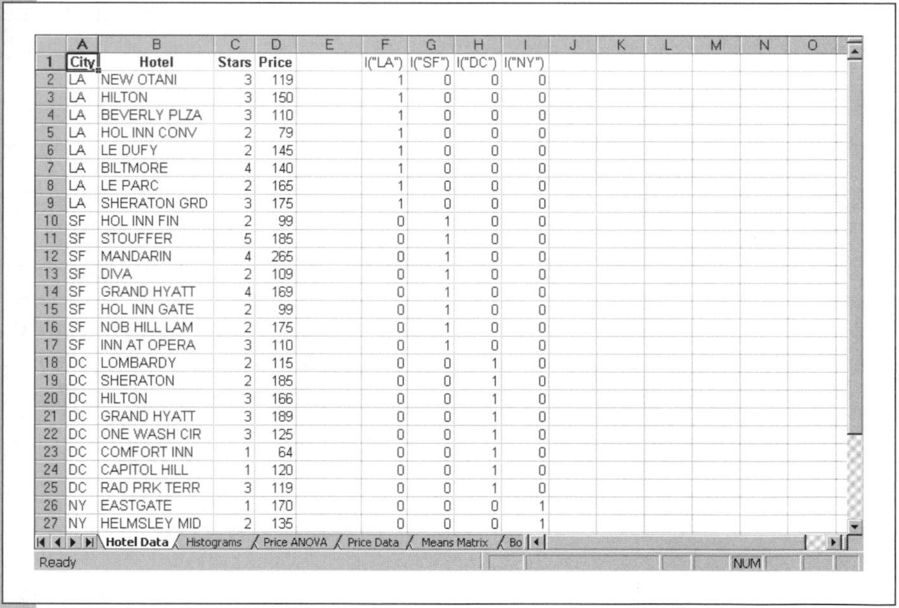

	A	B	C	D	E	F	G	H	I	J	K	L	M	N	O
1	City	Hotel	Stars	Price		I("LA")	I("SF")	I("DC")	I("NY")						
2	LA	NEW OTANI	3	119		1	0	0	0						
3	LA	HILTON	3	150		1	0	0	0						
4	LA	BEVERLY PLZA	3	110		1	0	0	0						
5	LA	HOL INN CONV	2	79		1	0	0	0						
6	LA	LE DUFY	2	145		1	0	0	0						
7	LA	BILTMORE	4	140		1	0	0	0						
8	LA	LE PARC	2	165		1	0	0	0						
9	LA	SHERATON GRD	3	175		1	0	0	0						
10	SF	HOL INN FIN	2	99		0	1	0	0						
11	SF	STOUFFER	5	185		0	1	0	0						
12	SF	MANDARIN	4	265		0	1	0	0						
13	SF	DIVA	2	109		0	1	0	0						
14	SF	GRAND HYATT	4	169		0	1	0	0						
15	SF	HOL INN GATE	2	99		0	1	0	0						
16	SF	NOB HILL LAM	2	175		0	1	0	0						
17	SF	INN AT OPERA	3	110		0	1	0	0						
18	DC	LOMBARDY	2	115		0	0	1	0						
19	DC	SHERATON	2	185		0	0	1	0						
20	DC	HILTON	3	166		0	0	1	0						
21	DC	GRAND HYATT	3	189		0	0	1	0						
22	DC	ONE WASH CIR	3	125		0	0	1	0						
23	DC	COMFORT INN	1	64		0	0	1	0						
24	DC	CAPITOL HILL	1	120		0	0	1	0						
25	DC	RAD PRK TERR	3	119		0	0	1	0						
26	NY	EASTGATE	1	170		0	0	0	1						
27	NY	HELMSLEY MID	2	135		0	0	0	1						

Figure 10-11 Indicator variables in columns F:I

The values in column F, labeled I ("LA"), are equal to 1 if the values in the row come from a hotel in Los Angeles, and 0 if they do not. Similarly, the values for the next three columns are 1 if the observations come from San Francisco, Washington, D.C., and New York City, respectively, and 0 otherwise.

Fitting the Effects Model

With these columns of indicator variables, you can now fit the effects model to the hotel pricing data.

To fit the effects model using regression analysis:

1 Click **Tools > Data Analysis**, click **Regression** in the Analysis Tools list box, and then click **OK**.

2 Type **D1:D33** in the Input Y Range text box, press **[Tab]**, and then type **F1:H33** in the Input X Range text box.

Recall that you have to remove one of the effect terms to keep from overparametrizing the model. For this example, remove the New York effect term. (You could have removed any one of the four city effect terms.)

3 Click the **Labels** checkbox to select it because the range includes a header row.

4 Click the **New Worksheet Ply** option button; then type **Effects Model** in the corresponding text box.

5 Verify that all four Residuals checkboxes are deselected; then click **OK**.

The regression output appears as in Figure 10-12. (The columns are resized to show the labels.)

**Figure 10-12
Created
effects
model
with the
Regression
command**

F-ratio is equal to
the one shown in
the analysis of
variance output

	A	B	C	D	E	F	G
1	SUMMARY OUTPUT						
2							
3	*Regression Statistics*						
4	Multiple R	0.528					
5	R Square	0.279					
6	Adjusted R Square	0.201					
7	Standard Error	44.205					
8	Observations	32					
9							
10	ANOVA						
11		*df*	*SS*	*MS*	*F*	*Significance F*	
12	Regression	3	21,145.375	7048.458	3.607	0.025	
13	Residual	28	54,714.500	1954.089			
14	Total	31	75,859.875				
15							
16		*Coefficients*	*Standard Error*	*t Stat*	*P-value*	*Lower 95%*	*Upper 95%*
17	Intercept	198.125	15.629	12.677	0.000	166.111	230.139
18	I("LA")	-62.750	22.103	-2.839	0.008	-108.025	-17.475
19	I("SF")	-46.750	22.103	-2.115	0.043	-92.025	-1.475
20	I("DC")	-62.750	22.103	-2.839	0.008	-108.025	-17.475

New York City
average

difference from the
New York City
average hotel price

The analysis of variance table produced by the regression (cells A10:F14) and shown in Figure 10-12 should appear familiar to you because it is equivalent to the ANOVA table created earlier and shown in Figure 10-7. There are two differences: the Between Groups row from the earlier ANOVA table is the Regression row in this table, and the Within Groups row is now termed the Residual row.

The parameter values of the regression are also familiar. The intercept coefficient, 198.125 (cell B17), is the same as the mean price in New York. The values of the LA, SF, and DC effect terms now represent the difference between the mean hotel price in these cities and the price in New York. Note that this is exactly what you calculated in the matrix of paired mean differences shown in Figure 10-9. The *p*-values for these coefficients are the uncorrected *p*-values for comparing the paired mean differences between these cities and New York. If you multiplied these *p*-values by 6 (the number of paired comparisons in the paired-mean-differences matrix), you would have the same *p*-values shown in Figure 10-9.

Can you see how the use of indicator variables allowed you to create the effects model? Consider the values for I ("LA"). For any non–Los Angeles hotel, the value of the indicator variable is 0, so the effect term is multiplied by 0, and therefore has no impact on the estimate of the hotel price. It is only for Los Angeles hotels that the effect term is present.

As you can see, using regression analysis to fit the effects model gives you much of the same information as the one-way analysis of variance.

The model you've considered suggests that the average price for a single room at a hotel in New York City is significantly higher than that for a single room in either Los Angeles or Washington, D.C., but not necessarily higher than that for such a room in San Francisco. You can expect to pay about an average of $198 for a single room in New York City, and $63 less than this in LA or DC. The San Francisco single-room average cost is about $47 less than the cost in New York City.

To save and close the Hotel2 workbook:

1   Click **File > Save** to save your file as Hotel2.

2   Click **File > Close**.

EXCEL TIPS

- You can use the Regression command to calculate the means model instead of the effects model. To do this, run the Analysis ToolPak's Regression command, choose *all* of the indicator variables in the Input X Range text box, and also select the "Constant Is Zero" checkbox. This will remove the constant term from the model. The parameter estimates will correspond to mean values of the different groups.

Two-Way Analysis of Variance

One-way analysis of variance compares several groups corresponding to a single categorical variable, or factor. A **two-way analysis of variance** uses two factors. In agriculture, for example, you might be interested in the effects of both potassium and nitrogen on the growth of potatoes. In medicine you might want to study the effects of two factors—medication and dose—on the duration of headaches. In education you might want to study the effects of grade level and gender on the time required to learn a skill. A marketing experiment might consider the effects of advertising dollars and advertising medium (television, magazines, and so on) on sales.

Recall that earlier in the chapter you looked at the means model for a one-way analysis of variance. Two-way analysis of variance can also be expressed as a means model:

$$y_{ijk} = \mu_{ij} + \varepsilon_{ijk}$$

where y is the response variable and μ_{ij} is the mean for the ith level of one factor and the jth level of the second factor. Within each combination of the two factors, you might have multiple observations called **replicates**. Here ε_{ijk} is the error for the ith level of the first factor, the jth level of the second factor, and the kth replicate, following a normal distribution with mean 0 and variance σ^2.

The model is more commonly presented as an effects model where

$$y_{ijk} = \mu + \alpha_i + \beta_j + \alpha\beta_{ij} + \varepsilon_{ijk}$$

Here y is the response variable, μ is the overall mean, α_i is the effect of the ith treatment for the first factor, and β_j is the effect of the jth treatment for the second factor. The term $\alpha\beta_{ij}$ represents the interaction between the two factors—that is, the effect that the two factors have on each other. For example, in an experiment where the two factors are advertising dollars and advertising medium, the effect of an increase in sales might be the same regardless of what advertising medium (radio, newspaper, or television) is used, or it might vary depending on the medium. When the increase is the same regardless of the medium, the interaction is 0; otherwise, there is an interaction between advertising dollars and medium.

A Two-Factor Example

To see how different factors affect the value of a response variable, consider an example of the effects of four different assembly lines (A, B, C, or D) and two shifts (a.m. or p.m.) on the production of microwave ovens for an appliance manufacturer. Assembly line and shift are the two factors; the assembly line factor has four levels, and the shift factor has two levels. Each combination of the factors line and shift is called a **cell**, so there are $4 \times 2 = 8$ cells. The response variable is the total number of microwaves assembled in a week for one assembly line operating on one particular shift. For each of the eight combinations of assembly line and shift, six separate weeks' worth of data are collected.

You can describe the mean number of microwaves created per week with the effects model where

Mean number of microwaves =
overall mean + assembly line effect + shift effect + interaction + error

Now let's examine a possible model of how the mean number of microwaves produced could vary between shifts and assembly lines. Let the overall mean number of microwaves produced for all shifts and assembly lines be 240 per week. Now let the four assembly line effects be A, +66 (that is, assembly line A produces on average 66 more microwaves than the overall mean); B, –2; C, –100; D, +36. Let the two shift effects be p.m., –6; and a.m., +6. Notice that the four assembly line effects add up to zero, as do the two shift effects. This follows from the need to constrain the values of the effect terms to avoid overparametrization, as was discussed with the one-way effects model earlier in this chapter.

If you exclude the interaction term from the model, the population cell means (the mean number of microwaves produced) look like this:

	A	B	C	D
p.m.	300	232	134	270
a.m.	312	244	146	282

These values are obtained by adding the overall mean + the assembly line effect + the shift effect for each of the eight cells. For example, the mean for the p.m. shift on assembly line A is

Overall mean + assembly line effect + shift effect = 240 + 66 − 6 = 300

Without interaction, the difference between the a.m. and the p.m. shifts is the same (12) for each assembly line. You can say that the difference between a.m. and p.m. is 12 no matter which assembly line you are talking about. This works the other way, too. For example, the difference between line A and line C is the same (166) for both the p.m. shift (300 − 134) and the a.m. shift (312 − 146). You might understand these relationships better from a graph. Figure 10-13 shows a plot of the eight means with no interaction (you don't have to produce this plot).

Figure 10-13
Means plot
without an
interaction
effect

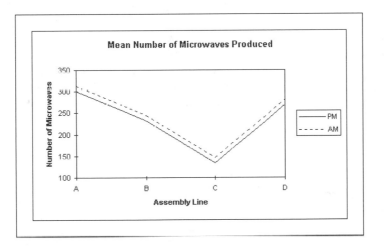

The cell means are plotted against the assembly line factor using separate lines for the shift factor. This is called an interaction plot; you'll create one later in this chapter.

Because there is a constant spacing of 12 between the two shifts, the lines are parallel. The pattern of ups and downs for the p.m. shift is the same as the pattern of ups and downs for the a.m. shift.

What if interaction is allowed? Suppose that the eight cell population means are as follows:

	A	B	C	D
p.m.	295	235	175	200
a.m.	317	241	142	220

In this situation, the difference between the shifts varies from assembly line to assembly line, as shown in Figure 10-14. This means that any inference on the shift effect must take into account the assembly line. You might claim that the a.m. shift generally produces more microwaves, but this is not true for assembly line C.

**Figure 10-14
Means plot
with an
interaction
effect**

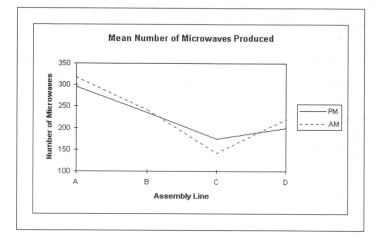

The assumptions for two-way ANOVA are essentially the same as those for one-way ANOVA. For one-way ANOVA, all observations on a treatment were assumed to have the same mean, but here all observations in a cell are assumed to have the same mean. The two-way ANOVA assumes independence, constant variance, and normality, just as in one-way ANOVA (and regression).

Two-Way Analysis Example: Comparing Soft Drinks

The workbook COLA.XLS contains data describing the effects of cola (Coke, Pepsi, Shasta, or generic) and type (diet or regular) on the foam volume of cola soft drinks. Cola and type are the factors; cola has four levels, and type has two levels. There are, therefore, eight combinations, or cells, of cola brand and soft drink type. For each of the eight combinations, the experimenter purchased and cooled a six-pack, so there are 48 different cans of soda. Then the experimenter chose a can at random, poured it in a standard way into a standard glass, and measured the volume of foam.

Why would it be wrong to test all of the regular Coke first, then the diet Coke, and so on? Although the experimenter might make every effort to keep everything standardized, trends that influence the outcome could appear. For example, the temperature in the room or the conditions in the refrigerator might change during the experiment. There could be subtle trends in the way the experimenter poured and measured the cola. If there were such trends, it would make a difference which brand was poured first, so it is best to pour the 48 cans in random order.

The COLA.XLS workbook contains the variables shown in Table 10-2.

Table 10-2 Data for Cola workbook

Range Name	Range	Description
Can_No	A2:A49	The number of the can (1–6) in the six-pack
Cola	B2:B49	The cola brand
Type	C2:C49	Type of cola: regular or diet
Foam	D2:D49	The foam content of the cola
Cola_Type	E2:E49	The brand and type of the cola

To open the Cola workbook:

1. Open the **Cola** workbook from the folder containing your student files.
2. Click **File > Save As** and save the workbook as **Cola2**. The workbook appears as shown in Figure 10-15.

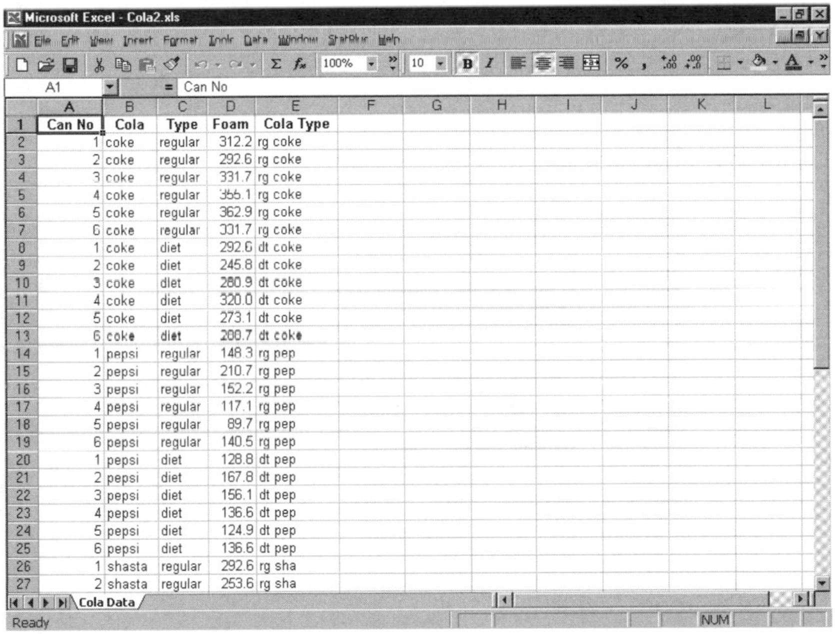

	Can No	Cola	Type	Foam	Cola Type
2	1	coke	regular	312.2	rg coke
3	2	coke	regular	292.6	rg coke
4	3	coke	regular	331.7	rg coke
5	4	coke	regular	355.1	rg coke
6	5	coke	regular	362.9	rg coke
7	6	coke	regular	331.7	rg coke
8	1	coke	diet	292.6	dt coke
9	2	coke	diet	245.8	dt coke
10	3	coke	diet	280.9	dt coke
11	4	coke	diet	320.0	dt coke
12	5	coke	diet	273.1	dt coke
13	6	coke	diet	288.7	dt coke
14	1	pepsi	regular	148.3	rg pep
15	2	pepsi	regular	210.7	rg pep
16	3	pepsi	regular	152.2	rg pep
17	4	pepsi	regular	117.1	rg pep
18	5	pepsi	regular	89.7	rg pep
19	6	pepsi	regular	140.5	rg pep
20	1	pepsi	diet	128.8	dt pep
21	2	pepsi	diet	167.8	dt pep
22	3	pepsi	diet	156.1	dt pep
23	4	pepsi	diet	136.6	dt pep
24	5	pepsi	diet	124.9	dt pep
25	6	pepsi	diet	136.6	dt pep
26	1	shasta	regular	292.6	rg sha
27	2	shasta	regular	253.6	rg sha

Figure 10-15 Cola2 workbook

Graphing the Data to Verify Assumptions

Before performing a two-way analysis of variance on the data, you should plot the data values to see whether there are any major violations of the assumptions of equal variability in the different cells. Note that you can use the Cola_Type variable to identify the eight cells.

To create multiple histograms of the foam data:

1 Click **StatPlus > Multi-variable Charts > Multiple Histograms**.

2 Click the **Data Values** button, select **Foam** from the range names list, and click **OK**.

3 Click the **Categories** button, select **Cola_Type** from the range names list, and click **OK**.

4 Click the **Display normal curve** checkbox.

5 Click the **Output** button, send the charts to a new worksheet named **Histograms**, and click **OK**.

6 Click **OK** to generate the histograms shown in Figure 10-16.

Figure 10-16
Multiple
histograms
of the
cola type

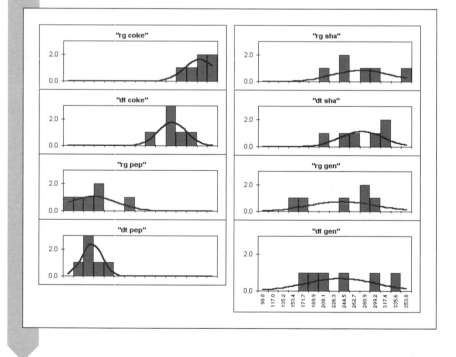

Because of the number of charts, you must either reduce the zoom factor on your worksheet or scroll vertically through the worksheet to see all the plots. Do you see major differences in spread among the eight groups? If so, it would suggest a violation of the equal-variances assumption, because all of the groups are supposed to have the same population variance. The histograms seem to indicate a greater variability in the generic colas and the Shasta brand, whereas less variability exists for the Coke and Pepsi brands. Once again, the two-way ANOVA is fairly robust with respect to the constant variance assumption, so this might not invalidate the analysis.

You should also look for outliers because extreme observations can make a big difference in the results. An outlier could be the result of a strange can of cola, a wrong observation, a recording error, or an error in entering the data. To gain further insight into the distribution of the data, create a boxplot of each of the eight combinations of brand and type.

To create boxplots of the foam data:

1 Click **StatPlus > Single Variable Charts > Boxplots**.

2 Click the **Data Values** button, select **Foam** from the range names list, and click **OK**.

3 Click the **Categories** button, select **Cola_Type** from the range names list, and click **OK**.

4 Click the **Output** button, and send the output to a new chart sheet named **Boxplots**. Click **OK**.

5 Click **OK** to create the boxplots.

6 You improve the chart by editing the labels at the bottom of the box-plot, removing the text string "Cola_Type=" from each label, and increasing the font size. See Figure 10-17.

**Figure 10-17
Boxplots of
foam vs.
cola type**

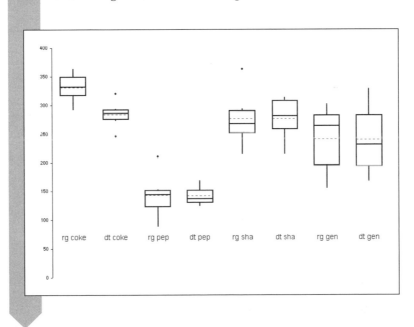

From the boxplots, you can see that there are no extreme outliers evident in the data, but there are several moderate outliers; perhaps most noteworthy are the outliers for regular Pepsi and regular Shasta. An advantage of the boxplot over the multiple histograms is that it is easier to view the relative change in foam volume from diet to regular for each brand of cola. The first two boxplots represent the range of foam values for regular and diet Coke, respectively, after which come the Pepsi values, Shasta values, and, finally, the generic values. Notice in the plot that the same pattern occurs for both the diet and the regular colas. Coke is the highest, Pepsi is the lowest, and Shasta and generic are in the middle. The difference in the foam between the diet and the regular sodas does not depend much on the cola brand. This suggests that there is no interaction between the cola effect and the type effect.

On the basis of this plot, can you draw preliminary conclusions regarding the effect of type (diet or regular) on foam volume? Does it appear that there is much difference due to cola type (diet or regular)? Because the foam levels do not appear to differ much between the two types, you can expect that the test for the type effect in a two-way analysis of variance will not be significant. However, look at the differences among the four brands of colas. The foamiest can of Pepsi is below the least foamy can of Coke, so you might expect that there will be a significant cola effect.

The Interaction Plot

The histograms and boxplots give us an idea of the influence of cola type and cola brand on foam volume. How do we graphically examine the interaction between the two factors? We can do so by creating an **interaction plot**, which displays the average foam volume for each combination of factors. To do this, we take advantage of Excel's pivot table feature.

To set up the pivot table:

1 Click the **Cola Data** sheet tab to return to the data.

2 Click **Data > PivotTable and PivotChart Report** to start the PivotTable Wizard.

3 Click the **PivtChart (with PivotTable)** option button and click **Next**.

4 Verify that the Range text box contains the range A1:E49; then click **Next**.

5 Click the **Layout** button.

6 Drag the **Type** field button to the Column area of the table.

7 Drag the **Cola** field button to the Row area of the table.

8 Drag the **Foam** field button to the Data area of the table so that Sum of Foam appears in the Data area.

9 Double-click the **Sum of Foam** button to open the PivotTable Field dialog box.

10 Click **Average** in the Summarize by List box; then click **OK** twice.

11 Click the **Options** button.

12 Deselect the **Grand Totals for Columns** and **Grand Totals for Rows** checkboxes; then click **OK**.

13 Click **Finish**.

By default, Excel produces a bar chart of the various cell averages. It would be more instructive to change this to a line chart.

To create a line chart of the cell average:

1 Click **Chart > Chart Type**.

2 Click **Line** from the list of chart types.

3 Click the first chart sub-type in the list and click **OK**.

Figure 10-18 displays the line chart of the cell means.

Figure 10-18 Interaction plot of cola type vs. cola brand

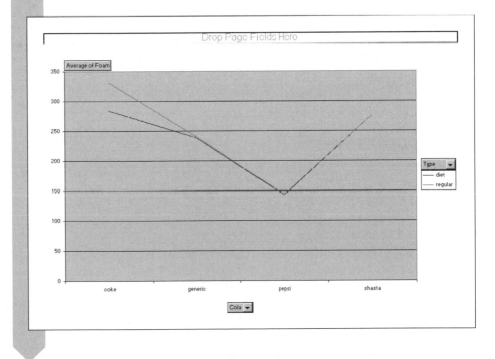

The plot shows that the foam volumes of diet and regular colas are very close, except for Coke. If there is no interaction between cola brand and cola type, the difference in foam volume for diet and regular should be the same for each cola brand. This means that the lines would move in parallel, always with the same vertical distance. Of course, there is a certain amount of random variation, so the lines will usually not be perfectly parallel. The plot would seem to indicate that there is no interaction between cola brand and cola type. To confirm our visual impression, we'll have to perform a two-way analysis of variance.

Using Excel to Perform a Two-Way Analysis of Variance

The Analysis ToolPak provides two versions of the two-way analysis of variance. One is for situations in which there is no replication of combination of factor levels. That would be the case in this example if the experimenter had tested only one can of soda for each cola brand and type. However, the experiment has been done with six cans, so you should perform a two-way analysis of variance with replication.

Note that the number of cans for each cell of brand and type must be the same. Specifically, you cannot use data that have five cans of diet Coke and six cans of regular Coke. Data with the same number of replications per cell are called **balanced data**. If the number of replicates is different for different combinations of brand and type, you cannot use the Analysis ToolPak's two-way analysis of variance command.

Finally, to use the Analysis ToolPak on this data set, it must be organized in a two-way table. Figure 10-19 shows this table for the cola data. The data are formatted so that the first factor (the four cola brands) is displayed in the columns, and the second factor (diet or regular) is shown in the rows of the table. Replications (the six cans in each pack) occupy six successive rows. Each cell in the two-way table is the value of the foam volume for a particular can. You can create this table using the Create Two-Way Table command included with StatPlus.

**Figure 10-19
Two-way
table of the
foam data**

Foam	Cola			
Type	coke	generic	pepsi	shasta
diet	292.6	167.8	128.8	292.6
	245.8	249.7	167.8	253.6
	280.9	187.3	156.1	214.6
	320.0	210.7	136.6	269.2
	273.1	292.6	124.9	312.2
	288.7	327.8	136.6	312.2
regular	312.2	156.1	148.3	292.6
	292.6	253.6	210.7	253.6
	331.7	273.1	152.2	362.9
	355.1	175.6	117.1	280.9
	362.9	284.8	89.7	249.7
	331.7	300.4	140.5	214.6

To create a two-way table:

1 Return to the Cola Data worksheet.

2 Click **StatPlus > Manipulate Columns > Create Two-Way Table**.

3 Click the **Data Values** button, select **Foam** from the range name list, and click **OK**.

4 Click the **Column Levels** button, select **Cola** from the list of range names, and click **OK**.

5 Click the **Row Levels** button, and select **Type** from the range names. Click **OK**.

6 Click the **Output** button, and direct the output to a new worksheet named **Two-Way Table**. Figure 10-20 shows the completed dialog box.

7 Click **OK**.

The structure of the data on the Two-Way Table worksheet now resembles Figure 10-19, and you can now use the Analysis ToolPak to compute the two-way ANOVA.

To calculate the two-way analysis of variance:

1 Click **Tools > Data Analysis**, click **Anova: Two-Factor With Replication** in the Analysis Tools list box, and then click **OK**.

2 Type **A2:E14** in the Input Range text box, press **[Tab]**.

You have to indicate the number of replicates in the two-way table for this command.

3 Type **6** in the Rows per sample text box.

4 Click the **New Worksheet Ply** option button, and type **Two-Way ANOVA** in the corresponding text box.

Your dialog box should look like Figure 10-21.

Figure 10-21
Anova:
Two-Factor
With
Replication
dialog box

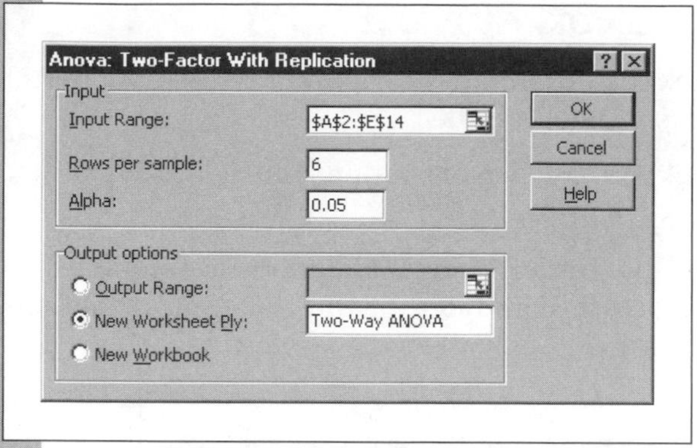

Anova: Two-Factor With Replication		? ✕
Input		
Input Range:	$A$2:$E$14	OK
Rows per sample:	6	Cancel
Alpha:	0.05	Help
Output options		
○ Output Range:		
● New Worksheet Ply:	Two-Way ANOVA	
○ New Workbook		

5 Click **OK**.

EXCEL TIPS

- If there is only one observation for each combination of the two factors, use Anova: Two-Factor Without Replication.

- If there is more than one observation for each combination of the two factors, use Anova: Two-Factor With Replication.

- If there are blanks for one or more of the factor combinations, you cannot use the Analysis ToolPak to perform two-way ANOVA.

- You can calculate the p-value for the F-distribution using Excel's FDIST(F, $df1$, $df2$), where F is the value of the F-statistic, $df1$ is the degrees of freedom for the factor, and $df2$ is the degrees of freedom for the error term.

**Figure 10-22
Two-Way
ANOVA table**

Type effect (diet
or regular)

Cola effect (Coke,
Pepsi, Shasta,
or generic)

Interaction of
the type and
cola effects

Error term

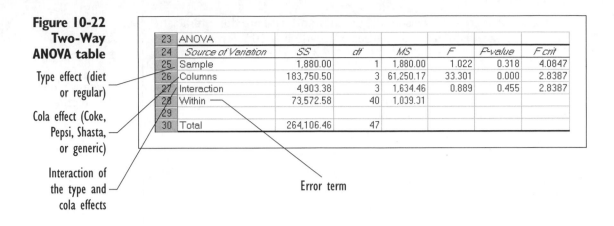

23	ANOVA						
24	Source of Variation	SS	df	MS	F	P-value	F crit
25	Sample	1,880.00	1	1,880.00	1.022	0.318	4.0847
26	Columns	183,750.50	3	61,250.17	33.301	0.000	2.8387
27	Interaction	4,903.38	3	1,634.46	0.889	0.455	2.8387
28	Within	73,572.58	40	1,039.31			
29							
30	Total	264,106.46	47				

Interpreting the Analysis of Variance Table

The analysis of variance table appears as in Figure 10-22, with the columns resized to show the labels (you might have to scroll to see this part of the output).

There are three effects now, whereas the one-way analysis had just one. The three effects are Sample for the type effect (row 25), Columns for the cola effect (row 26), and Interaction for the interaction between type and cola (row 27). The Within row (row 28) displays the Within sum of squares, also known as the error sum of squares.

As we saw earlier with the one-way ANOVA, the two-way ANOVA breaks the total sum of squares into different parts. If we designate SST as the sum of squares for the cola type, SSC as the sum of squares for cola brand, SSI for the interaction between brand and type, and SSE for random error, then

$$\text{Total} = \text{SST} + \text{SSC} + \text{SSI} + \text{SSE}$$

In this data set, the values for the various sums of squares are

SST	1,880.00
SSC	183,750.50
SSI	4,903.38
SSE	73,572.58

The degrees of freedom for each factor are equal to the number of levels in the factor minus 1. There are two cola types, diet and regular, so the degrees of freedom are 1. There are 3 degrees of freedom in the four cola brands (Coke, Pepsi, Shasta, and generic). The degrees of freedom for the interaction term are equal to the product of the degrees of freedom for the two factors. In this case, that would be $1 \times 3 = 3$. Finally, there are $n - 1$, or 47 degrees of freedom for the total sum of squares, leaving $47 - (1 + 3 + 3) = 40$ degrees of freedom for the error sum of squares. Note that the total degrees of freedom are equal to the sum of the degrees of freedom for each term in the model. In other words, if DFT are the degrees of freedom for the cola type, DFC are the degrees for freedom for cola brand, DFI are the interaction degrees of freedom, and DFE are the degrees of freedom for the error term, then

$$\text{Total degrees of freedom} = \text{DFT} + \text{DFC} + \text{DFI} + \text{DFE}$$

The next column of the two-way ANOVA table displays the mean square of each of the factors (equal to the sum of squares divided by the degrees of freedom). These are

Type	1,880.00
Cola	61,250.17
Interaction	1,634.46
Error	1,839.31

These values are the variances in foam volume within the various factors. The largest variance is displayed in the cola factor, which would indicate that this is where the greatest difference in foam volume lies. The mean square value for the error term, 1839.31, is an estimate of σ^2—the variance in foam volume after accounting for the factors of cola brand, type, and the interaction between the two. In other words, after accounting for these effects in your model, the typical deviation—or standard deviation—in foam volume is about $\sqrt{1840} = 42.9$.

As with one-way ANOVA, the next column of the table displays the ratio of each mean square to the mean square of the error term. These ratios follow a $F(m, n)$ distribution, where m is the degrees of freedom of the factor (type, cola or interaction) and n is the degrees of freedom of the error term. By comparing these values to the F-distribution, Excel calculates the p-values (cells F25:F27) for each of the three effects in the model. Examine first the interaction p-value, which is 0.455 (cell F27)—much greater than 0.05 and not even close to indicating significance at the 5% level. This confirms what we suspected from viewing the interaction plot. Now let's look at the type and cola factors.

The column or cola effect is highly significant, with a p-value given as 5.84×10^{-11} (cell F26). This is less than 0.05, so there is a significant difference among colas at the 5% level (because the p-value is less than 0.001, there is significance at the 0.1% level, too). However, the p-value is 0.318 for the sample or type effect (cell F25), so there is no significant difference between diet and regular.

These quantitative conclusions from the analysis of variance are in agreement with the qualitative conclusions drawn from the boxplot: There is a significant difference in foam volume between cola brands, but not between cola types. Nor does there appear to be an interaction between cola brand and type in how they influence foam volume.

Finally, how much of the total variation in foam volume has been explained by the two-way ANOVA model? Recall that the coefficient of determination (R^2-value) is equal to the fraction of the total sum of squares that is explained by the sums of squares of the various factors. In this case that value is

$$(1880.00 + 183{,}750.50 + 4903.38) / 264{,}106.46 = 0.721$$

Thus about 72% of the total variation in foam volume can be attributed to differences in cola brand, cola type, and the interaction between cola brand and type. Only about 28% of the total variation can be attributed to random causes.

Summary

To summarize the results from the plots and the analysis of variance, we conclude the following:

1. There is no reason to reject the hypothesis that foam volume is the same regardless of cola type (diet or regular).

2. There is a significant difference among the four cola brands (Coke, Pepsi, Shasta, and generic) with respect to foam volume. Coke has the highest volume of foam, Pepsi has the lowest, and the other two brands fall in the middle.

3. There is no significant interaction between cola type and cola brand. In other words, we don't reject the null hypothesis that the difference in foam volume between diet and regular is the same for all four brands.

You can save and close the Cola2 workbook now.

To save and close the Cola2 workbook:

1 Click **File > Save**.

2 Click **File > Exit** to quit Excel.

Exercises

1. You're performing a two-way ANOVA on an education study to evaluate a new teaching method. The two factors are region (East, Midwest, South, or West) and teaching method (standard or experimental). Schools are entered into the study, and their average test scores are recorded. There are five replicates for each combination of the region and method factors.
   a. Using the information about the design of the study, complete the following ANOVA table:

Term	SS	df	MS	F
Region	9,305	?	?	?
Method	12,204	?	?	?
Interaction	6,023	?	?	?
Error	?	?	?	
Total SS	60,341	?		

b. What is the R^2 value of the ANOVA model?

c. Use Excel's FDIST function to calculate the p-values for each of the factors and the interaction term in the model.

d. State your conclusions. What factors have a significant impact on the test scores? Is there an interaction between region and teaching method?

2. In analyzing the hotel data, you noticed that the Mandarin Hotel was an outlier. Reopen the HOTEL.XLS workbook.

   a. Remove the data about the Mandarin Hotel from the data set.

   b. Repeat the one-way ANOVA for the data (remember, you will have to unstack the data to use the Analysis ToolPak).

   c. Recalculate the matrix of paired differences (use Bonferroni correction in calculating the p-values).

   d. Summarize your results. Do your conclusions change with the removal of this one value?

   e. Save your workbook as E10HOTEL.XLS.

3. The HOTELTWOWAY.XLS workbook is taken from the same source as the HOTEL.XLS workbook, except that an effort was made to keep the data balanced for a two-way ANOVA. This means that the random sample was forced to have the same number of hotels in each of 12 cells of city and rating (four levels of city and three levels of stars). Hotels were excluded unless they had two, three, or four stars, and only two hotels are included per cell. Therefore, the sample has 24 hotels. Included in the file is a variable, city stars, which indicates the combination of city and stars. Open the workbook and perform the following analysis.

   a. Using Excel's pivot table feature, create an interaction plot of the average hotel price for the different combinations of city and stars. Is there evidence of an interaction apparent in the plot?

   b. Do a two-way ANOVA for price vs. stars and city. (You will have to create a two-way table that has stars as the row variable and city as the column variable.) Is there a significant interaction? Are the main effects significant?

   c. Based on the means for the three levels of stars, give an approximate figure for the additional cost per star.

   d. Compare the city effect in this model to the one-way analysis, which did not take into account the rating for each hotel.

   You notice that the gap between two-star hotels and three-star hotels is nearly the same as the gap between three-star and four-star hotels. This suggests a linear relationship between price and the star rating.

   e. Graph price vs. stars. Break down the chart into categories based on the city variable and then add trend lines to each of the four cities. Include the four regression equations on the chart. Do the slopes appear to be the same for the different cities?

   f. Save your results as E10HOTELTWOWAY.XLS.

4. Reopen the COLA.XLS workbook from this chapter. Because the two-way analysis of variance showed that the interaction term and the type effect were not significant, remove these factors from the model.

   a. Create boxplots and multiple histograms of the foam variable for the different cola brands.

   b. Redo your analysis as a one-way ANOVA with cola as the single factor.

   c. Create a matrix of paired differences, using Bonferroni correction. Which pairs of colas are different in terms of their foam volume?

   d. Summarize your results and save the workbook as E10COLA.XLS.

5. The FOURYR.XLS workbook has information on 24 colleges from the 1992 edition of *U.S. News and World Report*'s "America's Best Colleges," which lists 140 national liberal arts colleges, 35 in each of four quartiles. A random sample of six colleges was taken from each quartile. The data set includes College, the name of the college; Group, the quartile ranging from 1 to 4; Top10, the percentage of freshmen who were in the top 10% of their high school classes; Spending, the dollars spent by the college per pupil; and Tuition, a year's tuition at the college.

   a. Create a multiple histogram of the tuition for different group levels. Does the variability of the tuition values appear constant across the different levels?

   b. Perform a one-way ANOVA to compare tuition in the four quartiles. Does group appear to have a significant effect?

   c. Create a matrix of paired mean differences. Is the first quartile significantly more expensive than the others? Is the bottom quartile significantly less expensive than the others?

   d. Regress tuition on the group variable. Interpret the group regression coefficient in terms of the drop in tuition when you move to a higher group number. Conversely, how much extra does it cost to attend a college in a more prestigious group (with a lower group number)?

   e. To test the adequacy of the regression model as compared to the ANOVA model, perform an F-test. For testing the null hypothesis that the regression model fits, the F-ratio is

$$\frac{(\text{SSE regression} - \text{SSE ANOVA})/2}{\text{MSE ANOVA}}$$

   In the formula, SSE is the error sum of squares and MSE is the mean square value for error. What is the value of the F-ratio?

   Note: In the Analysis ToolPak regression output, the error sum of squares is the Residual sum of squares in cell C13. In the Analysis of Variance table, it is the Within Groups sum of squares in cell B14; the mean square value for error is the Within Group Mean Square in cell D14.

   The numerator is divided by 2 because the SSE for regression has two more degrees of freedom than the SSE for ANOVA. You should wind up with an F-ratio that is less than 1, which says that you can accept the smaller model because you would never reject the null hypothesis if F is less than 1. In other words, you can accept the regression model as a good fit. This is convenient because the regression model has a much simpler interpretation.

   f. Summarize your results, stating whether it is more expensive to attend a highly rated college and, if so, how the cost is related to the rating. Save your workbook as E10FOURYR.XLS.

6. The FOURYR.XLS workbook of Exercise 5 also includes the variable Computer, taken from the *Computer Industry Almanac 1991*. This represents the number of students per computer station, which includes microcomputers, mainframe terminals, and workstations. A low value indicates that the college offers ready access to computers. Unfortunately, data are not available for all colleges in the sample.

   a. Reopen the FOURYR.XLS workbook, and delete from the worksheet the rows in which there are no values for the computer variable.

   b. Create a boxplot of the computer variable broken down by group. There is an outlier present. Which college is it?

   c. Perform a one-way analysis of variance to compare quartiles. Are there significant differences among the four groups in terms of the availability of computers?

   d. Redo the analysis of variance, but this time do not include the outlier. Does removal of this observation make much difference in the results? If you find no significant difference, does this mean that access to computers is independent of college quartile? Would you make such an assertion based on data from only 14 colleges?

   e. Save your workbook as E10FOURYR2.XLS.

7. The BASEINFD.XLS workbook data set has statistics on 106 major league baseball infielders at the start of the 1988 season. The data include Salary, LN Salary (the logarithm of salary), and Position.

   a. Create multiple histograms and boxplots to see the distribution of Salary for each position. How would you describe the shape of the distribution?

   b. Make the same plots for LN Salary. How does the shape of the distribution change with the logarithm of the salary?

   c. Perform a one-way ANOVA of LN Salary on Position to see whether there is any significant difference of salary among positions.

   d. Report your results, saving your workbook as E10BASE.XLS.

8. The BASEINFD.XLS workbook also contains RBI Aver, the average runs batted in per time at bat. Reopen the BASEINFD workbook.

   a. Create multiple histograms and boxplots of the RBI variable against Position. Describe the shape of the distributions. Is there any reason to doubt the validity of the ANOVA assumptions?

   b. Perform a one-way ANOVA of RBI Aver against Position.

c. Create a matrix of paired mean differences to compare infield positions (use the Bonferroni correction factor). Which positions differ significantly? Can you explain why?

d. Summarize your results, and save your workbook as E10BASE2.XLS.

9. The HONDA25.XLS workbook contains the prices of used Hondas and indicates the age (in years) and whether the transmission is 5-speed or automatic.

a. Open the workbook and perform a two-sample t-test for the price data based on the transmission type.

b. Perform a one-way ANOVA with price as the dependent variable and transmission as the grouping variable.

c. Compare the value of the t-statistic in the t-test to the value of the F-ratio in the F-test. Do you find that the F-ratios for ANOVA are the same as the squares of the t-values from the t-test and that the p-values are the same?

d. Use one-way ANOVA to compare the ages of the Hondas for the two types of transmissions. Does this explain why the difference in price is so large?

e. Perform two regressions of price vs. age—the first for automatic transmissions and the second for 5-speed transmissions. Compare the two linear regression lines. Do they appear to be the same? What problems do you see with this approach?

f. Summarize your results. Save your workbook as E10HONDA.XLS.

10. The workbook HONDA12.XLS contains a subset of the HONDA25.XLS workbook in which the age variable is made categorical and has the values 1–3, 4–5, and 6 or more. Some observations of HONDA25.XLS have been removed to balance the data. The variable trans indicates the transmission, and the variable Trans Age indicates the combination of transmission and age class.

a. Create a multiple histogram and boxplot of price vs. trans age. Does the constant variance assumption for a two-way analysis of variance appear justified?

b. Create an interaction plot of price vs. trans and age (you will need to create a pivot table of means for this). Does the plot give evidence for an interaction between the trans and age factors?

c. Perform a two-way analysis of variance of price on age and trans (you will have to create a two-way table using age as the row variable and trans as the column variable).

d. Report your conclusions and save your workbook as E10HON12.XLS.

11. At the Olympics, competitors in the 100-meter dash go through several rounds of races, called "heats," before reaching the finals. The first round of heats involves over a hundred runners from countries all over the globe. The heats are evenly divided among the premier runners so that

one particular heat does not have an overabundance of top racers. You decide to test this assumption by analyzing data from the 1996 Summer Olympics in Atlanta, Georgia. To do this, open the workbook RACE.XLS on your student disk.

a. Create a boxplot of the race times broken down by heats. Note any large outliers in the plot and then rescale the plot to show times from 9 to 13 seconds. Is there any reason not to believe, based on the boxplot, that the variation of race times is consistent between heats?

b. Perform a one-way ANOVA to test whether the mean race times among the 12 heats are significantly different.

c. Create a pairwise means matrix of the race times by heat.

d. Summarize your conclusions. Are the race times different between the heats? What is the significance level of the analysis of variance? Save the workbook as E10RACE.XLS.

12. Repeat Exercise 11, this time looking at the reaction times among the 12 heats and deciding whether these reaction times vary. Write your conclusions and save your workbook as E10RACE2.XLS.

13. Another question of interest to race observers is whether reaction times increase as the level of competition increases. Try to answer this question by analyzing the reaction times for the 14 athletes who competed in the first three rounds of heats of the men's 100-meter dash at the 1996 Summer Olympics. Open the workbook RACEPAIR.XLS in your Student disk.

a. Use the Analysis ToolPak's ANOVA: Two-Factor Without Replication command to perform a two-way analysis of variance on the data in the Reaction Times worksheet. What are the two factors in the ANOVA table?

b. Examine the ANOVA table. What factors are significant in the analysis of variance? What percentage of the total variance in reaction time can be explained by the two factors? What is the R^2-value?

c. Examine the means and standard deviations of the reaction times for each of the three heats. Using these values, form a hypothesis for how you think reaction times vary with rounds.

d. Test your hypothesis by performing a paired t-test on the difference in reaction times between each pair of rounds (1 vs. 2, 2 vs. 3, and 1 vs. 3). Which pairs show significant differences at the 5% level? Does this confirm your hypothesis from the previous step?

e. Because there is no replication of a racer's reaction time within a round, you cannot add an interaction term to the analysis of variance. You can still create an interaction plot, however. Create an interaction plot with round on the x-axis and the reaction times for each racer as a separate line in the chart. Based on the appearance of the chart, do you believe there is an interaction between reaction time and the racer involved? What impact does this have on your overall conclusions as to whether reaction times vary with round?

f. Report your results and save your workbook as E10REACT.XLS.

14. Researchers are examining the effect of exercise on heart rate. They've asked volunteers to exercise by going up and down a set of stairs. The experiment has two factors: step height and rate of stepping. The step heights are 5.75 inches (coded as 0) and 11.5 inches (coded as 1). The stepping rates are 14 steps/min (coded as 0), 21 steps/min (coded as 1), and 28 steps/min (coded as 2). The experimenters recorded both the resting heart rate (before the exercise) and the heart rate afterward. Their results are stored in the HEART.XLS workbook.

   a. Create a two-way table using StatPlus. Place frequency in the row area of the table, place height in the column area of the table, and use heart rate after the exercise as the response variable.

   b. Analyze the values in the two-way table with a two-way ANOVA (with replication). Is there a significant interaction between the frequency at which subjects climb the stairs and the height of the stairs as it affects the subject's heart rate?

   c. Create an interaction plot. Discuss why the interaction plot supports your findings from the previous step.

   d. Create a new variable named Change, which is the change in heart rate due to the exercise. Repeat steps a–c for this new variable and answer the question of whether there is an interaction between frequency and height in affecting the change in heart rate.

   e. Summarize your findings and save your workbook as E10HEART.XLS.

15. The NOISE.XLS workbook contains data from a statement by Texaco, Inc. to the Air and Water Pollution Subcommittee of the Senate Public Works Committee on June 26, 1973. Mr. John McKinley, president of Texaco, cited an automobile filter developed by Associated Octel Company as effective in reducing pollution. However, questions had been raised about the effects of filters on vehicle performance, fuel consumption, exhaust gas back-pressure, and silencing. On the last question, he referred to the data included here as evidence that the silencing properties of the Octel filter were at least equal to those of standard silencers.

   a. Create boxplots and histograms of the Noise variable, broken down by the Size_Type variable (you should edit the labels in the boxplot to make the plot easier to read).

   b. Create an interaction plot of the Noise variable for different levels of the Size and Type factors. Is there evidence of an interaction from the plot?

   c. Create a two-way table of the Noise data for the Size and Type factors.

   d. Using the two-way table, perform a two-way ANOVA on the data. What factors are significant?

   e. Summarize your findings and save the workbook as E10NOISE.XLS.

16. The WASTE.XLS workbook contains data from a clothing manufacturer. The firm's quality control department collects weekly data on percent waste, relative to what can be achieved by computer layouts of patterns

on cloth. A negative value indicates that the plant employees beat the computer in controlling waste. Your job is to determine whether there is a significant difference among the five plants in their percent waste values.

a. Open the WASTE.XLS workbook and create boxplots of the waste value for the five plants. Are there any extreme outliers in the data that you should be concerned about?

b. Perform a one-way analysis of variance on the data.

c. Create a matrix of paired mean differences for the data. State your tentative conclusions.

d. Copy the waste data to another worksheet in the workbook, and delete any observations that were identified as extreme outliers on the boxplots.

e. Redo your one-way ANOVA and means matrix on the revised data. Have your conclusions changed?

f. Summarize your findings and save the workbook as E10WASTE.XLS.

Time Series

Objectives

In this chapter you will learn to:

- Plot a time series

- Compare a time series to lagged values of the series

- Use the autocorrelation function to determine the relationship between past and current values

- Use moving averages to smooth out variability

- Use simple exponential smoothing and two-parameter exponential smoothing

- Recognize seasonality and adjust data for seasonal effects

- Use three-parameter exponential smoothing to forecast future values of a time series

- Optimize the simple exponential smoothing constant

Time Series Concepts

A **time series** is a sequence of observations taken at evenly spaced time intervals. The sequence could be daily temperature measurements, weekly sales figures, monthly stock market prices, quarterly profits, or yearly power-consumption data. Time series analysis involves looking for patterns that help us understand what is happening with the data and help us predict future observations. For some time series data (for example, monthly sales figures), you can identify patterns that change with the seasons. This seasonal behavior is important in forecasting.

Usually the best way to start analyzing a time series is by plotting the data against time to show trends, seasonal patterns, and outliers. If the variability of the series changes with time, the series might benefit from a transformation that stabilizes the variance. Constant variance is assumed in much of time series analysis, just as in regression and analysis of variance, so it pays to see first whether a transformation is needed. The logarithmic transformation is one such example that is especially useful for economic data. For example, if there is growth in power consumption over the years, then the month-to-month variation might also increase proportionally. In this case, it might be useful to analyze either the log or the percentage change, which should have a variance that changes little over time.

Time Series Example: The Dow in the 1980s

To illustrate these ideas, consider the DOW.XLS workbook, which contains the monthly closing Dow Jones averages for 1981 through 1990. Time series analysis can shed light on how the Dow Jones has changed, for better or for worse, over this 10-year period. The DOW.XLS workbook contains the variables and reference names shown in Table 11-1.

Table 11-1 Dow.xls Workbook

Range Name	Range	Description
Year	A2:A121	The year
Month	B2:B121	The month
Year_Month	C2:C121	The year and the month
Dow_Jones	D2:D121	The monthly Dow Jones average

To open the Dow.xls workbook:

1 Open the **Dow** workbook from the folder containing your student files.

2 Click **File > Save As** and save the workbook as **Dow2**. The workbook appears as shown in Figure 11-1.

**Figure 11-1
The Dow
workbook**

	A	B	C	D	E	F	G	H	I	J	K
1	Year	Month	Year_Month	Dow Jones							
2	1981	Jan	Jan 1981	947.27							
3	1981	Feb	Feb 1981	974.58							
4	1981	Mar	Mar 1981	1003.87							
5	1981	Apr	Apr 1981	997.75							
6	1981	May	May 1981	991.75							
7	1981	Jun	Jun 1981	976.88							
8	1981	Jul	Jul 1981	952.34							
9	1981	Aug	Aug 1981	881.47							
10	1981	Sep	Sep 1981	849.98							
11	1981	Oct	Oct 1981	852.55							
12	1981	Nuv	Nuv 1981	000.90							
13	1981	Dec	Dec 1981	875.00							
14	1982	Jan	Jan 1982	871.10							
15	1982	Feb	Feb 1982	824.39							
16	1982	Mar	Mar 1982	822.77							
17	1982	Apr	Apr 1982	848.36							
18	1982	May	May 1982	819.54							
19	1982	Jun	Jun 1982	811.93							
20	1982	Jul	Jul 1982	808.60							
21	1982	Aug	Aug 1982	901.31							
22	1982	Sep	Sep 1982	896.25							
23	1982	Oct	Oct 1982	991.72							
24	1982	Nuv	Nuv 1982	1039.20							
25	1982	Dec	Dec 1982	1046.54							
26	1983	Jan	Jan 1983	1075.70							
27	1983	Feb	Feb 1983	1112.62							

Plotting the Dow Jones Time Series

Before doing any computations, it is best to explore the time series graphically.

To plot the Dow Jones average:

1 Select the range **C1:D121**.

2 Click **Insert > Chart** to open the Chart Wizard.

3 Click the **Line** Chart type and then click the first chart sub-type (lines without symbols). Click **Next** twice.

4 Type **Dow Jones Average 1981-1990** in the Chart title text box, **Year** in the Category (X) axis text box, and **Dow Jones** in the Value (Y) axis text box.

5 Remove the legend and gridlines from the chart and then click **Next**.

6 Click the **As new sheet** option button and type **Dow Jones Chart** in the accompanying text box.

7 Click **Finish**.

Because the labels crowd each other, the x-axis of the chart is difficult to read. You can format the chart so that it shows tick marks and labels at 12-month intervals (which makes sense because each category level on the x-axis represents a month).

To reduce the number of categories on the x-axis:

1 Double-click the **x-axis** to open the Format Axis dialog box; then click the **Scale** tab.

2 Type **12** in the Number of Categories between Tick-Mark Labels text box and **12** in the Number of Categories between Tick Marks text box; then click **OK**.

3 The plot of the Dow Jones average in the 1980s is shown in Figure 11-2.

**Figure 11-2
Time plot
of the
Dow Jones
average**

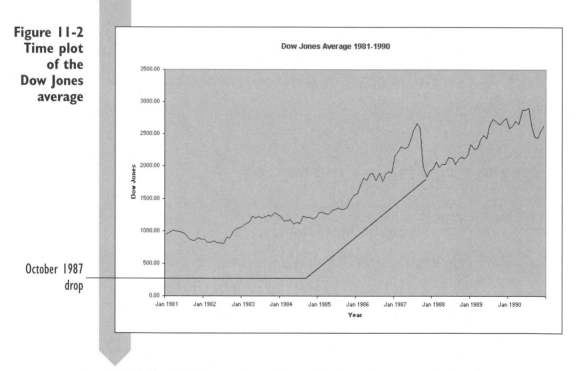

October 1987
drop

The plot in Figure 11-2 shows the amazing growth from near 1,000 to near 3,000 in just ten years. Notice also the big drop that occurred in

October 1987, when the Dow dropped 600 points. Throughout the 1980s there were prophets of doom advising everyone to get out of the market because the sky was about to fall. Yet in spite of the crash of October 1987, average stockholders profited by just retaining their holdings for these ten years.

How variable are the average Dow values from one month to another? Let's find out for each of the ten years of the 1980s.

To calculate the statistics about the monthly Dow averages:

1 Click the **Dow Jones Avg** sheet tab.

2 Click **StatPlus > Univariate Statistics**.

3 Click the **Summary** tab and select the **Count** and **Average** checkboxes.

4 Click the **Variability** tab and select the **Range** and **Std. Deviation** checkboxes.

5 Click the **General** tab and click the **Columns** option button to display the statistics by columns rather than by rows.

6 Click the **Input** button and select **Dow_Jones** from the list of range names. Click **OK**.

Now you'll break down the statistics year by year.

7 Click the **By** button and select **Year** from the range names list. Click **OK**.

8 Click the **Output** button and direct the output to a new worksheet named **Dow Statistics**.

9 Click **OK**.

Excel generates the table shown in Figure 11-3.

**Figure 11-3
Descriptive statistics of the 1980s Dow Jones monthly average**

	A	B	C	D	E	F
1				Univariate Statistics		
2			Count	Average	Range	Standard Deviation
3	Dow Jones	Year = 1981	12	932.7017	153.89	50.94470
4		Year = 1982	12	890.1492	237.94	88.37117
5		Year = 1983	12	1,197.9050	200.32	60.53523
6		Year = 1984	12	1,175.1967	119.53	41.49327
7		Year = 1985	12	1,345.8075	288.61	85.19115
8		Year = 1986	12	1,815.1075	343.24	101.22069
9		Year = 1987	12	2,273.3658	829.40	264.09811
10		Year = 1988	12	2,077.3400	210.35	68.27902
11		Year = 1989	12	2,534.3758	494.81	184.91840
12		Year = 1990	12	2,662.2325	462.87	155.83707
13		Overall	120	1,690.4182	2,096.60	647.39347

As you examine the table, notice that the average of the monthly figures in 1981 is about 930 and that this value is tripled by 1990 to over 2600. However, at the same time that the Dow is increasing in value, the standard deviation of the monthly averages also triples, from about 59 in 1981 to over 155 in 1990. Generally, as the Dow increases in value, it also increases in variability. There are some exceptions to this. In 1988 the average of the monthly Dow values was high at 2077, but the variability was comparatively low at around 68 points.

Analyzing the Change in the Dow

For people who own stocks, the changes in the Dow Jones average from month to month are important. Your next step in examining these data is to analyze the month-to-month change.

To calculate the change in the Dow:

1 Click the **Dow Jones Avg** sheet tab to return to the data.

2 Click cell **E1**, type **Diff**, and then press **Enter**.

3 Select the range **E3:E121** (*not* E2:E121).

4 Type **=D3-D2** in cell **E3**; then press **Enter**.

5 Click **Edit > Fill > Down** to enter the rest of the differences in the range E3:E121.

Now that you have calculated the differences in the Dow from one month to the next, you can plot those differences.

To plot the change in the Dow vs. time:

1 Select the range **C1:C121**, press and hold the **CTRL** key, and then select the range **E1:E121**.

2 Click **Insert > Chart**, click the **Line** chart type, click the fourth chart sub-type (lines with symbol markers), and click **Next** twice.

3 Enter **Dow Changes 1981-1990** in the Chart title box, **Year** in the Value (X) axis box, and **Monthly Differences** in the Value (Y) axis box.

4 Remove the legend and gridlines from the plot and click **Next**.

5 Send the chart to a new chart sheet named **Difference Chart** and click **Finish**.

6 Once again reformat the x-axis scale as you did for the plot of the Dow Jones average so that tick marks and labels appear at 12-month intervals.

7 Double-click the line on the chart to display the Format Data Series dialog box.

8 Click the **Patterns** tab, and click the **None** option button in the list of line options to remove the line from the plot (leaving only the symbol markers). Click **OK**.

Your final chart should look like Figure 11-4.

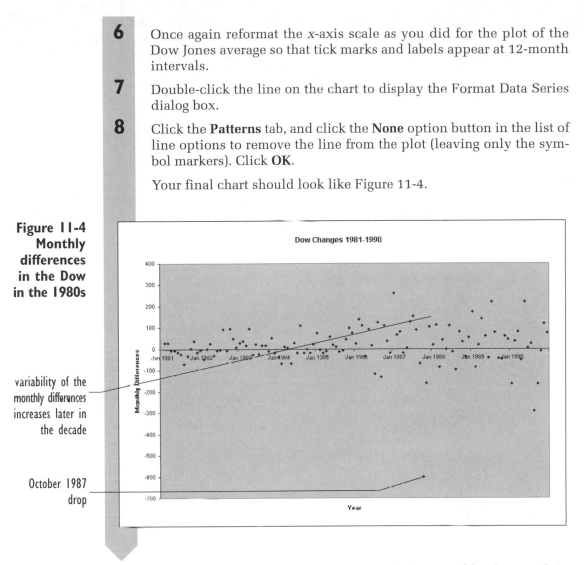

Figure 11-4 Monthly differences in the Dow in the 1980s

variability of the monthly differences increases later in the decade

October 1987 drop

The plot shows steadily increasing variance in the monthly change of the Dow over the 10 years. Notice the outlier, the drop of 600 points in October 1987.

Plotting the Percentage Change in the Dow

Because the variance in the monthly change in the Dow increased through the 1980s, it might be useful to transform the values to stabilize the variance, because many statistical procedures require constant variance. A logarithmic transformation is often used in situations where the variance of a data set changes in proportion to the size of the data values. Another transformation you might consider is the percentage change in the Dow by month. Let's calculate this value and see whether it stabilizes the variance.

To calculate the percent change in the Dow and then chart the data:

1 Click the **Dow Jones Avg** tab to return to the data.

2 Click cell F1, type **PDiff**, and then press **Enter**.

3 Select the range **F3:F121**(not **F2:F121**).

4 Type **=100*E3/D2** in cell F3 (this is the percent change from the previous month); then press **Enter**.

5 Click **Edit > Fill > Down** to enter the rest of the percent changes in the range F3:F121.

Now plot the percent change on a new chart sheet.

6 Select the range **C1:C121**, press and hold the **CTRL** key, and then select the range **F1:F121**.

7 Click **Insert > Chart**.

8 As you did for the Difference chart, select the **Line** chart type and the four chart sub-type (lines with symbol markers). Click the **Next** button twice.

9 Remove the legend and gridlines from the plot. Enter **Dow % Changes 1981-1990** for the chart title, **Year** for the Category (X) axis title, and **Percent Changes** for the Value (Y) axis title. Click **Next**.

10 Save the chart to a new chart sheet named **%Change Chart** and click **Finish**.

11 Reformat the x-axis to show tick marks and labels at 12-month intervals. Remove the line from the chart series, showing only the markers on the plot.

Your final plot should look like Figure 11-5.

Figure 11-5
Plot of
percent
change in
the Dow per
month

± 10% range ——

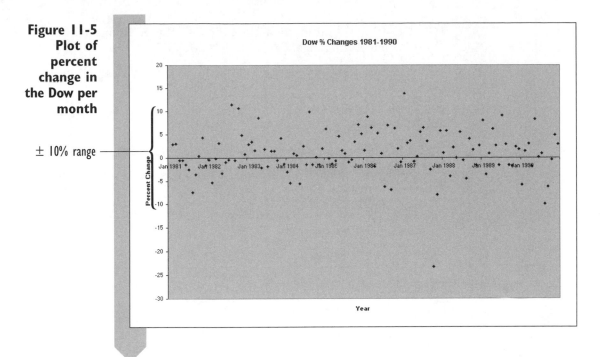

The variance is nearly stable across the plot. The month-to-month changes are mostly in the range of ±10%, and this fact does not change from 1981 to 1990, except for the drop occurring in October 1987. You've discovered that the percentage change in the Dow per month has a more stable variance across time.

Because the variance is more stable, you should feel more comfortable about applying statistical analysis to the monthly percentage change in the Dow Jones average. As a first step, calculate the descriptive statistics for the average monthly percent change in the Dow during the 1980s.

To calculate descriptive statistics for the percent change in the Dow:

1 Click the **Dow Jones Avg** tab to return to the data.

2 Click **StatPlus > Univariate Statistics**.

3 Click the **Summary** tab and select the **Count**, **Average**, and **Median** checkboxes.

4 Click the **Variability** tab and select the **Minimum**, **Maximum**, **Range**, and **Std. Deviation** checkboxes.

5 Click the **General** tab and click the **Columns** option button to display the statistics in different columns.

6 Click the **Input** button; then click the **Use Range References** option button, select the range, **F1:F121**, and click **OK**.

7 Click the **By** button, select **Year** from the range names list, and click **OK**.

8 Click the **Output** button, direct the output to a new sheet named **PDiff Statistics**, and click **OK**.

9 Click **OK** to start generating the descriptive statistics table.

10 To format the table values, select the range **D3:I13** and click the **Decrease Decimal** button several times, reducing the number of decimal places to three.

Figure 11-6 shows the revised table.

Figure 11-6 Descriptive statistics of the percent change in the Dow per month

	A	B	Count	Average	Median	Minimum	Maximum	Range	Standard Deviation
1					Univariate Statistics				
2			Count	Average	Median	Minimum	Maximum	Range	Standard Deviation
3	Pdiff	Year = 1981	11	-0.668	-0.610	-7.442	4.273	11.715	3.327
4		Year = 1982	12	1.618	-0.303	-5.362	11.465	16.828	5.122
5		Year = 1983	12	1.590	1.489	-2.138	8.510	10.649	2.998
6		Year = 1984	12	-0.243	-0.694	-5.629	9.782	15.411	4.074
7		Year = 1985	12	2.097	1.211	-1.342	7.118	8.460	3.034
8		Year = 1986	12	1.831	1.756	-6.888	8.789	15.677	5.130
9		Year = 1987	12	0.613	3.295	-23.216	13.824	37.040	9.224
10		Year = 1988	12	0.988	1.346	-4.560	5.791	10.351	3.341
11		Year = 1989	12	2.081	2.024	-3.583	9.041	12.624	4.017
12		Year = 1990	12	-0.247	0.495	-10.011	8.277	18.288	5.120
13		Overall	119	0.980	0.853	-23.216	13.824	37.040	4.793

From the table, you see that the average percentage change in the Dow during the 1980s is 0.98%, or slightly less than a 1% increase per month. This indicates that, on average, you would expect the Dow to increase about 1% each month over the previous month's level. There are some large departures from this assumption. The greatest percent drop in the monthly average of the Dow occurred in 1987, when the average dropped over 23% from the previous month. The highest percent increase in the monthly average also occurred in 1987 as the Dow climbed over 13 percent.

However, because the average percent change is positive, the long-run changes are upward. Would you be willing to invest long-term in the market if you could expect an upward trend of about 1% per month? Does it make sense that stock market investors tended to be happy with the Reagan and Bush administrations? However, would you be cautious about investing in the Dow over the short term if the average percentage change has the possibility of being strongly negative?

Looking at Lagged Values

Often in time series you will want to compare a value observed at one time point to the value observed one or more time points earlier. In the Dow Jones data, for example, you might be interested in comparing the monthly value of the Dow with the value it had the previous month. Such prior values are known as **lagged values**. Lagged values are an important concept in time series analysis. You can lag observations for one or more time points. In the example of the Dow Jones average, the lag 1 value is the value of the Dow one month prior, the lag 2 value is the Dow Jones average two months prior, and so forth.

You can calculate lagged values by letting the values in rows of the lagged column be equal to values one or more rows above in the unlagged column. Let's add a new column to the Dow Jones Avg worksheet, consisting of Dow Jones averages lagged one month.

To create a column of lag 1 values for the Dow Jones average:

1 Click the **Dow Jones Avg** sheet tab to return to the data.

2 Right-click the **D** column header so that the entire column is selected and the shortcut menu opens. Click **Insert** in the shortcut menu.

3 Click cell **D1**, type **Lag1 Dow**, and press **Enter**.

4 Select the range **D3:D121** (not D2:D121).

5 Type =**E2** in cell D3 (this is the value from the previous month); then press **Enter**.

6 Click **Edit > Fill > Down** to enter the rest of the values in the lagged column.

Each row of the lagged Dow Jones average is equal to the value of the Dow one row, or one month, prior. You could have created a column of lag 2 values by selecting the range D4:D121 and letting D4 be equal to E2 and so on. Note that for the lag 2 values you have to start two rows down, as compared to one row down for the lag 1 values. The lag 3 values would have been put into the range D5:D121.

How do the values of the Dow compare to its values one month prior? To see the relationship between the Dow and the lag 1 value of the Dow, create a scatterplot.

To create a scatterplot of the Dow vs. the lagged Dow:

1 Select the range **D1:E121**.

2 Click **Insert > Chart**.

3 Select the **XY (Scatter)** chart type along with the first sub-type (points and no lines). Click **Next** twice.

4 Remove the gridlines and legends from the plot. Name the chart title **Lagged Dow Values**, the Value (X) axis title **Prior Month Value**, and the Value (Y) axis title **Dow Jones Average**. Click **Next**.

5 Send the chart to a new chart sheet named **Lag 1 Chart**. Click **Finish**.

The Lag 1 Chart appears in Figure 11-7.

Figure 11-7 Scatterplot of the lag 1 values for the Dow Jones average

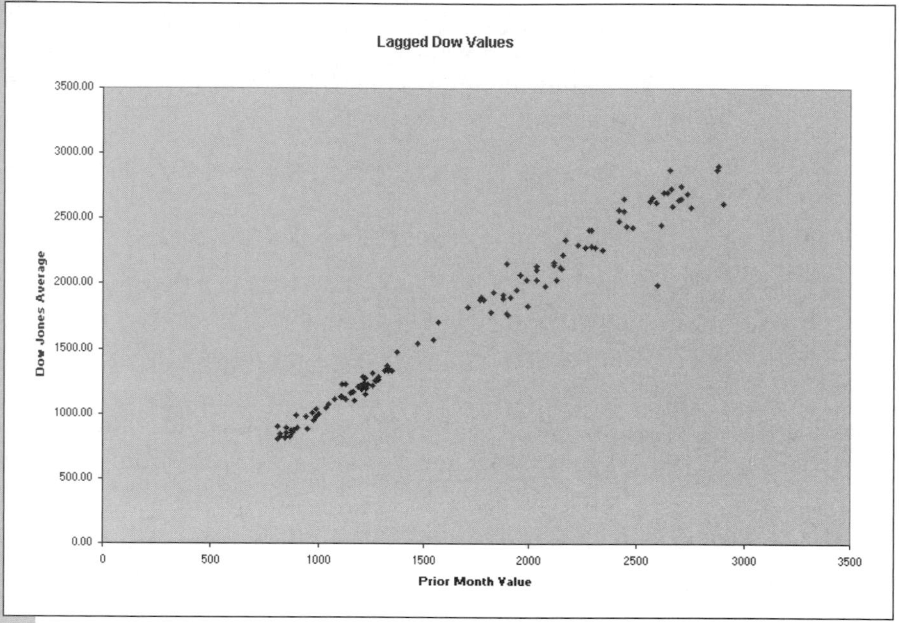

As shown in the chart, there is a strong positive relationship between the value of the Dow and the value of the Dow in the previous month. This means that a high value for the Dow in one month implies a high (or above average) value in the following month; a low value in one month indicates a low (or below average) value in the next month. In time series analysis, we study the correlations among observations, and these relationships are sometimes helpful in predicting future observations. In this example, the Dow is strongly correlated with the value of the Dow in the previous month. The Dow Jones might also be correlated with observations two, three, or more months earlier. To discover the relationship between a time series and other lagged values of the series, statisticians calculate the autocorrelation function.

The Autocorrelation Function

If there is some pattern in how the values of your time series change from observation to observation, you could use it to your advantage. Perhaps a below-average value in one month means it is more likely that the series will be high in the next month. Or maybe the opposite is true—a month in which the series is low makes it more likely that the series will continue to stay low for a while.

The **autocorrelation function** (**ACF**) is useful in finding such patterns. It is similar to a correlation of a data series with its lagged values. The ACF value for lag 1 (denoted by r_1) calculates the relationship between the data series and its lagged values. The formula for r_1 is

$$r_1 = \frac{(y_2 - \overline{y})(y_1 - \overline{y}) + (y_3 - \overline{y})(y_2 - \overline{y}) + \cdots + (y_n - \overline{y})(y_{n-1} - \overline{y})}{(y_1 - \overline{y})^2 + (y_2 - \overline{y})^2 + \cdots + (y_n - \overline{y})^2}$$

Here, y_1 represents the first observation, y_2 the second observation and so forth. Finally, Y_n represents the last observation in the data set. Similarly, the formula for r_2, the ACF value for lag 2, is

$$r_2 = \frac{(y_3 - \overline{y})(y_1 - \overline{y}) + (y_4 - \overline{y})(y_2 - \overline{y}) + \cdots + (y_n - \overline{y})(y_{n-2} - \overline{y})}{(y_1 - \overline{y})^2 + (y_2 - \overline{y})^2 + \cdots + (y_n - \overline{y})^2}$$

The general formula for calculating the autocorrelation for lag k is

$$r_k = \frac{(y_{k+1} - \overline{y})(y_1 - \overline{y}) + (y_{k+2} - \overline{y})(y_2 - \overline{y}) + \cdots + (y_n - \overline{y})(y_{n-k} - \overline{y})}{(y_1 - \overline{y})^2 + (y_2 - \overline{y})^2 + \cdots + (y_n - \overline{y})^2}$$

Before considering the autocorrelation of the Dow Jones data, let's apply these formulas to a smaller data set, as shown in Table 11-2:

Table 11-2 Sample Autocorrelation data

Observation	Values	Lag 1 Values	Lag 2 Values
1	6		
2	4	6	
3	8	4	6
4	5	8	4
5	0	5	8
6	7	0	5

The average of the values is 5, y_1 is 6, y_2 is 4, y_3 is 8, and so forth through y_n, which is equal to 7. To find the lag 1 autocorrelation, use the formula for r_1 so that

$$r_1 = \frac{(4-5)(6-5) + (8-5)(4-5) + \cdots + (7-5)(0-5)}{(6-5)^2 + (4-5)^2 + \cdots + (7-5)^2}$$

$$= \frac{-14}{40} = -0.35$$

In the same way, the value for r_2, the lag 2 ACF value, is

$$r_2 = \frac{(8-5)(6-5) + (5-5)(4-5) + \cdots + (7-5)(5-5)}{(6-5)^2 + (4-5)^2 + \cdots + (7-5)^2}$$

$$= \frac{-12}{40} = -0.30$$

The values for r_1 and r_2 imply a negative correlation between the current observation and its lag 1 and lag 2 values (that is, the previous two values). So a low value at one time point indicates high values for the next two time points. Now that you've seen how to compute r_1 and r_2, you should be able to compute r_3, the lag 3 autocorrelation. Your answer should be 0.275, a positive correlation, indicating that values of this series are positively correlated with observations three time points earlier.

Recall from earlier chapters that a constant variance is needed for statistical inference in simple regression and also for correlation. The same holds true for the autocorrelation function. The ACF can be misleading for a series with unstable variance, so it might first be necessary to transform for a constant variance before using the ACF.

Applying the ACF to the Dow Jones Average

Now apply the ACF to the Dow Jones average in the 1980s. You can use StatPlus to compute and plot the autocorrelation values for you.

To compute the autocorrelation function for the Dow Jones average:

1 Click **the Dow Jones Avg** tab.

2 Click **StatPlus > Time Series > ACF Plot**.

3 Click the **Data Values** button and select **Dow_Jones** from the range names list. Click **OK**.

4 Enter **18** in the Calculate ACF up through lag spin box to calculate the autocorrelations between the Dow and values of the Dow up to 18 months earlier.

5 Click the **Output** button, and send the output to a new sheet named **Dow Jones ACF**. Click **OK**.

6 The completed dialog box should appear as shown in Figure 11-8.

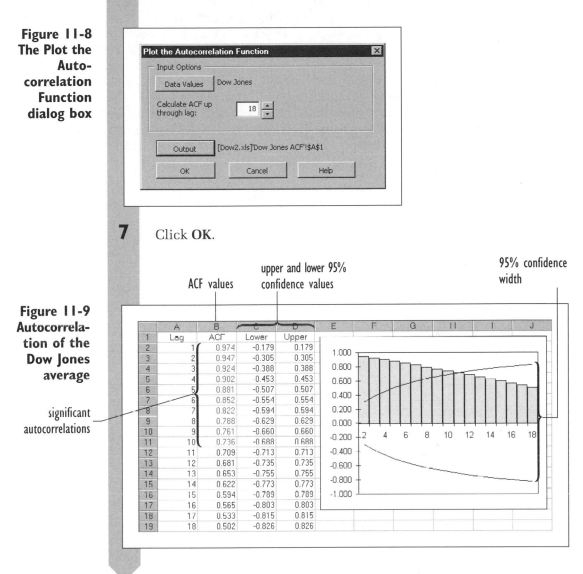

**Figure 11-8
The Plot the Auto-correlation Function dialog box**

7 Click **OK**.

**Figure 11-9
Autocorrela-tion of the Dow Jones average**

The output shown in Figure 11-9 lists the lags from 1 to 18 and gives the corresponding autocorrelations in the next column.

The lower and upper ranges of the autocorrelations are shown in the next two columns and indicate how low or high the correlation needs to be for statistical significance at the 5% level. Autocorrelations that lie outside this range are shown in red. The plot of the ACF values and confidence widths gives a visual picture of the patterns in the data. The two curves indicate the width of the 95% confidence interval of the autocorrelations.

The autocorrelations are very high for the lower lag numbers, and they remain significant (that is, they lie outside the 95% confidence width boundaries) through lag 10. Specifically, the correlation between the Dow Jones average and the value for the Dow in the previous month is 0.974 (cell B2). The correlation between the Dow and the Dow's value 12 months prior is the correlation for lag 12, 0.681 (cell B13). This is typical for a series that has a strong trend upward or downward. Given the increase in the Dow during the 1980s, it shouldn't be surprising that high values of the Dow are correlated with previous high values. In such a series, if an observation is above the mean, then its neighboring observations are also likely to be above the mean, and the autocorrelations with nearby observations are high. In fact, when there is a trend, the autocorrelations tend to remain high even for high lag numbers.

STATPLUS TIPS

- You can use StatPlus's ACF(*range*, *lag*) function to compute autocorrelations for specific lag values. Here, *range* is the range of cells containing the time series data, and *lag* is the number of observations to lag. Note that values must be placed within a single column.

Other ACF Patterns

Other time series show different types of autocorrelation patterns. Figure 11-10 shows four examples of time series (trend, cyclical, oscillating, and random), along with their associated autocorrelation functions.

Figure 11-10 Four sample time series with the ACF patterns

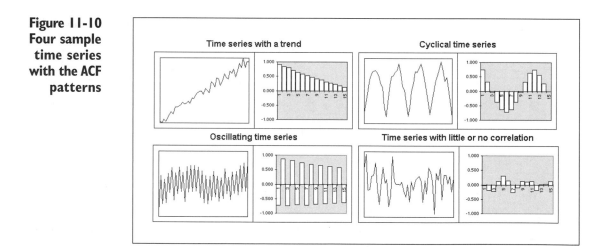

You have already seen the first example with the Dow Jones data. The trend need not be increasing; a decreasing trend also produces the type of ACF pattern shown in the first example in Figure 11-10.

The seasonal or cyclical pattern shown in the second example is common in sales data that follows a seasonal pattern (such as sales of winter apparel). The length of the cycle in this example is 12, indicated in the ACF by the large positive autocorrelation for lag 12. Because the data in the time series follows a cycle of length 12, you would expect that values 12 units apart would be highly correlated with each other. Seasonal time series models are covered more thoroughly later in this chapter.

The third example shows an oscillating time series. In this case, a large value is followed by a low value and then by another large value. An example of this might be winter and summer sales over the years for a large retail toy company. Winter sales might always be above average because of the holiday season, whereas summer sales might always be below average. This pattern of oscillating sales might continue and could follow the pattern shown in Figure 11-10. The ACF for this time series has an alternating pattern of positive and negative autocorrelations.

Finally, if the observations in the time series are independent or nearly independent, there is no discernible pattern in the ACF and all the autocorrelations should be small, as shown in the fourth example.

There are many other possible patterns of behavior for time series data besides the four examples shown here.

Applying the ACF to the Percent Change in the Dow

Having looked at the autocorrelation function for the Dow, let's look at the ACF for the percent change in the monthly average. You could argue that the percent change is what really matters in the study of the stock market. The ACF tells you whether the percent change in the Dow can be related to percent changes observed in previous months.

To calculate the autocorrelation for the percent change in the Dow:

1 Click the **Dow Jones Avg** sheet tab.

2 Click **StatPlus > Time Series > ACF Plot**.

3 Click the **Data Values** button, click the **Use Range References** option button, and select the range **G3:G121**.

4 Deselect the **Range includes a row of column labels** checkbox and click **OK**.

You want to *deselect* this checkbox because this selection does not include a header row.

5 Enter **18** in the Calculate ACF up through lag spin box.

6 Click the **Output** button, and send the output to a new sheet named **PDiff ACF**. Click **OK** twice.

	A	B	C	D	E	F	G	H	I	J
1	Lag	ACF	Lower	Upper						
2	1	0.087	-0.180	0.180						
3	2	0.003	-0.181	0.181						
4	3	-0.060	-0.181	0.181						
5	4	-0.085	-0.182	0.182						
6	5	0.105	-0.183	0.183						
7	6	-0.002	-0.185	0.185						
8	7	0.032	-0.185	0.185						
9	8	-0.110	-0.185	0.185						
10	9	-0.115	-0.187	0.187						
11	10	0.085	-0.189	0.189						
12	11	0.017	-0.191	0.191						
13	12	-0.122	-0.191	0.191						
14	13	0.032	-0.193	0.193						
15	14	-0.098	-0.193	0.193						
16	15	-0.045	-0.195	0.195						
17	16	0.034	-0.195	0.195						
18	17	-0.042	-0.195	0.195						
19	18	-0.049	-0.196	0.196						

The autocorrelations of Figure 11-11 show nothing significant; in fact, none of the autocorrelations is even close to being significant. This suggests that there is not much correlation between past and future changes in the market. This is characteristic of the **random walk model** for price movements, which says that market changes in different time periods are independent random variables with mean = 0. Past changes are useless in predicting future values.

However, the ten years of raw Dow Jones averages showed a definite upward trend, with changes in different time periods not averaging to 0. Thus the random walk model is not valid for the monthly Dow Jones averages. These ten years of past changes suggest that the future is predictable—that changes will be generally positive and that it pays to invest in the market. Although the recent past and the long-term trend over the history of the Dow Jones average have been kind to investors, the upward trend is not necessarily permanent. If the market loses its upward trend over an extended period, then it might require a change of terminology. Stock market "investors" might need to be called stock market "gamblers"!

Moving Averages

As you saw with the percent change in the Dow Jones average, time series data can fluctuate unpredictably from one observation to the next. To smooth the unpredictable ups and downs of a time series, you can form averages of consecutive observations. If you wanted to see whether the present observation is an improvement over the recent past, you can compute the average of the previous observations and compare that value to the present observation.

For example, if you calculate the average of the last six months, and you do this every month, you are forming a **moving average**. Specifically, to calculate the 6-month moving average for values prior to the observation, y_n, you define the moving average, $y_{ma(6)}$, such that

$$y_{ma(6)} = \frac{(y_{n-1} + y_{n-2} + y_{n-3} + y_{n-4} + y_{n-5} + y_{n-6})}{6}$$

The number of observations used in the moving average is called the **period**. Here the period is 6.

Excel provides the ability to add a moving average to a scatterplot using the Insert Trendline command. Let's add a 10-month moving average to the percent change in the Dow in the 1980s.

To add a moving average to a scatterplot:

1 Click the **% Change Chart** sheet tab.

2 Right-click the data series (any data value) on the chart to select it and open the shortcut menu.

3 Click **Add Trendline** in the shortcut menu.

4 Click the **Type** tab if it is not already in the forefront, click the **Moving average** trend type, and then click the **Period** up spin arrow until **10** appears as the period. Your dialog box should look like Figure 11-12.

Figure 11-12
The completed Add Trendline dialog box

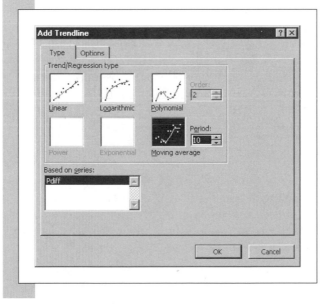

5 Click **OK**; then click outside the chart to deselect the data series. The moving-average curve appears as in Figure 11-13.

Figure 11-13
The
10-month
moving
average of
the percent
change in
the Dow

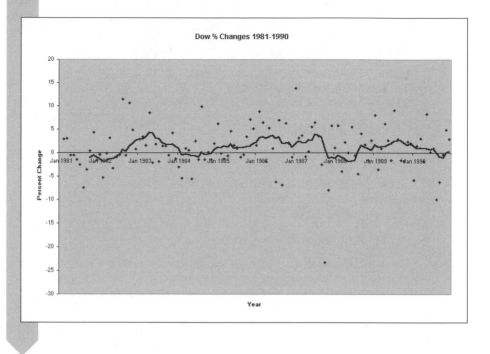

The addition of the 10-month moving average to the scatterplot smoothes the ups and downs of the data. It's clear that for much of the 1980s, the 10-month moving average of the percentage change in the Dow remained above zero. Moreover, there appear to be three periods, or phases, of sustained above-average growth: from July 1982 through January 1984, from January 1985 to September 1987, and from July 1988 to July 1990. In each of these three periods, the 10-month moving average remained above zero. The only period of sustained decline, as characterized by the moving average, is from October 1987 to October 1988, in which the moving average was constantly below zero. This is because you included the October 1987 value in the average.

EXCEL TIPS

- Excel's Analysis ToolPak also includes a command to calculate a moving-average and display the moving-average values in a chart. To run the command, click "Tools > Data Analysis" from the Excel menu and then select "Moving Average" from the list of analysis tools.

Simple Exponential Smoothing

Why choose 10 months for a moving average of the Dow? A strategy frequently used by some stock market investors is to buy if the current price of the stock exceeds the average of its ten most recent prices. Conversely, if the present price is lower than the average of the last ten, it is time to sell. In his book *Stock Market Logic*, Norman Fosback devotes Chapter 41 to the topic of moving averages. Using 10-week moving averages applied to the Standard and Poor's 500 Index, he found that the market did better when the price was above the moving average than when it was below the moving average. He compared prices three months later and also six months later. Although the differences were not great, this does lend support to the strategy of buying when the market goes above its moving average.

Although Fosback's research showed some effectiveness for 10-period moving averages, he was also critical of this method. An important point to remember with regard to using moving averages is that all prior observations within the moving-average interval are weighted equally in the calculation. That is a large part of the reason why there was a negative value for the moving average of the percentage change in the Dow after October 1987. Even in August 1988, the October 1987 drop was weighted as heavily in calculating the average as the more recent July 1988 values.

Many analysts advocate an average that gives greater weight to more recent prices, in which the value of the weights drops off exponentially. This moving average does not include only a certain number of observations but, rather, gives some weight to all observations in the past. The most recent observation gets weight w, where the value of w ranges from 0 to 1. The next most recent observation gets weight $w(1 - w)$, and the one before that gets weight $w(1 - w)^2$, and so on. In general, the weight assigned to an observation k units prior to the current observation is equal to $w(1 - w)^{k-1}$. The exponentially weighted moving average is therefore

$$\text{Exponentially weighted average } = wy_{n-1} + w(1 - w)y_{n-2} + w(1 - w)^2 y_{n-3} + \cdots$$

Here w is called a **smoothing factor** or **smoothing constant**. This technique is called **exponential smoothing** or, specifically, **one-parameter exponential smoothing**. Table 11-3 gives the weights for prior observations under different values of w.

Table 11-3 Exponential Weights

y_{n-1}	y_{n-2}	y_{n-3}	y_{n-4}	y_{n-5}	y_{n-6}
w	$w(1-w)$	$w(1-w)^2$	$w(1-w)^3$	$w(1-w)^4$	$w(1-w)^5$
0.01	0.0099	0.0098	0.0097	0.0096	0.0095
0.15	0.1275	0.1084	0.0921	0.0783	0.0666
0.45	0.2475	0.1361	0.0749	0.0412	0.0226
0.75	0.1875	0.0469	0.0117	0.0029	0.0007

As the table indicates, different values of w cause the weights assigned to previous observations to change. For example, when w equals 0.01, approximately equal weight is given to a value from the most recent observation and to values observed six units earlier. However, when w has the value of 0.75, the weight assigned to previous observations quickly drops, so that values collected six units prior to the current time receive essentially no weight. In a sense, you could say that as the value of w approaches zero, the smoothed average has a longer memory, whereas as w approaches 1, the memory of prior values becomes shorter and shorter.

Forecasting with Exponential Smoothing

Exponential smoothing is often used to forecast the value of the next observation, given the current and prior values. In this situation, you already know the value of y_n and are trying to forecast the next value, y_{n+1}. Call the forecast S_n. The formula for S_n is similar to the one we derived for the exponentially weighted moving average; it is

$$S_n = wy_n + w(1-w)y_{n-1} + w(1-w)^2 y_{n-2} + \cdots + w(1-w)^{n-1}y_1 + (1-w)^n S_0$$

S_n is more commonly written in an equivalent recursive formula, where

$$S_n = wy_n + (1-w)S_{n-1}$$

so that S_n is equal to the sum of the weighted values of the current observation and the previous forecast. Therefore, to create the forecasted value, an initial forecasted value, S_0, is required. One option is to let S_0 equal y_1, the initial observation. Another choice is to let S_0 equal the average of the first few values in the series. The examples in this chapter will use the first option, setting S_0 equal to the first value in the time series.

Once you determine the value of S_0, you can generate the exponentially smoothed values as follows:

$$S_1 = wy_1 + (1 - w)S_0$$
$$S_2 = wy_2 + (1 - w)S_1$$
$$\vdots$$
$$S_n = wy_n + (1 - w)S_{n-1}$$

and then S_n becomes the value you predict for the next observation in the time series.

Assessing the Accuracy of the Forecast

Once you generate the smoothed values, how do you measure their accuracy in forecasting values of the time series? One way is to use exponential smoothing to calculate $\hat{y}_t$, the predicted value of the time series at time t. Then, for each value in the time series, compare $\hat{y}_t$ to the observed value, y_t. The **mean square error**, or **MSE**, gives the sum of the squared differences between the forecasted values and the observed values. The formula for the MSE is

$$\text{MSE} = \frac{\sum_{t=1}^{n}(y_t - \hat{y}_t)^2}{n}$$

By comparing the MSE of one set of smoothed values to another, one can determine which set does a better job of forecasting the data.

The square root of the MSE gives us the **standard error**, which indicates the magnitude of the typical forecasted error. A standard error of 5 would indicate that the forecasts are typically off by about 5 points.

Another way of measuring the magnitude is to take the sum of the absolute values of the differences between the forecasted and observed values. This measure, called the **mean absolute deviation**, or **MAD**, has the formula

$$\text{MAD} = \frac{\sum_{t=1}^{n}\left|y_t - \hat{y}_t\right|}{n}$$

One of the differences between the MAD and the MSE is that the MAD does not penalize a forecast as much for very large errors. Because the MSE squares the deviations, large errors become even more prominent.

Another measure is the **mean absolute percent error**, or **MAPE**, which expresses the accuracy as a percentage of the observed value. The formula for the MAPE is

$$\text{MAPE} = \frac{\sum_{t=1}^{n} \left| (y_t - \hat{y}_t)/y_t \right|}{n} \times 100$$

To help you get a visual image of the impact that differing values of w have on smoothing the data and forecasting the next value in the series, you can open the instructional workbook Exponential Smoothing.

CONCEPT TUTORIALS:
One-Parameter Exponential Smoothing

To use the Exponential Smoothing workbook:

1   Open the **Exponential Smoothing** file in the Explore folder. Enable the macros in the workbook.

2   Review the contents of the workbook up to the section entitled "Explore One-Parameter Exponential Smoothing." The worksheet is shown in Figure 11-14.

**Figure 11-14
Exploring one-parameter exponential smoothing**

observed values

forecasted values

value of *w*

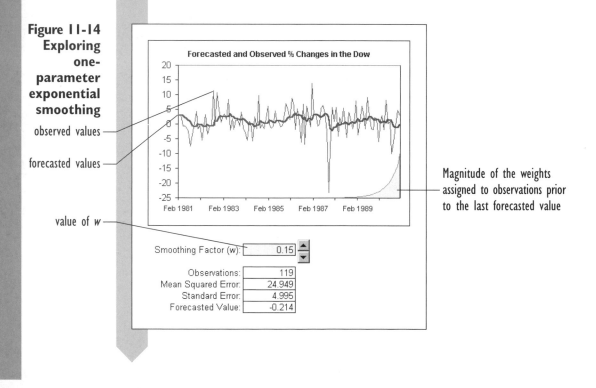

Magnitude of the weights assigned to observations prior to the last forecasted value

This worksheet shows the observed percent changes in the Dow overlaid with the one-parameter exponentially smoothed values. The smoothing factor w is set at 0.15. In the lower-right corner, the worksheet contains an area curve indicating the magnitude of the weights assigned to the observations prior to the last value in the series.

The final forecasted value is −0.214. The most recent observation has the most weight in calculating this result, with observations decreasing exponentially in importance. Comparing the curve to the time series tells you that the large drop the Dow experienced in October 1987 has little weight in estimating the final value. In fact, observations prior to February 1989 have negligible impact.

The mean square error is 24.949 and the standard error is 4.995, showing that if you had used exponential smoothing on these data throughout the 1980s, your typical error in forecasting would have been about 5 percentage points.

One way of choosing a value for the smoothing constant is to pick the value that results in the lowest mean square error. Let's see what happens to the mean square error when you decrease the value of the smoothing constant.

To decrease the value of the smoothing constant:

I Click the **down spin button** repeatedly to reduce the value of w to 0.03. The forecasted values and the weight assigned to prior observations change dramatically. See Figure 11-15.

Figure 11-15
Reducing the value of w to 0.03

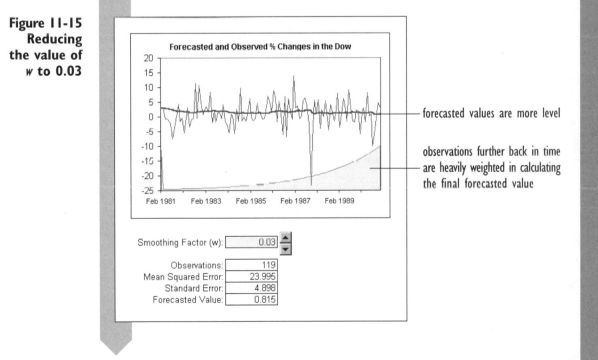

With such a small value for w, the smoothed value has a long memory. In fact, the final forecasted value, 0.815, is based in some part on observations spanning the entire 1980s. A consequence of having such a small value for w is that individual events, such as the crash of 1987, have a minor impact on the smoothed values. The line of forecasted values is practically straight. Note as well that the standard error has declined from 5.022 to 4.898. This decline shouldn't be too surprising. Recall that earlier in the chapter, you showed that there were no significant autocorrelations for the percentage change data and that, in fact, you could regard the time series values as independent. In that case, the percentage change is best estimated by the overall average or smoothed value that has a long memory.

Now increase the value of the smoothing factor to make the forecasts more susceptible to month-by-month change.

To increase the smoothing factor:

▎ Click the **up spin button** repeatedly to increase the value of w to 0.60. See Figure 11-16.

Figure 11-16 Increasing the value of w to 0.60

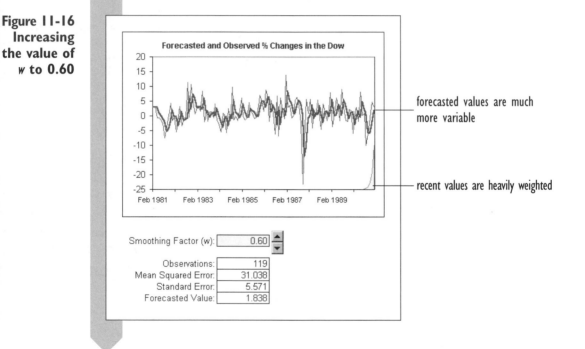

With a larger value for w, the forecasted values are much more variable—almost as variable as the observations themselves. This is because so much weight is assigned to the value in the previous month. If one month has a large upward swing, then the forecasted value for the next month tends to be high. As w approaches 1, the forecasted values appear more and more like lag 1 values.

Continue trying different values for w to see how it affects the smoothed curve and the standard error of the forecasts. Can you find a value for w that results in forecasts with the smallest standard error?

Close the Exponential Smoothing workbook without saving your changes. You'll return to the workbook later in this chapter.

Choosing a Value for w

As you saw in the Exponential Smoothing workbook, you have to choose the value of w with care. When choosing a value, keep several factors in mind. Generally, you want the standard error of the forecasts to be low, but this is not the only consideration. The value of w that gives the lowest standard error might be very high (such as 0.9) so that the exponential smoothing does not result in very smooth forecasts. If your goal is to simplify the appearance of the data or to spot general trends, you would not want to use such a high value for w, even if it produced forecasts with a low standard error. Analysts generally favor values for w ranging from 0.01 to 0.3. Fosback advocates using a value of 0.18 for w. Choosing appropriate parameter values for exponential smoothing is often based on intuition and experience. Nevertheless, exponential smoothing has proved valuable in forecasting time series data.

Using Forecasting to Guide Decisions

Recall that a strategy sometimes used by investors is to purchase a stock if the current price exceeds the average of its ten most recent prices. Let's recast this idea using exponential smoothing so that it is time to buy if the stock's price is higher than was forecasted based on one-parameter exponential smoothing, and it is time to sell if the stock's price is lower than was forecasted. The strength of the buy signal or the sell signal is based on how large a difference exists between the stock's observed value and its forecasted value. Note that the buy signal is not a forecasted value but an indicator of whether the stock's value is expected to increase in the next month. A large positive difference would strongly indicate (under this strategy) that you should buy, whereas a large negative difference would indicate that you should sell. Let's see how well this strategy would have performed in the 1980s.

The ability to perform exponential smoothing on time series data has been provided for you with StatPlus. Let's smooth the Dow Jones average from the 1980s.

To create exponentially smoothed forecasts of the Dow Jones data:

1 If necessary, reopen the Dow2 workbook, and go to the Dow Jones Avg worksheet.

2 Click **StatPlus > Time Series > Exponential Smoothing**.

3 Click the **Data Values** button and select **Dow_Jones** from the list of range names. Click **OK**.

4 Type **0.18** in the Weight box under General Options.

5 Click the **Output** button and direct the output to a new worksheet named **Smoothed Dow**. Click **OK**.

The completed dialog box appears in Figure 11-17.

**Figure 11-17
The Perform
Exponential
Smoothing
dialog box**

the value of *w* ———

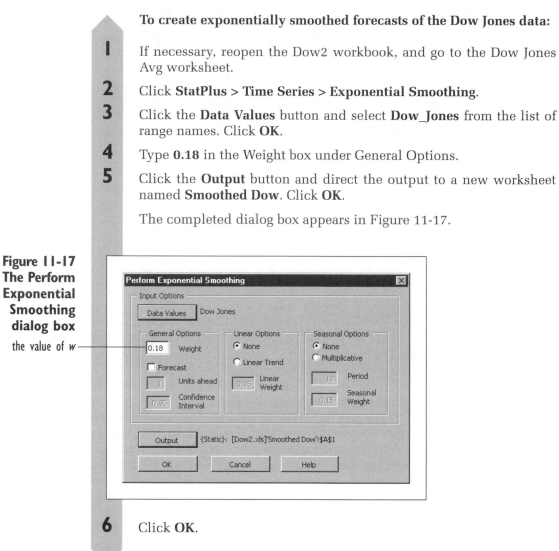

6 Click **OK**.

**Figure 11-18
Smoothed
Dow values**

monthly Dow Jones averages   smoothed forecasts   descriptive statistics

chart of observed
and forecasted
values

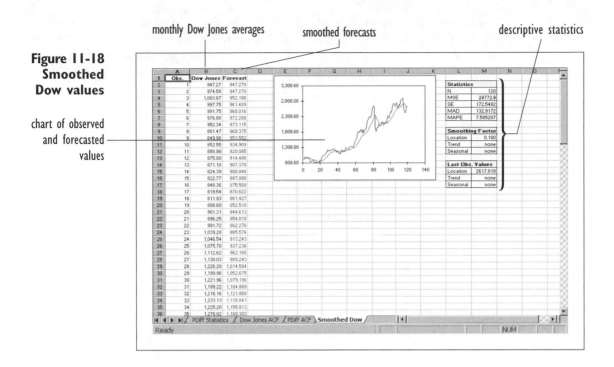

The output shown in Figure 11-18 consists of three columns: the observation number, the observed Dow Jones average, and the Dow Jones value forecasted for each month. The values are then plotted on the chart. It appears that the forecasted values generally underestimated the value of the Dow Jones throughout the 1980s. Of course, this is caused by the extraordinary growth. The standard error of the forecasts, 172.5482, indicates that the typical forecasting error was about 170 points per month. Based on the MAPE value, this represents a typical mean percent error of about 7.6%.

Now let's use the forecasted values to ascertain the reliability of our stock-purchasing rule of thumb: Buy when the Dow exceeds the forecasted value. First you'll have to copy the forecasted values into the Dow Jones data worksheet.

To copy the data:

1. Highlight the range **C1:C121** on the Smoothed Dow sheet; then click **Edit > Copy**.

2. Click the **Dow Jones Avg** sheet tab (you might have to scroll to see it).

3. Click cell **H1**; then click **Edit > Paste**.

To calculate the buy or sell signal for each month, calculate the percent difference between the forecasted value and the observed value. The larger the percent difference between the two, the stronger the signal to buy. Then you want to compare this value to the percent change in the Dow's value. You've already calculated this and placed it in column G. You will want to lag the buy signal values 1 month in order to compare the buy signal of the previous month to the actual change that was witnessed in the market.

To calculate the percent difference between the observed and forecasted values:

1 Click cell **I1** and type **Buy Signal**.

2 Select the range **I3:I121** (*not* I2:I121).

3 Type **=(E2-H2)/E2*100** and press **Enter**.

4 Click **Edit > Fill Down** to fill in the rest of the values in the column. See Figure 11-19.

Figure 11-19
The first few
values of
the buy
signal

	A	B	C	D	E	F	G	H	I	J
1	Year	Month	Year_Month	Lag1 Dow	Dow Jones	Diff	Pdiff	Forecast	Buy Signal	
2	1981	Jan	Jan 1981		947.27			947.270		
3	1981	Feb	Feb 1981	947.27	974.58	27.31	2.883022	947.270	-2.0618E-06	
4	1981	Mar	Mar 1981	974.58	1003.87	29.29	3.005397	952.186	2.80223067	
5	1981	Apr	Apr 1981	1003.87	997.75	-6.12	-0.60964	961.489	5.14849328	
6	1981	May	May 1981	997.75	991.75	-6.00	-0.60135	968.016	3.63428057	
7	1981	Jun	Jun 1981	991.75	976.88	-14.87	-1.49937	972.288	2.3931468	
8	1981	Jul	Jul 1981	976.88	952.34	-24.54	-2.51208	973.115	0.4700537	

Now let's answer the following question: Are the values of the buy signal positively correlated with the observed changes in the Dow?

To calculate the correlation:

1 Click **StatPlus > Multivariate > Correlation Matrix**.

2 Click the **Data Values** button; then click the **Use Range References** option button. Select the range **G3:G121** and then, holding down the **CTRL** key, select the range **I3:I121**.

3 Deselect the Range Include a Row of Column Labels checkbox, because this range does not include the label.

4 Click **OK**.

5 Click the **Output** button and direct the output to a new worksheet named **Buy Signal Correlation**. Click **OK** twice.

Excel generates the correlation matrix shown in Figure 11-20.

Figure 11-20
Correlation
of the buy
signal with
the percent
change in
the Dow

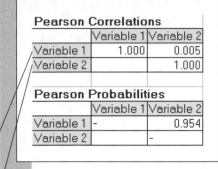

Pearson Correlations		
	Variable 1	Variable 2
Variable 1	1.000	0.005
Variable 2		1.000

Pearson Probabilities		
	Variable 1	Variable 2
Variable 1	-	0.954
Variable 2		-

percent change in
the Dow

buy signal ─

The correlation matrix indicates that there is no relationship between the buy signal and the subsequent behavior of the market in the following month. The correlation is 0.005 with a p-value of 0.954. The lack of significance is not surprising, considering the low value for the lag 1 autocorrelation shown earlier. Because there is no indication of significant correlation between the previous month's percent change value and the current percent change in the Dow, it is reasonable that a previous month's buy signal would not correlate well with the percent change either. To see this visually, create a scatterplot of the buy signal vs. the subsequent percent change in the Dow.

To create a scatterplot of percent change vs. the buy signal:

1 Return to the Dow Jones Avg worksheet.

2 Click **StatPlus > Single Variable Charts > Fast Scatterplot**.

Note: Excel's Chart Wizard expects that the variable you want to plot for the y-axis (the percent change in the Dow) is located to the left of the variable you want to plot on the x-axis (the buy signal). Rather than move the columns around or negotiate a long series of dialog boxes, we'll use the Fast Scatterplot command from StatPlus. Using this command, we can generate a quick scatterplot without worrying about the order of the columns on the worksheet.

3 Click the **x-axis** button, click the **Use Range References** option button, and select the range **I3:I121**. Deselect the Range Include a Row of Column Labels checkbox. Click **OK**.

4 Click the **y-axis** button, click the **Use Range References** option button, and select the range **G3:G121**. Again, deselect the Range Include a Row of Column Labels checkbox and click **OK**.

5 Click the **Chart Options** button and enter **Percent Change vs. Buy Signal** in the Chart Title box, **Buy Signal** in the X-axis Title box, and **Percent Change in the Dow** in the Y-axis title box. Click **OK**.

6 Click the **Output** button and direct the scatterplot to a new chart sheet named **Buy Signal Chart**.

7 Click **OK** twice.

Excel generates the chart shown in Figure 11-21.

Figure 11-21 Scatterplot of the percent change in the Dow vs. the buy signal

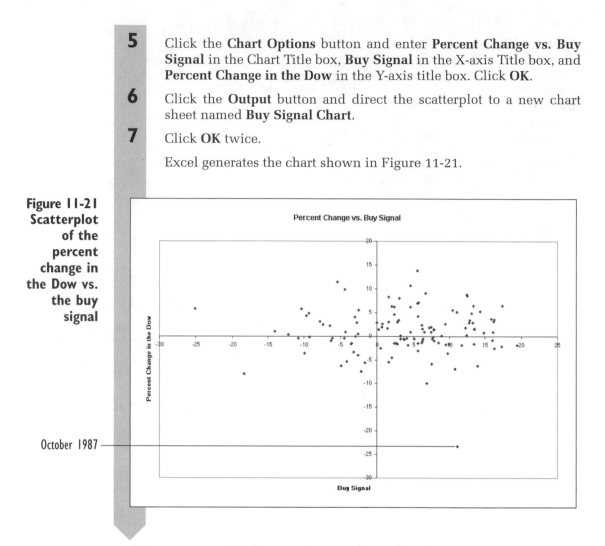

In agreement with the correlation values, the plot shows no trend at all. The correlation says there is no apparent linear relationship, and the plot says there is no apparent relationship of any kind. Notice the outlying point at the lower right, which comes from the October 1987 market crash.

The crash illustrates that buy signals based on forecasted values can be bad advice. The two most negative signals were given by the moving average in October and November of 1987, after the market had plunged. At the end of October the Dow was 15.5% below its moving average, and at the end of November the Dow was 20.1% below its moving average. Had an investor followed these sell signals and sold all holdings, it would have been a big mistake. As shown earlier in Figure 11-2, the market gained rapidly after November 1987.

Given the failure of exponential smoothing to predict short-term market activity, you might expect that no simple investment formula exists. That might indeed be the case. If it were a simple matter to predict short-term market activity, we would all be doing it!

To close the DOW2.XLS workbook and save it:

1 Click **File > Close**.

2 Click **Yes** when prompted to save your changes.

EXCEL TIPS

- Excel's Analysis ToolPak also includes a command to perform one-parameter exponential smoothing. To run the command, click "Tools > Data Analysis" from the Excel menu and then "Exponential Smoothing" from the list of analysis tools.

Two-Parameter Exponential Smoothing

The two-parameter exponential smoothing method builds on one-parameter exponential smoothing by adding a new factor—the trend, or slope, of the time series. To see how this works, let's express one-parameter exponential smoothing in terms of the following equation, where y_t is the value of the y-variable at time t:

$$y_t = \beta_0 + \varepsilon_t$$

where β_0 is the **location parameter** that changes slowly over time, and ε_t is the random error at time t. If β_0 were constant throughout time, you could estimate its value by taking the average of all the observations. Using that estimate, you would forecast values that would always be equal to your estimate of β_0. However, if β_0 varies with time, you weight the more recent observations more heavily than distant observations in any forecasts you make. Such a weighting scheme could involve exponential smoothing. How could such a situation occur in real life? Consider tracking crop yields over time. The average yield could slowly change over time as equipment or soil science technology improves. An additional factor in changing the average yield would be the weather, because a region of the country might go through several years of drought or good weather.

Now suppose the values in the time series follow a linear trend so that the series is better represented by this equation:

$$y_t = \beta_0 + \beta_1 t + \varepsilon_t$$

where β_1 is the **trend parameter**, whose value can also change over time. If β_0 and β_1 were constant throughout time, you could estimate their values using simple linear regression. However, when the values of these parameters change, you can try to estimate their values using the same smoothing techniques you used with one-parameter exponential smoothing (this approach is known as **Holt's method**). This type of smoothing estimates a line fitting the time series, with more weight given to recent data and less weight given to distant data. A county planner might use this method to forecast the growth of a suburb. The planner would not expect the rate of growth to be constant over time. When the suburb is new, it could have a very high growth rate, which might change as the area becomes saturated with people, as property taxes change, or as new community services are added. In forecasting the probable growth of the community, the planner tends to weight recent growth rates much more heavily than older ones.

Calculating the Smoothed Values

The formulas for two-parameter smoothing are very similar in form to the simple one-parameter equations. Define S_n to be the value of the location parameter for the n_{th} observation and T_n to be the trend parameter. Because we have two parameters, we also need two smoothing constants. We'll use the familiar w constant for smoothing the estimates of S_n, and we call t the smoothing constant for T_n. Using the same recursive form as was discussed with one-parameter exponential smoothing, we calculate S_n and T_n as follows:

$$S_n = wy_n + (1 - w)(S_{n-1} + T_{n-1})$$

$$T_n = t(S_n - S_{n-1}) + (1-t)T_{n-1}$$

and the formula for the forecasted value of y_{n+1} is

$$y_{n+1} = S_n + T_n$$

The values of the parameters need not be equal. Although the equations might seem complicated, the idea is fairly straightforward. The value of S_n is a weighted average of the current observation and the previous forecasted value. The value of T_n is a weighted average of the change in S_n and the previous estimate of the trend parameter. As with simple exponential smoothing, you must determine the initial values, S_0 and T_0. One method is to fit a linear regression line to the entire series and use the intercept and slope of the regression equation as initial estimates for the location and trend parameters.

CONCEPT TUTORIALS:
Two-Parameter Exponential Smoothing

The Exponential Smoothing workbook that you used earlier also contains an interactive tutorial on two-parameter exponential smoothing.

To view the Exponential Smoothing workbook:

1   Open the **Exponential Smoothing** file in the Explore folder. Enable the macros in the workbook.

2   Scroll through the workbook until you reach "Explore Two-Parameter Exponential Smoothing." See Figure 11-22.

**Figure 11-22
Exploring
two-
parameter
exponential
smoothing**

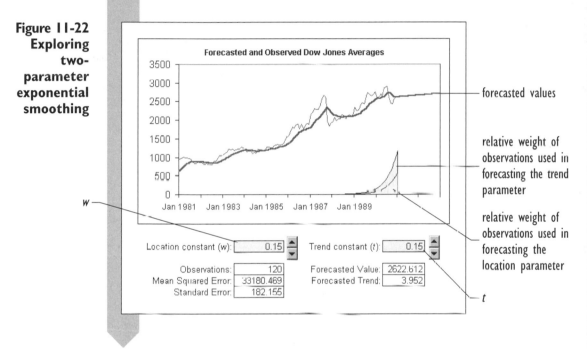

The worksheet shows the now-familiar Dow Jones average from the 1980s. The smoothing factor for the location is equal to 0.15, as is the smoothing factor for trend. The area curves at the bottom of the chart indicate the relative weights assigned to previous values in calculating the final forecast for the location and trend parameters. For t and w equal to 0.15, the most prominent observations range from January 1991 back to January 1989. Weights for observations prior to 1989 are too small to be visible on the chart. Based on the two-parameter exponential smoothing estimates, the next value of the Dow is projected to be about 2622, increasing at a rate of 3.95 points per month.

The values chosen for *w* and *t* are important in determining what the forecasted value for the Dow Jones average will be. If an investor assumes that the market will behave pretty much as it did during the previous ten years, smaller values for *w* and *t* might be used, because those would result in an estimate that has a longer "memory" of previous values. Let's see what kind of difference this would make by reducing the value of *t* from 0.15 to 0.05.

To reduce the value of *t*:

1. Repeatedly click the **down spin arrow** next to the Trend constant until the value of *t* equals **0.05**. See Figure 11-23.

**Figure 11-23
Decreasing
the value of
t to 0.05**

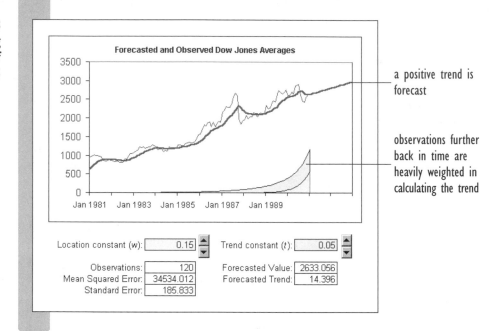

With this value of *t*, the expected increase in the Dow rises to 14.40 points per month, reflecting the belief that there will be an increase in the Dow similar to what was observed during much of the previous ten years. Note that the weights for the trend factor, as shown by the area curve, indicate that observations as far back as January 1985 are well represented in the forecast. Now let's see what would happen if we increased the value of *t*, focusing more on short-term trends.

To increase the value of *t*:

Repeatedly click the **up spin arrow** next to the Trend constant until the value of *t* equals **0.40**. See Figure 11-24.

**Figure 11-24
Increasing
the value of
t to 0.40**

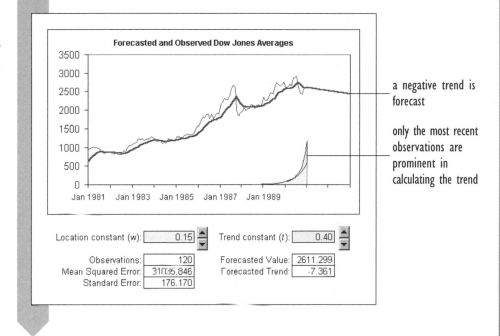

Forecasted and Observed Dow Jones Averages

a negative trend is forecast

only the most recent observations are prominent in calculating the trend

Location constant (w):	0.15		Trend constant (*t*):	0.40
Observations:	120		Forecasted Value:	2611.299
Mean Squared Error:	31035.846		Forecasted Trend:	-7.361
Standard Error:	176.170			

With a higher smoothing constant, the forecasted trend of the Dow shows a decline of 7.36 points per month. The area curve indicates that the smoothed trend estimate has a shorter memory—only the previous 12 months seem relevant in estimating the trend.

Which smoothing factor value you use might depend on what type of investment you're considering. If you are a long-term investor, then averaging trend estimates over a longer period of time might be more indicative of future performance. However, for short-term investments, the behavior of the market eight years earlier might not be indicative of future short-term prospects. Of course, as you've seen in this chapter, it is very difficult to forecast short-term changes in any case.

Using this worksheet, you can change the values of the smoothing constant for the location and trend parameters. What combinations result in the lowest values for the standard error? When you are finished with your investigations, close the workbook. You do not have to save any of your changes.

Seasonality

Often time series are measured on a seasonal basis, such as monthly or quarterly. If the data are sales of ice cream, toys, or electric power, there is a pattern that repeats each year. Ice cream and electric power sales are high in the summer, and toy sales are high in December.

Multiplicative Seasonality

If the sales of some of your products are seasonal, you might want to adjust your sales for the seasonal effect, in order to compare figures from month to month. To compare November and December sales, should you use the difference of the values or the ratio? Recall that the Dow Jones data had a month-to-month variation that grew as the series itself grew, but the variation of the percent change per month was relatively stable. Similarly, in many cases seasonal changes are best expressed in ratios, especially if there is substantive growth in yearly sales.

As annual sales increase, the difference between the November and December values should also increase, but the ratio of sales between the two months might remain nearly constant. This is called **multiplicative seasonality**. To quantify the effect of the season on each month's value, we need to assign a multiplicative factor to each month. If the month's sales are equal to the expected yearly average, we'll give it a multiplicative factor of 1. Consequently, months with higher-than-average sales have multiplicative factors greater than 1, and months with lower-than-average sales have multiplicative factors less than 1.

As an example, consider Table 11-4, which shows seasonal sales and multiplicative factors.

Table 11-4 Multiplicative Seasonality

	Jan	Feb	Mar	Apr	May	Jun	Jul	Aug	Sep	Oct	Nov	Dec
Sales	220	310	359	443	374	660	1030	1320	1594	1093	950	610
Factor	0.48	0.58	0.60	0.69	0.59	1.00	1.48	1.69	1.99	1.29	1.02	0.59
Adjusted Sales	458.3	534.5	598.3	642.0	633.9	660.0	695.9	781.1	801.0	847.3	931.4	1033.9

The monthly sales figures are shown in the first row of the table. The multiplicative factors based on previous years' sales are shown in the second row. Dividing the sales in each month by the multiplicative factor yields the adjusted sales. Plotting the sales values and adjusted sales values in Figure 11-25 reveals that sales have been steadily increasing throughout the year. This information is masked in the raw sales data by the seasonal effects.

Figure 11-25
Plot of
adjusted
sales data

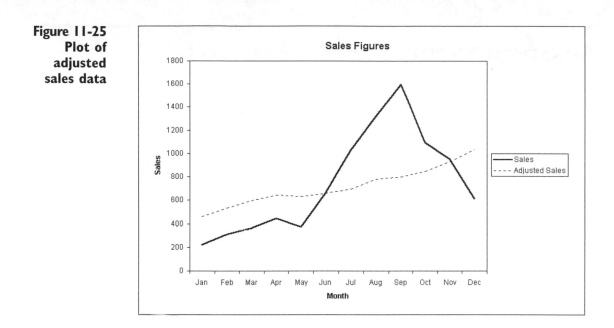

Additive Seasonality

Sometimes the seasonal variation is expressed in additive terms, especially if there is not much growth. If the highest annual sales total is no more than twice the lowest annual sales total, it probably does not matter whether you use differences or ratios. If you can express the month-to-month changes in additive terms, the seasonal variation is called **additive seasonality**. Additive seasonality is expressed in terms of differences from the expected average for the year. In Table 11-5, the seasonal adjustment for December sales is −240, resulting in an adjusted sales for that month of 681. After adjustment for the time of the year, December turned out to be one of the most successful months, at least in terms of exceeding goals.

Table 11-5 Additive Seasonality

	Jan	Feb	Mar	Apr	May	Jun	Jul	Aug	Sep	Oct	Nov	Dec
Sales	298	378	373	443	374	660	1004	1153	1388	904	715	441
Factor	−325	−270	−270	−200	−280	−55	350	450	550	220	70	−240
Adjusted Sales	623	648	643	643	654	715	654	703	838	684	645	681

In this chapter you'll work with multiplicative seasonality only, but you should be aware of the principles of additive seasonality.

Seasonal Example: Beer Production

Is beer production seasonal? The BEER.XLS workbook has U.S. production figures for each month, 1980 through 1991, in thousands of barrels. The workbook contains the variables and reference names shown in Table 11-6.

Table 11-6 The Beer Workbook

Range Name	Range	Description
Year	A2:A145	The year
Month	B2:B145	The month
Year_Month	C2:C145	The year and the month
Barrels	D2:D145	The monthly production of beer in thousands of barrels

To open the Beer workbook:

1. Open the **Beer** workbook from the folder containing your student files.
2. Click **File > Save As** and save the workbook as **Beer2**. The workbook appears as shown in Figure 11-26.

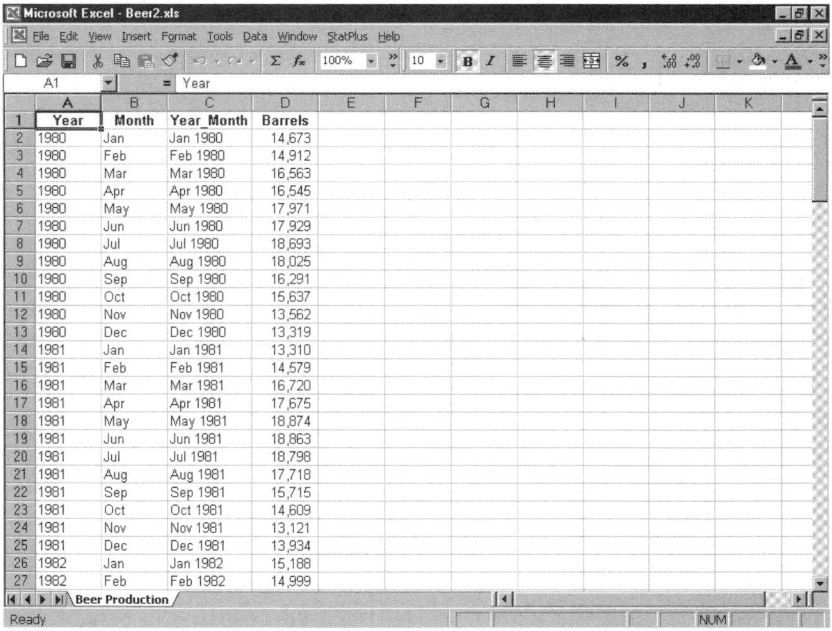

	Year	Month	Year_Month	Barrels
2	1980	Jan	Jan 1980	14,673
3	1980	Feb	Feb 1980	14,912
4	1980	Mar	Mar 1980	16,563
5	1980	Apr	Apr 1980	16,545
6	1980	May	May 1980	17,971
7	1980	Jun	Jun 1980	17,929
8	1980	Jul	Jul 1980	18,693
9	1980	Aug	Aug 1980	18,025
10	1980	Sep	Sep 1980	16,291
11	1980	Oct	Oct 1980	15,637
12	1980	Nov	Nov 1980	13,562
13	1980	Dec	Dec 1980	13,319
14	1981	Jan	Jan 1981	13,310
15	1981	Feb	Feb 1981	14,579
16	1981	Mar	Mar 1981	16,720
17	1981	Apr	Apr 1981	17,675
18	1981	May	May 1981	18,874
19	1981	Jun	Jun 1981	18,863
20	1981	Jul	Jul 1981	18,798
21	1981	Aug	Aug 1981	17,718
22	1981	Sep	Sep 1981	15,715
23	1981	Oct	Oct 1981	14,609
24	1981	Nov	Nov 1981	13,121
25	1981	Dec	Dec 1981	13,934
26	1982	Jan	Jan 1982	15,188
27	1982	Feb	Feb 1982	14,999

Figure 11-26 The Beer workbook

The figures for the number of barrels produced come from pages 15–17 of the *Brewers Almanac 1992*, published by the Beer Institute in Washington, D.C. As a first step in analyzing these data, create a line plot of barrels vs. year and month.

To create a time series plot:

1 Select the range **C1:D145** and click **Insert > Chart**.

2 Select the **Line** chart type and the first chart sub-type (lines with no symbols) and click **Next** twice.

3 Enter the chart title **Beer Production 1980-1991**, enter **Year** for the Category (X) axis title, and enter **Barrels** for the Value (Y) axis title. Remove the gridlines and legend from the plot and click **Next**.

4 Save the chart to a new chart sheet named **Beer Chart** and click **Finish**.

5 Rescale the *y*-axis to cover a range from 12,000 to 20,000 barrels.

6 Rescale the *x*-axis so that there are 12 categories between the tick marks and the tick-mark labels.

Figure 11-27 shows the final plot.

**Figure 11-27
Beer
production
in the 1980s**

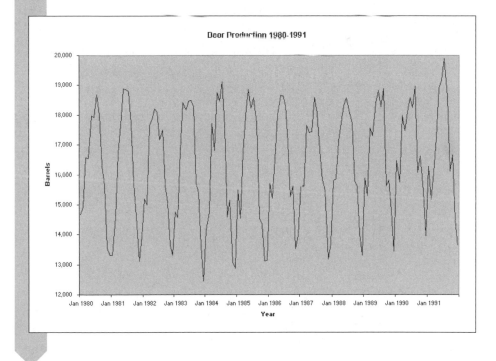

The plot shows that production is seasonal, with peaks occurring each summer. Winter production is much lower. Usually, the minimum each year is about 13,000 barrels in the winter, and the maximum is about 18,000 or 19,000 barrels in the summer. That is, production is about 40% higher in the summer.

Examining Seasonality with a Boxplot

One way to see the seasonal variation is to make a boxplot, with a box for each of the 12 months. This gives you a picture of the month-to-month variation in beer production. The shape of each box tells you how production for that month varied throughout the 1980s.

To create the boxplot:

1 Click the **Beer Production** sheet tab.

2 Click **StatPlus > Single Variable Charts > Boxplots**.

3 Click the **Connect Medians between Boxes** checkbox.

4 Click the **Data Values** button and select **Barrels** from the range names list. Click **OK**.

5 Click the **Categories** button and select **Month** from the list of range names. Click **OK**.

6 Click the **Output** button and direct the plot to a new chart sheet named **Beer Boxplot**. Click **OK** twice.

7 Rescale the y-axis to go from 12,000 to 20,000 barrels.

8 Edit the labels at the bottom of the boxplot, removing the "Month =" text from each.

Figure 11-28
Boxplot of
monthly
barrel
production

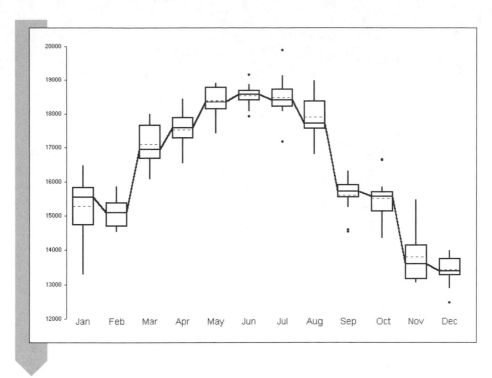

The boxplot in Figure 11-28 clearly shows the effect of seasonality. The only month that seems to depart from the trend is February, which appears to be about 10% lower relative to the trend from January to March. However, February has about 10% fewer days than January and March, so it is reasonable for February to be low. The boxplot also indicates the range of production levels for each month. January, for example, is one of the most variable months in terms of beer production. How might this fact affect any forecasts you would make regarding beer production in January?

Examining Seasonality with a Line Plot

You can also take advantage of the two-way table to create a line plot of barrels vs. month for each year of the data set. This is another way to get insight into the monthly distribution of barrel production during the 1980s. You will first have to create a two-way table of the data in the Beer Production worksheet.

To create a line plot of beer production:

1 Return to the Beer Production worksheet.

2 Click **StatPlus > Manipulate Columns > Create Two-way Table**.

3   Select **Barrels** for the Data Values variable, **Month** for the Column Levels, and **Year** for the Row Levels.

4   Deselect the **Sort the Column Levels** checkbox.

Note: You want to deselect this checkbox to prevent the two-way table from sorting the columns in alphabetical order, rather than leaving them in time order.

5   Send the two-way table to a new worksheet named **Production by Year**.

6   Click **OK**.

7   Go to the Production by Year worksheet.

8   Select the range **B2:L14** and click **Insert > Chart**. Select the **Line** chart type and the first sub-type (lines without symbols). Click the **Next** button twice.

9   Enter **Beer Production vs. Months** for the chart title, **Months** for the Category title, and **Barrels** for the Value title. Remove the legend and gridlines from the plot. Click **Next**.

10   Rescale the *y*-axis to range from 12,000 to 20,000 barrels.

11   Send the plot to a new chart sheet named **Beer Line Plot** and click **Finish**.

**Figure 11-29
Line plot of
monthly
barrel
production**

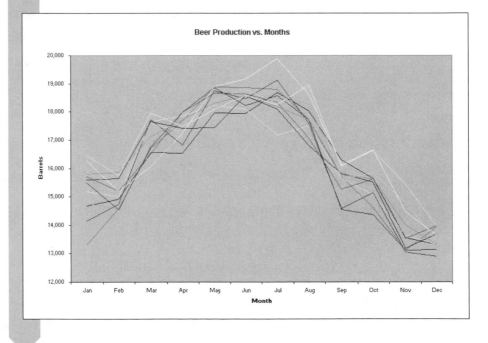

The line plot in Figure 11-29 demonstrates the seasonal nature of the data and also allows you to observe individual values. Plots like this are sometimes called **spaghetti plots**, for obvious reasons.

Applying the ACF to Seasonal Data

You can also use the autocorrelation function to display the seasonality of the data. For a seasonal monthly series, the ACF should be very high at lag 12, because the current value should be strongly correlated with the value from the same month in the previous year.

To calculate the ACF:

1. Return to the Beer Production worksheet.
2. Click **StatPlus > Time Series > ACF Plot**.
3. Select **Barrels** for the Data Values variable.
4. Click the **up spin arrow** to calculate the ACF up through a lag of **24**.
5. Send the output to a new worksheet named **Beer ACF**.
6. Click **OK**.

Figure 11-30 ACF for the beer production data

	A	B	C	D
1	Lag	ACF	Lower	Upper
2	1	0.723	-0.163	0.163
3	2	0.433	-0.234	0.234
4	3	0.013	-0.254	0.254
5	4	-0.460	-0.254	0.254
6	5	-0.693	-0.275	0.275
7	6	-0.828	-0.319	0.319
8	7	-0.659	-0.372	0.372
9	8	-0.402	-0.402	0.402
10	9	0.045	-0.412	0.412
11	10	0.438	-0.412	0.412
12	11	0.670	-0.425	0.425
13	12	0.858	-0.452	0.452
14	13	0.630	-0.493	0.493
15	14	0.376	-0.514	0.514
16	15	0.001	-0.522	0.522
17	16	-0.409	-0.522	0.522
18	17	-0.609	-0.530	0.530
19	18	-0.731	-0.549	0.549
20	19	-0.597	-0.574	0.574
21	20	-0.374	-0.590	0.590
22	21	0.018	-0.597	0.597
23	22	0.379	-0.597	0.597
24	23	0.597	-0.603	0.603
25	24	0.765	-0.619	0.619

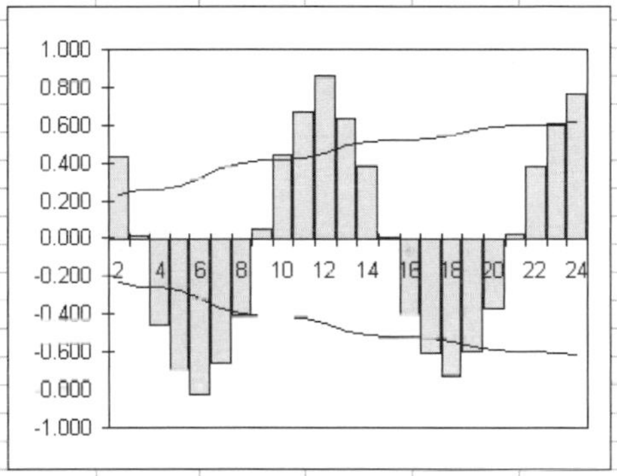

The plot shown in Figure 11-30 indicates strong seasonality, because the lag 6 autocorrelation is strongly negative and the lag 12 autocorrelation is strongly positive. Notice that these correlations are both bigger in absolute value than the lag 1 autocorrelation. Thus, to predict this month's figure, you would do better to use the value from a year ago rather than last month's value.

Adjusting for Seasonality

Because the beer production data have a strong seasonal component, it would be useful to adjust the values for the seasonal effect. In this way you can determine whether a drop in production during one month is due to seasonal effects or is a true decline. Adjusting the production data for seasonality also gives you a better indication of whether a trend exists for the data. You can use StatPlus to adjust time series data for multiplicative seasonality.

To adjust the beer production data:

1 Return to the Beer Production worksheet.

2 Click **StatPlus > Time Series > Seasonal Adjustment**.

3 Click the **Data Values** button and select **Barrels** from the range names list. Click **OK**.

4 Verify that a period of length "12" is entered into the Length of Period box.

5 Click the **Output** button and send the output to a new worksheet named **Adjusted Production**. Click **OK**.

Your completed dialog box should look like Figure 11-31.

**Figure 11-31
The Perform
Seasonal
Adjustment
dialog box**

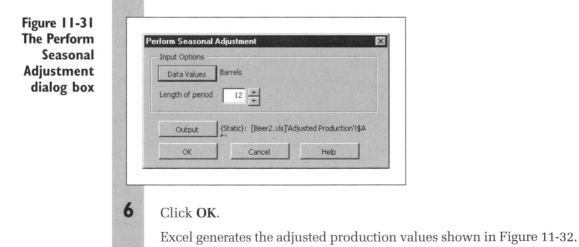

6 Click **OK**.

Excel generates the adjusted production values shown in Figure 11-32.

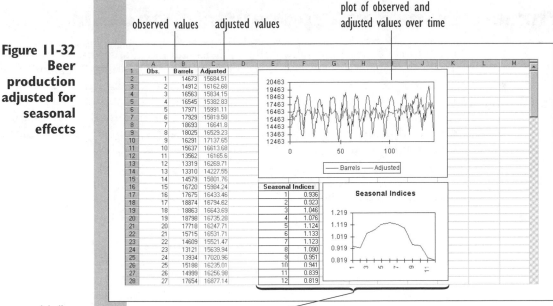

Figure 11-32 Beer production adjusted for seasonal effects

observed values   adjusted values   plot of observed and adjusted values over time

seasonal indices

The observed production levels are shown in column B, and the seasonally adjusted values are shown in column C. Using the adjusted values can give you some insight into the production levels. For example, between observations 8 and 9 (corresponding to August and September 1980) the production level drops 1,734 units from 18,025 units to 16,291. However, when you adjust these figures for the time of the year, the production level actually increases 608 units from 16,529.23 units to 17,137.65. From this you conclude that although production dropped in September 1980, the decline in production was less than would have been predicted for September on the basis of the usual seasonal levels.

You can get some idea of the relative sales for different months of the year from the table of seasonal indices. For example, the seasonal index for June is 1.133 and for May it is 1.124. This indicates that you can expect a percent increase in beer production of $(1.133 - 1.124) / 1.124 = 0.00848$, or 0.8%, going from May to June each year. Seasonal indices for the multiplicative model must add up to the length of the period—in this case, 12. You can use this information to tell you that 9.45% of the beer produced each year is produced in June (because $1.133 / 12 = 0.0945$). A line plot of the seasonal indices is provided and shows a profile very similar to the one you saw earlier with the boxplot.

A scatterplot is also included, showing both the production data and the adjusted production values. The adjusted values do not appear to show any kind of trend, but the chart is not completely clear. To explore this question further, you can use three-parameter exponential smoothing.

Three-Parameter Exponential Smoothing

You perform exponential smoothing on seasonal data using three smoothing constants. This process is known as **three-parameter exponential smoothing** or **Winters' method**. The smoothing constants in the Winters' method involve location, trend, and seasonality. Winters' method can be used for either multiplicative or additive seasonality, though in this text, we'll assume only multiplicative seasonality. The equation for a time series variable y_t with a multiplicative seasonality adjustment is

$$y_t = (\beta_0 + \beta_1 t) \times I_p + \varepsilon_t$$

and for additive seasonality adjustment the equation is

$$y_t = (\beta_0 + \beta_1 t) + I_p + \varepsilon_t$$

In these equations β_0, β_1, and ε_t once again represent the location, trend, and error parameters of the model, and I_p represents the seasonal index at point p in the season. For example, if we used the multiplicative seasonal indices shown in Figure 11-32, I_5 would equal 1.124. Once again, these parameters are not considered to be constant but can vary with time. The beer production data might be an example of such a series. The production values are seasonal, but there might also be a time trend to the data such that production levels increase from year to year after adjusting for seasonality.

Let's concentrate on smoothing with a multiplicative seasonality factor. The smoothing equations used in three-parameter exponential smoothing are similar to equations you've already seen. For the smoothed location value S_n and the smoothed trend value T_n, from a time series where the length of the period is p, the recursive equations are

$$S_n = w \frac{y_n}{I_{n-p}} + (1 - w)(S_{n-1} + T_{n-1})$$
$$T_n = t(S_n - S_{n-1}) + (1 - t)T_{n-1}$$

Note that the recursive equation for S_n is identical to the equation used in two-parameter exponential smoothing except that the current observation y_n must be seasonally adjusted. Here, I_{n-p} is the seasonal index taken from the index values of the previous period. The recursive equation for T_n is identical to the recursive equation for the two-parameter model.

As you would expect, three-parameter smoothing also smooths out the values of the seasonal indices, because these might also change over time. We use a different smoothing constant, c, for these indices. The recursive equation for a seasonal index I_p is

$$I_p = c \frac{y_n}{S_n} + (1 - c)I_{n-p}$$

The value of the smoothed seasonal index is a weighted average of the current seasonal index based on the values of y_n and S_n, and the index value from the previous period. Calculating initial estimates of S_0, T_0, and I_p is beyond the scope of this book.

Forecasting Beer Production

Let's use exponential smoothing to predict beer production. For the purposes of demonstration, we'll assume a multiplicative model. You will have to decide on values for each of the three smoothing constants. The values need not be the same. For example, seasonal adjustments often change more slowly than the trend and location factors, so you might want to choose a low value for the seasonal smoothing constant, say about 0.05. However, if you feel that the trend factor or location factor will change more rapidly over the course of time, you will want a higher value for the smoothing constant, such as 0.15. As you have seen in this chapter, the values you choose for these smoothing constants depend in part on your experience with the data. Excel does not provide a feature to do smoothing with the Winters' method. One has been provided for you with the Exponential Smoothing command found in StatPlus.

To forecast values of beer production:

1 Return to the Beer Production worksheet.

2 Click **StatPlus > Time Series > Exponential Smoothing**.

3 Select **Barrels** for your Data Values variable.

4 Enter **0.15** in the General Options Weight box. This is your value for w.

5 Click the **Linear Trend** option button, and enter **0.15** in the Linear Weight box. This is your value for t.

6 Click the **Multiplicative** option button, and enter **0.05** in the Seasonal Weight box. This is the value of c. Verify that the length of the period is set to **12**.

7 Click the **Forecast** checkbox and enter **12** in the Units ahead box. This will forecast the next year of beer production.

8 Verify that **0.95** is entered in the Confidence Interval box. This will produce a 95% confidence region around your forecasted values.

9 Click the **Output** button and direct your output to a new sheet named **Forecasted Values**. Click **OK**.

Your dialog box should look like Figure 11-33.

Figure 11-33
The Perform
Exponential
Smoothing
dialog box

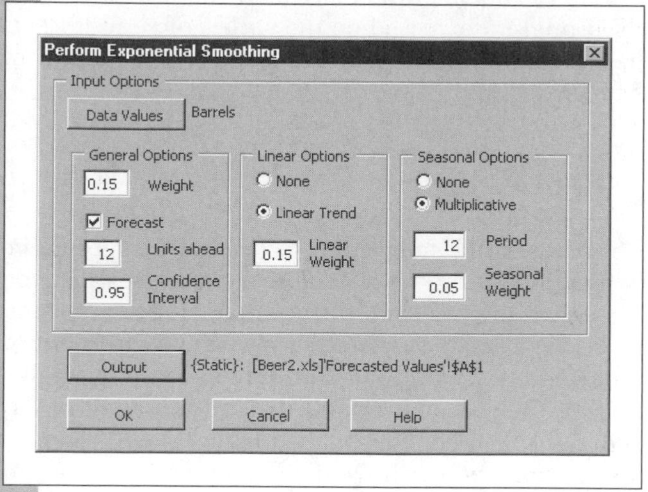

10 Click **OK**.

Output from the command appears on the Forecasted Values work-sheet. To view the forecasted values, drag the vertical scroll bar down. See Figure 11-34.

Figure 11-34
Forecasted
production
values with
the 95%
confidence
region

	A	B	C	D
147	Obs.	Forecast	Lower	Upper
148	145	15,936.9	14,860.1	17,013.6
149	146	15,732.3	14,643.0	16,821.5
150	147	17,794.9	16,692.3	18,897.4
151	148	18,263.5	17,146.9	19,380.1
152	149	19,164.0	18,032.7	20,295.4
153	150	19,337.6	18,190.8	20,484.4
154	151	19,291.2	18,128.3	20,454.2
155	152	18,695.2	17,515.5	19,874.9
156	153	16,278.8	15,081.6	17,475.9
157	154	16,207.3	14,992.2	17,422.5
158	155	14,423.5	13,189.9	15,657.2
159	156	14,020.1	12,767.4	15,272.8

The output does not give the month for each forecast, but you can easily confirm that observation 145 in column A is January 1992 because observation 144 on the Beer Production worksheet is December 1991. On the basis of the values in column B, you forecast that in the next year, beer production will reach a peak in June 1992 (observation 150) with about 19,338 units (a unit being 1,000 barrels). The 95% prediction interval for this estimate is about 18,191 to 20,484 units. In other words, in June 1992 you would not expect to produce fewer than 18,191 or more than 20,484 units. You could use these estimates to plan your production strategy for the upcoming year. Before putting much faith in the prediction intervals, you should verify the assumptions for the smoothed forecasts. If the smoothing model is correct, the residuals should be independent (show no discernible ACF pattern) and follow a normal distribution with mean 0. You would find for the beer data that these assumptions are met.

Scrolling back up the worksheet, you can view how well the smoothing method forecasted the beer production in the previous 12 years, as shown in Figure 11-35.

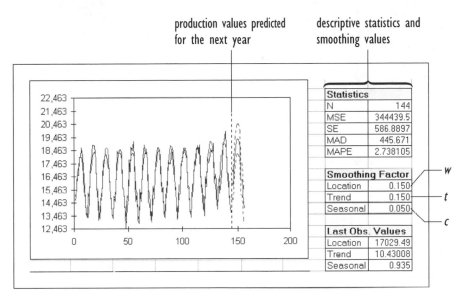

Figure 11-35 Output from the three-parameter exponential smoothing equation

The standard error of the forecast is 586.8897, indicating that the typical forecasting error in the time series is about 600 units (the MAD value is lower with a value of about 446 units). According to the MAPE, this represents an average error of about 2.7%.

The final estimates for the location and trend values are 17,029.49 and 10.43, respectively. The location value represents the monthly production value after adjusting for seasonality effects. The trend estimate indicates that production is increasing at a rate of about 10.43 units per month—hardly a large increase given the magnitude of the monthly production.

The output also includes a scatterplot comparing the observed, smoothed, and forecasted production values. Because of the number of points in the time series, the seasonal curves are close together and difficult to interpret. To make it easier to view the comparison between the observed and forecasted values, rescale the x-axis to show only the current year and the forecasted year's values.

To rescale the x-axis:

1 Double-click the x-axis scale in the scatterplot.

2 Click the **Scale** tab, type **130** in the Minimum text box, and click **OK**.

The revised plot appears in Figure 11-36.

**Figure 11-36
Plot of
forecasted
and
observed
values for
the current
year and the
next year**

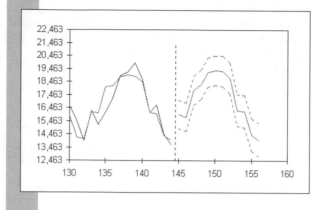

From the rescaled plot, you would conclude that exponential smoothing has done a good job of modeling the production data and that the resulting forecasts appear reasonable.

The final part of the exponential smoothing output is the final estimate of the seasonal indices, shown in Figure 11-37.

**Figure 11-37
Seasonal
indices**

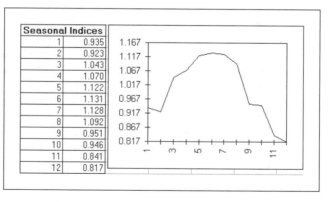

The values for the seasonal indices are very similar to those you calculated using the seasonal adjustment command. The difference is due to the fact that these seasonal indices are calculated using a smoothed average, whereas the earlier indices were calculated using an unweighted average.

You're finished with the workbook. You can close it now, saving your changes.

Optimizing the Exponential Smoothing Constant (optional)

As you've seen in this chapter, the choice for the value of the exponential smoothing constant depends partly on the analyst's experience and intuition. Many analysts advocate using the value that minimizes the mean square error. You can use an Excel add-in called the "Solver" to calculate this value. To demonstrate how this technique works, open the file EXPSOLVE.XLS, which contains the monthly percentage changes in the Dow that you worked with earlier in this chapter.

To open the EXPSOLVE workbook:

1 Open **Expsolve.xls** from the folder containing your student files.

2 Click **File > Save As** and save the workbook as **Exp2**.

The workbook displays the column of percent differences. Let's create a column of exponentially smoothed forecasts. First you must decide on an initial estimate for the smoothing constant w; you can start with any value you want, so let's start with 0.15. From this value, we'll calculate the mean squared error.

To calculate the mean squared error:

1 Click cell **F1**, type **0.15**, and then press **Enter**.

Next determine a value for S_0 to be put in cell C2. We'll use the first value in the time series.

2 Click cell **C2**, type **=B2**, and then press **Enter**.

Now create a column of smoothed forecasts, S_n, using the recursive smoothing equation.

3 Select the range **C3:C120**.

4 Type **=$F$1*B2+(1-$F$1)*C2**; then press **Enter**.

5 Click **Edit > Fill > Down** to fill in the rest of the forecasts.

Now create a column of squared errors [(forecast − observed)^2].

6 Select the range **D2:D120**.

7 Type **=(C2-B2)^2**, and then press **Enter**.

8 Click **Edit > Fill > Down**.

Finally, calculate the mean squared error for this particular value of *w*.

9 Click cell **F2**, type **=SUM(D2:D120)/119**, and then press **Enter**.

10 Verify that the values in your spreadsheet match the values in Figure 11-38.

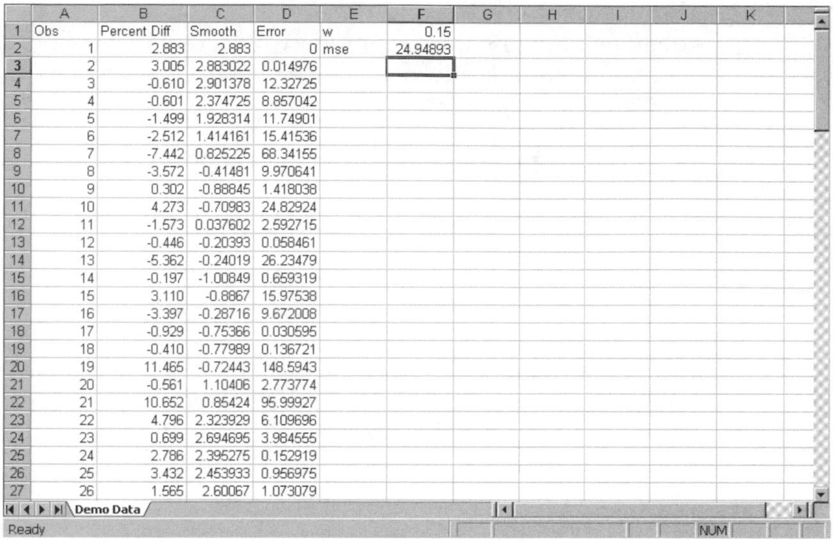

Figure 11-38
Exp2
workbook
values

	A	B	C	D	E	F	G	H	I	J	K
1	Obs	Percent Diff	Smooth	Error	w	0.15					
2	1	2.883	2.883	0	mse	24.94893					
3	2	3.005	2.883022	0.014976							
4	3	-0.610	2.901378	12.32725							
5	4	-0.601	2.374725	8.857042							
6	5	-1.499	1.928314	11.74901							
7	6	-2.512	1.414161	15.41536							
8	7	-7.442	0.825225	68.34155							
9	8	-3.572	-0.41481	9.970641							
10	9	0.302	-0.88845	1.418038							
11	10	4.273	-0.70983	24.82924							
12	11	-1.573	0.037602	2.592715							
13	12	-0.446	-0.20393	0.058461							
14	13	-5.362	-0.24019	26.23479							
15	14	-0.197	-1.00849	0.659319							
16	15	3.110	-0.8867	15.97538							
17	16	-3.397	-0.28716	9.672008							
18	17	-0.929	-0.75366	0.030595							
19	18	-0.410	-0.77989	0.136721							
20	19	11.465	-0.72443	148.5943							
21	20	-0.561	1.10406	2.773774							
22	21	10.652	0.85424	95.99927							
23	22	4.796	2.323929	6.109696							
24	23	0.699	2.694695	3.984555							
25	24	2.786	2.395275	0.152919							
26	25	3.432	2.453933	0.956975							
27	26	1.565	2.60067	1.073079							

Demo Data

Ready — NUM

You now have everything you need to use Solver.

To open Solver:

1 Click **Tools > Add-Ins**.

2 If the **Solver Add-In** checkbox appears in the list box, click it to select it if necessary; then click **OK**. If the checkbox does not appear, then you must install the Solver. Check your Excel User's Guide for information on installing the Solver from your original Excel installation disks.

Once the Solver is installed, you can determine the optimal value for the smoothing constant.

To determine the optimal value for the smoothing constant:

1 Click **Tools > Solver**.

2 Type **F2** in the Set Target Cell text box. This is the cell that you will use as a target for the Solver.

3 Click the **Min** option button to indicate that you want to minimize the value of the mean squared error (cell F2).

4 Type **F1** in the By Changing Cells text box to indicate that you want to change the value of F1, the smoothing constant, in order to minimize cell F2.

Because the exponential smoothing constant can take on only values between 0 and 1, you have to add some constraints to the values that the Solver will investigate.

5 Click the **Add** button.

6 Type **F1** in the Cell Reference text box, select **<=** from the Constraint drop-down list, type **1** in the Constraint text box, and then click **Add**.

7 Type **F1** in the Cell Reference text box, select **>=** from the Constraint drop-down list, type **0** in the Constraint text box, and then click **Add**.

8 Click **Cancel** to return to the Solver Parameters dialog text box. The completed Solver Parameters dialog box should look like Figure 11-39.

Figure 11-39
The Solver
Parameters
dialog box

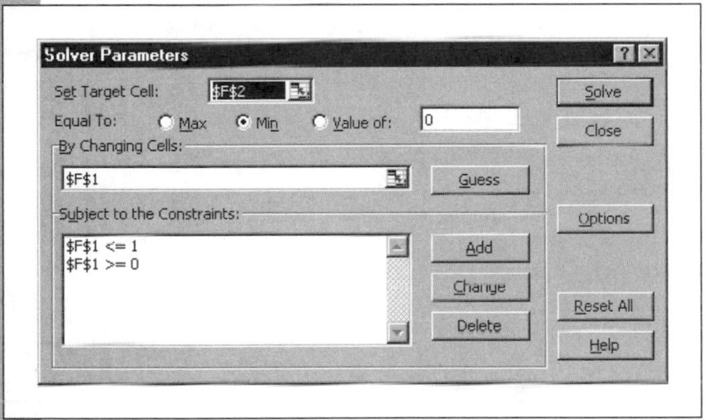

9 Click **Solve**.

The Solver now determines the optimal value for the smoothing constant (at least in terms of minimizing the mean squared error). When the Solver is finished, it will prompt you either to keep the Solver solution or to restore the original values.

10 Click **OK** to keep the solution.

The Solver returns a value of 0.028792 (cell F1) for the smoothing constant, resulting in a mean squared error of 23.99456 (cell F2). This is the optimal value for the smoothing constant.

It's possible to set up similar spreadsheets for two-parameter and three-parameter exponential smoothing, but that will not be demonstrated here. The main difficulty in setting up the spreadsheet to do these calculations is in determining the initial estimates of S_0, T_0, and the seasonal indices.

In the case of two-parameter exponential smoothing, you would use linear regression on the entire time series to derive initial estimates for the location and trend values. Once this is done, you would derive the forecasted values using the recursive equations described earlier in the chapter. You would then apply the Solver to minimize the mean squared error of the forecasts by modifying both the location and the trend smoothing constants. Using the Solver to derive the best smoothing constants for the three-parameter model is more complicated because you have to come up with initial estimates for all of the seasonal indices. The interested student can refer to more advanced texts for techniques to calculate the initial estimates.

You can now save and close the Exp2 workbook.

Exercises

1. Do the following calculations for one-parameter exponential smoothing, where $w = 0.10$.
   a. $S_4 = 23.4$, $y_5 = 29$. What is S_5?
   b. If the observed value of y_6 is 25, what is the value of S_6? Assume the same values from 1a.

2. Do the following calculation for two-parameter exponential smoothing, where $w = 0.10$ and $t = 0.20$.
   a. $S_4 = 23.4$, $T_4 = 1.1$, and $y_5 = 29$. What is S_5? What is T_5?
   b. If the observed value of y_6 is 25, what are the values of S_6 and T_6? Assume the same values from 2a.

3. If monthly sales are equal to 4,811 units and the seasonal index for that month is 0.85, what is the adjusted sales figure?

4. How can you tell whether a series is seasonal? Mention plots, including the ACF. What is the difference between additive and multiplicative seasonality?

5. A politician citing the latest raw monthly unemployment figures claimed that unemployment had fallen by 88,000 workers. The Bureau of Labor Statistics, however, using seasonally adjusted totals, claimed that unemployment had increased by 98,000. Discuss the two interpretations of the data. Which number gives a better indication of the state of the economy?

6. The BBAVER.XLS workbook contains data on the leading major league base-ball batting averages for the years 1901 to 1991. Open the workbook and:
   a. Create a line chart of the batting average vs. year. Do you see any apparent trends? Do you see any outliers? Does George Brett's average of 0.390 in 1980 stand out compared with other observations?
   b. Insert a trend line smoothing the batting average using a 10-year moving average.
   c. Calculate the ACF and state your conclusions (notice that the ACF does not drop off to zero right away, which suggests a trend component).
   d. Calculate the difference of the BBAVER.XLS averages from one year to the next. Plot the difference series and also compute its ACF. Does the plot show that the variance of the original series is reasonably stable? That is, are the changes roughly the same size at the beginning, middle, and end of the series?
   e. Looking at the ACF of the differenced series, do you see much correlation after the first few lags? If not, it suggests that the differenced series does not have a trend, and this is what you would expect. Interpret any lags that are significantly correlated.
   f. Perform one-parameter exponential smoothing forecasting one year ahead, using w values of 0.2, 0.3, 0.4, and 0.5. In each case, notice the value predicted for 1992 (observation 92). Which parameter gives the lowest standard error?
   g. An almanac shows that the actual highest batting average of 1992 is 0.343. Compare the predictions with the actual value. Does the parameter with the minimum standard error also give the best prediction for 1992?
   h. Report your results, and save your workbook as E11BBAVER.XLS.

7. The ELECTRIC.XLS workbook has monthly data on U.S. electric power production, 1978 through 1990. The variable called power is measured in billions of kilowatt hours. The figures come from the 1992 *CRB Commodity Year Book*, published by the Commodity Research Bureau in New York. Open the workbook and:
   a. Create a line chart of the power data. Is there any seasonality to the data?
   b. Fit a three-parameter exponential model with location, linear, and seasonal parameters. Use a smoothing constant of 0.05 for the location parameter, 0.15 for the linear parameter, and 0.05 for the seasonal parameter. What level of power production do you forecast over the next 12 months?
   c. Using the seasonal index, which are the three months of highest power production? Is this in accordance with the plots you have seen? Does it make sense to you as a consumer? By what percentage does the busiest month exceed the slowest month?

d. Repeat the exponential smoothing of part b of Exercise 6 with the smoothing constants shown in Table 11-7.

Table 11-7 Exponential Smoothing Constants

Location	Linear	Seasonal
0.05	0.30	0.05
0.15	0.15	0.05
0.15	0.30	0.05
0.30	0.15	0.05
0.30	0.30	0.05

e. Which forecasts give the smallest standard error?

f. Report your conclusions, saving the workbook as E11ELECTRIC.XLS.

8. The VISIT.XLS workbook contains monthly visitation data for two sites at the Kenai Fjords National Park in Alaska from January 1990 to June 1994. You'll analyze the visitation data for the Exit Glacier site.

a. Create a line plot of visitation for Exit Glacier vs. year and month. Summarize the pattern of visitation at Exit Glacier between 1990 and mid-1994.

b. Create two line plots, one showing the visitation at Exit Glacier plotted against year with different lines for different months, and the second showing visitation plotted against month with different lines for different years (you will have to create a two-way table for this). Are there any unusual values? How might the June 1994 data influence future visitation forecasts?

c. Calculate the seasonally adjusted values for visits to the park. Is there a particular month in which visits to the park jump to a new and higher level?

d. Smooth the visitation data using exponential smoothing. Use smoothing constants of 0.15 for both the location and the linear parameter, and use 0.05 for the seasonal parameter. Forecast the visitation 12 months into the future. What are the projected values for the next 12 months?

e. A lot of weight of the projected visitations for 1994–1995 is based on the jump in visitation in June 1994. Assume that this jump was an aberration, and refit two exponential smoothing models with 0.05 and 0.01 for the location parameter (to reduce the effect of the June 1994 increase), 0.15 for the linear parameter, and 0.05 for the seasonal parameter. Compare your results with your first forecasts. How do the standard errors compare? Which projections would you work with and why? What further information would you need to decide between these three projections?

f. What problems do you see with either forecasted value? (*Hint*: Look at the confidence intervals for the forecasts.)

g. Save your workbook as E11VISIT.XLS.

9. The visitation data in the VISIT.XLS workbook cover a wide range of values. It might be appropriate to analyze the $\log_{10}$ of the visitation counts instead of the raw counts.

   a. Reopen the VISIT.XLS workbook and create a new column in the workbook of the $\log_{10}$ counts of the Exit Glacier data (use the Excel function $\log_{10}$).

   b. Create a line plot of $\log_{10}$ (visitation) for the Exit Glacier site from 1990 to mid-1994. What seasonal values does this chart reveal that were hidden when you charted the raw counts?

   c. Use exponential smoothing to smooth the $\log_{10}$ (visitation) data. Use a value of 0.15 for the location and linear effects, and use 0.05 for the seasonal effect. Project $\log_{10}$ (visitation) 12 months into the future. Untransform the projections and the prediction intervals by raising 10 to the power of $\log_{10}$ (visitation) [that is, if $\log_{10}$ (visitation) = 1.6, then visitation = $10^{1.6}$ = 39.8]. What do you project for the next year at Exit Glacier? What are the 95% prediction intervals? Are the upper and lower limits reasonable?

   d. Redo your forecasts, using 0.01 and then 0.05 for the location parameter, 0.15 for the linear parameter, and 0.05 for the seasonal parameter. Which of the three projections results in the smallest standard error?

   e. Compare your chosen projections from Exercise 8, using the raw counts, with your chosen projections from this exercise, using the $\log_{10}$-transformed counts. Which would you use to project the 1994–1995 visitations? Which would you use to determine the amount of personnel you will need in the winter months and why?

   f. Save your workbook as E11VISIT2.XLS.

10. The workbook NFP.XLS contains daily body temperature data for 239 consecutive days for a woman in her twenties. Daily temperature readings are one component of natural family planning (NFP) in which a woman uses her monthly cycle with a number of biological signs to determine the onset of ovulation. The file has four columns: Observation, Period (the menstrual period), Day (the day of the menstrual period), and Waking Temperature. Day 1 is the first day of menstruation. Open the workbook and:

   a. Create a line plot of the daily body temperature values. Do you see any evidence of seasonality in the data?

   b. Create a boxplot of temperature vs. day. What can you determine about the relationship between body temperature and the onset of menstruation?

   c. Calculate the ACF for the temperature data up through lag 70. On the basis of the shape of the ACF, what would you estimate as the length of the period in days?

   d. Smooth the data using exponential smoothing. Use 0.15 as the location parameter, 0.01 for the linear parameter (it will not be important in this model), and 0.05 for the seasonal parameter. Use the period length that you estimated in part c of Exercise 9. What body temperature values do you forecast for the next cycle?

e. Repeat your forecast with values of 0.15 and 0.25 for the seasonal parameters. Which model has the lowest standard error?

f. Save your workbook as E11NFP.XLS.

11. The DRAFT.XLS workbook contains data from the 1970 Selective Service draft. Each birth date was given a draft number. Those eligible men with a low draft number were drafted first. One way of presenting the draft number data is through exponential smoothing. The draft numbers vary greatly from day to day, but by smoothing the data, one may be better able to spot trends in the draft numbers. In this exercise, you'll use exponential smoothing to examine the distribution of the draft numbers. Open the workbook and:

a. Create one-parameter exponential smoothed plots of the number variable on the Draft Numbers worksheet. Use values of 0.15, 0.085, and 0.05 for the location parameter. Which value results in the lowest mean square error?

b. Examine your plots. Does there appear to be any sort of pattern in the smoothed data?

c. Test to see whether any autocorrelation exists in the draft numbers. Test for autocorrelation up to a lag of 30. Is there any evidence for autocorrelation in the time series?

d. Summarize your conclusions and save your workbook as E11DRAFT.XLS.

12. The OIL.XLS workbook displays information on monthly production of crude cottonseed oil from 1992 to 1995. The production of cottonseed oil follows a seasonal pattern. Using the data in this workbook, project the monthly values for 1996. Open the OIL.XLS workbook.

a. Restructure the data in the worksheet into a two-way table. Create a line plot of the production values in the table using a separate line for each year. Describe the seasonal nature of cottonseed oil production.

b. Smooth the production data using a value of 0.15 for all three smoothing factors. Forecast the values 12 months into the future. What are your projections and your upper and lower limits for 1996?

c. Adjust the production data for the seasonal effects. Is there evidence that the adjusted production values have increased over the 4-year period? Test your assumption by performing a linear regression of the adjusted values on the month number (1–48). Is the regression significant at the 5% level?

d. Summarize your results, saving your workbook as E11OIL.XLS.

13. The Bureau of Labor Statistics maintains data on consumer prices. The HOUSE.XLS workbook on your Student disk displays the monthly values of the index for housing values from 1967 to 1996. Open this workbook and use the information it contains to project values for the index in 1997.

a. Use a three-parameter exponential smoothing model to smooth the data. Use 0.15 for all three smoothing parameters. Forecast index values 12 months into future with a 95% prediction interval.

b. What are the forecasted values for 1997? What are the lower and upper prediction boundaries?

c. Rescale the chart in the Exponential Smoothing output so that the x-axis goes from observation 348 to 372 and the house values in the y-axis go from 145 to 165.

d. Study the seasonal index values. Which month has the lowest index value? Which month has the highest? On the basis of the index, which time of the year seems to be better for buying a house?

e. Summarize your results and save the workbook as E11HOUSE.XLS.

14. The Bureau of Labor Statistics records the number of work stoppages each month that involve 1,000 or more workers in the period. Are such work stoppages seasonal in nature? Are there more work stoppages in summer than in winter? Information on work stoppages has been stored in the WORK.XLS on your Student disk. Open the workbook and complete the following.

a. Restructure the data in the Work Stoppage worksheet into a two-way table, with each year in a separate row and each month in a separate column.

b. Use the two-way table to create a boxplot and line plot of the work stoppage values. Which months have the highest work stoppage numbers? Do work stoppages occur more often in winter or in summer?

c. Adjust the work stoppage values assuming a 12-month cycle. Is there evidence in the scatterplot that the adjusted number of work stoppages has decreased over the past decade?

d. Smooth adjusted values using one-parameter exponential smoothing. Use a value of 0.15 for the smoothing parameter.

e. Summarize your findings regarding work stoppages of 1,000 or more workers. Are they seasonal? Have they declined in recent years? Use whatever charts and tables you created to support your conclusions. Save your workbook as E11WORK.XLS.

15. The JOBS.XLS workbook contains monthly youth unemployment rates from 1981 to 1996. Analyze the data in the workbook and try to determine whether unemployment rates are seasonal. Open the JOBS.XLS workbook and:

a. Restructure the data in the Youth Unemployment worksheet into a two-way table, with each year in a separate row and each month in a separate column.

b. Create a spaghetti plot of the unemployment values.

c. Create a boxplot of youth unemployment rates. Is any pattern apparent in the boxplot?

d. Adjust the unemployment rates assuming a 12-month cycle. Is there evidence in the chart that youth unemployment varies with the season?

e. Report your results, saving the workbook as E11JOBS.XLS.

Chapter 12

Quality Control

Objectives

In this chapter you will learn to:

- Distinguish between controlled and uncontrolled variation

- Distinguish between variables and attributes

- Determine control limits for several types of control charts

- Use graphics to create statistical control charts with Excel

- Interpret control charts

- Create a Pareto chart

In this chapter you will look at one of the statistical tools used in manufacturing and industry. The proper use of quality control can improve productivity, enhance quality, and reduce production costs. In this chapter, you'll learn about one such tool, the control chart, that is used to determine when a process is out of control and requires human intervention.

Statistical Quality Control

The immediately preceding chapters have been dedicated to the identification of relationships and patterns among variables. Such relationships are not immediately obvious, mainly because they are never exact for individual observations. There is always some sort of variation that obscures the true association. In some instances, once the relationship has been identified, an understanding of the types and sources of variation becomes critical. This is especially true in business, where people are interested in controlling the variation of a process. A **process** is any activity that takes a set of inputs and creates a product. The process for an industrial plant takes raw materials and creates a finished product. A process need not be industrial. For example, another type of process might be to take unorganized information and produce an organized analysis. Teaching could even be considered a process, because the teacher takes uninformed students and produces students capable of understanding a subject (such as statistics!). In all such processes, people are interested in controlling the procedure so as to improve quality. The analysis of processes for this purpose is called **statistical quality control** (**SQC**) or **statistical process control** (**SPC**).

Statistical process control originated in 1924 with Walter A. Shewhart, a researcher for Bell Telephone. A certain Bell product was being manufactured with great variation in quality, and the production managers could not seem to reduce the variation to an acceptable level. Dr. Shewhart developed the rudimentary tools of statistical process control to improve the homogeneity of Bell's output. Shewhart's ideas were later championed and refined by W. Edwards Deming, who tried unsuccessfully to persuade American firms to implement SPC as a methodology underlying all production processes. Having failed to convince U.S. executives of the merits of SPC, Deming took his cause to Japan, which, before World War II, was renowned for its shoddy goods. The Japanese adopted SPC wholeheartedly, and Japanese production became synonymous with high and uniform quality. In response, American firms jumped on the SPC bandwagon, and many of their products regained market share.

Controlled Variation

The reduction of variation in any process is beneficial. However, you can never eliminate all variation, even in the simplest process, because there are bound to be many small, unobservable, chance effects that influence the process outcome. Variation of this kind is called **controlled variation** and is analogous to the random-error effects in the ANOVA and regression models you studied earlier. As in those statistical models, many individually insignificant random factors interact to have some net effect on the process output. In quality-control terminology, this random variation is said to be "in control," not because the process operator is able to control the factors absolutely, but rather because the variation is the result of normal disturbances, called **common causes**, within the process. This type of variation can be predicted. In other words, given the limitations of the process, each of these common causes is controlled to the greatest extent possible.

Because controlled variation is the result of small variations in the normally functioning process, it cannot be reduced unless the entire process is redesigned. Furthermore, any attempts to reduce the controlled variation without redesigning the process will create more, not less, variation in the process. Endeavoring to reduce controlled variation is called **tampering**; this increases costs and must be avoided. Tampering might occur, for instance, when operators adjust machinery in response to normal variations in the production process. Because normal variations will always occur, adjusting the machine is more likely to harm the process, actually *increasing* the variation in the process, than to help it.

Uncontrolled Variation

The other type of variation that can occur within a process is called uncontrolled variation. **Uncontrolled variation** is due to **special causes**, which are sources of variation that arise sporadically and for reasons outside the normally functioning process. Variation induced by a special cause is usually significant in magnitude and occurs only occasionally. Examples of special causes include differences between machines, different skill or concentration levels of workers, changes in atmospheric conditions, and variation in the quality of inputs.

Unlike controlled variation, uncontrolled variation can be reduced by eliminating its special cause. The failure to bring uncontrolled variation into control is costly.

SPC is a methodology for distinguishing whether variation is controlled or uncontrolled. If variation is controlled, then only improvements in the process itself can reduce it. If variation is uncontrolled, then further analysis is needed to identify and eliminate the special cause.

Table 12-1 summarizes the two types of variation studied in SPC.

Table 12-1 Types of Variation

Variation	Descriptive	Remedy
Controlled	Variation that is native to the process, resulting from normal factors called "common causes"	Redesign the process to result in a new set of controlled variations with better properties.
Uncontrolled	Variation that is the result of "special causes" and need not be inherent in the process	Analyze the process to locate the source of the uncontrolled variation and then remove or fix that special cause.

Control Charts

The principal tool of SPC is the control chart. A **control chart** is a graph of the process values plotted in time order. Figure 12-1 shows a sample control chart.

Figure 12-1
A control chart

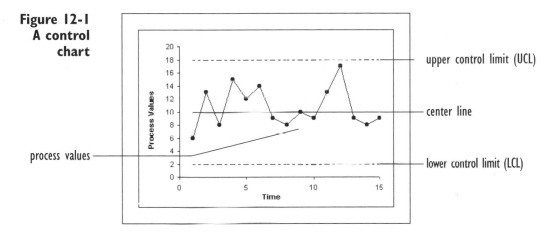

The chief features of the control chart are the **lower** and **upper control limits** (**LCL** and **UCL**, respectively), which appear as dotted horizontal lines. The solid line between the upper and lower control limits is the **center line** and indicates the expected values of the process.

As the process goes forward, values are added to the control chart. As long as the points remain between the lower and upper control limits, we assume that the observed variation is controlled variation and that the process is **in control** (there are a few exceptions to this rule, which we'll discuss shortly). Figure 12-1 shows a process that is in control. It is important to note that control limits do not represent specification limits or maximum variation targets. Rather, control limits illustrate the limits of normal controlled variation.

In contrast, the process depicted in Figure 12-2 is **out of control**. Both the fourth and the twelfth observations lie outside of the control limits, leading us to believe that their values are the result of uncontrolled variation. At this point a shop manager, or the person responsible for the process, might examine the conditions for those observations that resulted in such extreme values. An analysis of the causes could lead to a better, more efficient, and more stable process.

**Figure 12-2
A process
not in
control**

process value
exceeding control
limits

process value
below control
limits

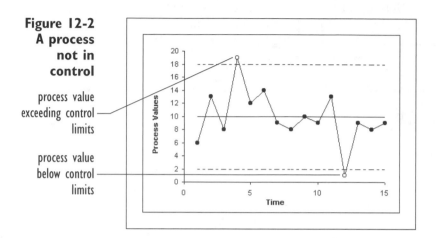

Even control charts in which all points lie between the control limits might suggest that a process is out of control. In particular, the existence of a pattern in eight or more consecutive points indicates a process out of control, because an obvious pattern violates the assumption of random variability. In Figure 12-3, for example, the last eight observations depict a steady upward trend. Even though all of the points lie within the control limits, you must conclude that this process is out of control because of the evident trend the data values exhibit.

**Figure 12-3
Process out
of control
because of
an upward
trend**

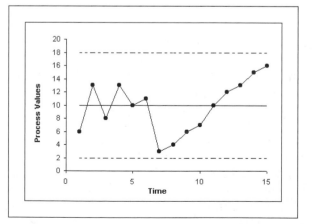

Another common example of a process that is out of control, even though all points lie between the control limits, appears in Figure 12-4. The first eight observations are below the center line, whereas the second seven observations all lie above the center line. Because of prolonged periods where values are either small or large, this process is out of control. One could use the Runs test, discussed in Chapter 8 in the context of examining residuals, to test whether the data values are clustered in a nonrandom way.

Figure 12-4
Process out of control because of a nonrandom pattern

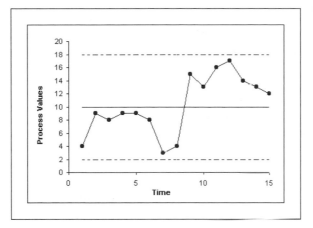

Here are two other situations that may show a process out of control, even though all values lie within the control limits:

- 9 points in a row, all on the same side of the center line
- 14 points in a row, alternating above and below the center line

Other suspicious patterns could appear in control charts. Unfortunately, we cannot discuss them all here. In general, though, any clear pattern in the process values indicates that a process is subject to uncontrolled variation and that it is not in control.

Statisticians usually highlight out-of-control points in control charts by circling them. As you can see, the control chart makes it very easy for you to identify visually points and processes that are out of control without using complicated statistical tests. This makes the control chart an ideal tool for the shop floor, where quick and easy methods are needed.

Control Charts and Hypothesis Testing

The idea underlying control charts should be familiar to you. It is closely related to confidence intervals and hypothesis testing. The associated null hypothesis is that the process is in control; you reject this null hypothesis if any point lies outside the control limits or if any clear pattern appears in the distribution of the process values. Another insight from this analogy is that the possibility of making errors exists, just as errors can occur in standard hypothesis testing. In other words, occasionally a point that lies outside the control limits does not have any special cause but occurs because of normal process variation. On the other hand, there could exist a special cause that is not big enough to move the point outside of the control limits. Statistical analysis can never be 100% certain.

Variable and Attribute Charts

There are two categories of control charts: those that monitor variables and those that monitor attributes. **Variable charts** display continuous measures, such as weight, diameter, thickness, purity, and temperature. As you have probably already noticed, much statistical analysis focuses on the mean values of such measures. In a process that is in control, you expect the mean output of the process to be stable over time.

    Attribute charts differ from variable charts in that they describe a feature of the process rather than a continuous variable such as a weight or volume. Attributes can be either discrete quantities, such as the number of defects in a sample, or proportions, such as the percentage of defects per lot. Accident and safety rates are also typical examples of attributes.

Using Subgroups

In order to compare process levels at various points in time, we usually group individual observations together into **subgroups**. The purpose of the subgroup is to create a set of observations in which the process is relatively stable with controlled variation. Thus the subgroup should represent a set of homogeneous conditions. For example, if we were measuring the results of a manufacturing process, we might create a subgroup consisting of values from the same machine closely spaced in time. Once we create the subgroups, we can calculate the subgroup averages and calculate the variance of the values. The variation of the process values within the subgroups is then used to calculate the control limits for the entire set of process values. A control chart might then answer the question "Do the averages *between* the subgroups vary more than expected, given the variation *within* the subgroups?"

The $\bar{x}$-Chart

One of the most common variable control charts is the **$\bar{x}$-chart** (the "x-bar chart"). Each point in the $\bar{x}$-chart displays the subgroup average against the subgroup number. Because observations usually are taken at regular time intervals, the subgroup number is typically a variable that measures time, with subgroup 2 occurring after subgroup 1 and before subgroup 3. As an example, consider a clothing store in which the owner monitors the length of time customers wait to be served. He decides to calculate the average wait-time in half-hour increments. The first half-hour (for instance, customers who were served between 9 a.m. and 9:30 a.m.) forms the first subgroup, and the owner records the average wait-time during this interval. The second subgroup covers the time from 9:30 a.m. to 10:00 a.m., and so forth.

The $\bar{x}$-chart is based on the standard normal distribution. The standard normal distribution underlies the mean chart, because the Central Limit Theorem (see Chapter 5) states that the subgroup averages approximately follow the normal distribution even when the underlying observations are not normally distributed.

The applicability of the normal distribution allows the control limits to be calculated very easily when the standard deviation of the process is known. You might recall from Chapter 5 that 99.74% of the observations in a normal distribution fall within 3 standard deviations of the mean (μ). In SPC, this means that points that fall more than 3 standard deviations from the mean occur only 0.26% of the time. Because this probability is so small, points outside the control limits are assumed to be the result of uncontrolled special causes. Why not narrow the control limits to ±2 standard deviations? The problem with this approach is that you might increase the **false-alarm rate**—that is, the number of times you stop a process that you incorrectly believed was out of control. Stopping a process can be expensive, and adjusting a process that doesn't need adjusting might increase the variability through tampering. For this reason, a 3-standard-deviation control limit was chosen as a balance between running an out-of-control process and incorrectly stopping a process when it doesn't need to be stopped.

You might also recall that the statistical tests you learned earlier in the book differed slightly depending on whether the population standard deviation was known or unknown. An analogous situation occurs with control charts. The two possibilities are considered in the following sections.

Calculating Control Limits When σ Is Known

If the true standard deviation of the process (σ) is known, then the control limits are

$$LCL = \mu - \frac{3\sigma}{\sqrt{n}}$$

$$UCL = \mu + \frac{3\sigma}{\sqrt{n}}$$

and 99.74% of the points should lie between the control limits if the process is in control. If σ is known, it usually derives from historical values. Here, n

is the number of observations in the subgroup. Note that in this control chart and the charts that follow, n need not be the same for all subgroups. Control charts are easier to interpret if this is the case, though.

The value for μ might also be known from past values. Alternatively, μ might represent the target mean of the process rather than the actual mean attained. In practice, though, μ might also be unknown. In that case, the mean of all of the subgroup averages, $\bar{\bar{x}}$, replaces μ as follows:

$$\text{LCL} = \bar{\bar{x}} - \frac{3\sigma}{\sqrt{n}}$$

$$\text{UCL} = \bar{\bar{x}} + \frac{3\sigma}{\sqrt{n}}$$

The interpretation of the mean chart is the same whether the true process mean is known or unknown.

Here is an example to help you understand the basic mean chart. Students are often concerned about getting into courses with "good" professors and staying out of courses taught by "bad" ones. In order to provide students with information about the quality of instruction provided by different instructors, many universities use end-of-semester surveys in which students rate various professors on a numeric scale. At some schools, such results are even posted and used by students to help them decide in which section of a course to enroll. Many faculty members object to such rankings on the grounds that although there is always some apparent variation among faculty members, there are seldom any significant differences. However, students often believe that variations in scores reflect the professors' relative aptitudes for teaching and are not simply random variations due to chance effects.

$\bar{x}$-Chart Example: Teaching Scores

One way to shed some light on the value of student evaluations of teaching is to examine the scores for one instructor over time. The workbook TEACH.XLS provides data ratings of one professor who has taught principles of economics at the same university for 20 consecutive semesters. The instruction in this course can be considered a process, because the instructor has used the same teaching methods and covered the same material over the entire period. Five student evaluation scores were recorded for each of the 20 courses. The five scores for each semester constitute a subgroup. Possible teacher scores run from 0 (terrible) to 100 (outstanding). The range names have been defined in Table 12-2 for the workbook.

Table 12-2 The Teach Workbook

Range Name	Range	Description
Semester	A2:A21	The semester of the evaluation
Score_1	B2:B21	First student evaluation
Score_2	C2:C21	Second student evaluation
Score_3	D2:D21	Third student evaluation
Score_4	E2:E21	Fourth student evaluation
Score_5	F2:F21	Fifth student evaluation

To open TEACH.XLS:

1 Open **TEACH.XLS** from the folder containing your student files.

2 Click **File > Save As** and save the workbook as **TEACH2.XLS**. Figure 12-5 displays the contents of the workbook.

**Figure 12-5
The Teach
workbook**

	Semester	Score 1	Score 2	Score 3	Score 4	Score 5
2	1	97	89	80	81	82
3	2	74	100	94	65	86
4	3	85	100	88	62	65
5	4	100	91	77	67	71
6	5	83	92	88	79	76
7	6	72	79	85	100	78
8	7	80	83	93	88	96
9	8	80	100	100	79	84
10	9	87	70	84	96	83
11	10	75	77	84	75	85
12	11	55	95	89	100	100
13	12	75	73	100	72	78
14	13	75	100	89	66	100
15	14	89	88	100	84	84
16	15	100	84	95	80	92
17	16	91	100	99	77	79
18	17	92	90	93	87	90
19	18	82	80	80	79	76
20	19	54	89	97	84	71
21	20	83	66	69	100	82

There is obviously some variation between scores across semesters, with scores varying from a low of 54.0 to a high of 100. Without further analysis, you and your friends might think that such a spread indicates that the professor's classroom performance has fluctuated widely over the course of 20 semesters. Is this interpretation valid?

If you consider teaching to be a process, with student evaluation scores as one of its products, you can use SPC to determine whether the process is in control. In other words, you can use SPC techniques to determine whether the variation in scores is due to identifiable differences in the quality of instruction that can be attributed to a particular semester's course (that is, special causes) or is due merely to chance (common causes).

Historical data from other sources show that σ for this professor is 5.0. Because there are five observations in each subgroup, $n = 5$. You can use StatPlus to calculate the mean scores for each semester and then the average of all 20 mean scores.

To create a control chart of the teacher's scores:

1 Click **StatPlus > QC Charts > Xbar Chart**.

2 Click the **Subgroups in rows across columns** option button.

3 Click the **Data Values** button and select the range names **Score_1** through **Score_5**. Click **OK**.

4 Click the **Sigma Known** checkbox and type **5** in the accompanying text box.

5 Click the **Output** button and send the control chart to a new chart sheet named **XBar Chart**. Click **OK**.

Figure 12-6 shows the completed dialog box.

**Figure 12-6
The Create
an XBAR
Control
Chart dialog
box**

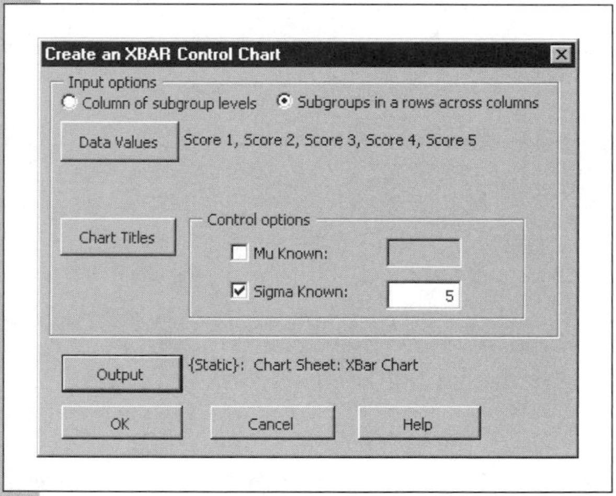

6 Click **OK**.

**Figure 12-7
Control
chart of the
teacher's
scores**

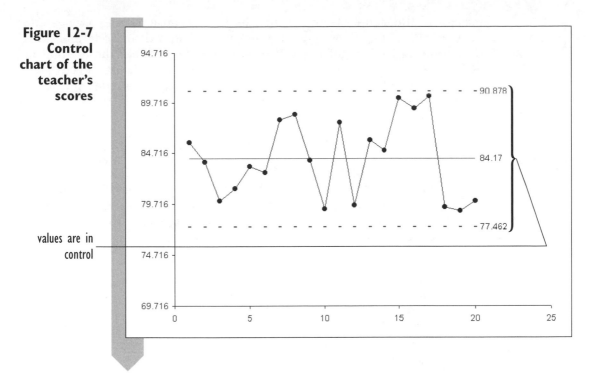

values are in
control

As you can see from Figure 12-7, no mean score falls outside the control limits. The lower control limit is 77.462, the mean subgroup average is 84.17, and the upper control limit is 90.878. There is no evident trend to the data or nonrandom pattern. You conclude that there is no reason to believe the teaching process is out of control.

Because we conclude that the process is in control, in contrast to what the typical student might conclude from the data, there is no evidence that this professor's performance was better or worse in one semester than in another. The raw scores from the last three semesters are misleading. A student might claim that using a historical value for σ is also misleading, because a smaller value for σ could lead one to conclude that the scores were not in control after all. The exercises at the end of this chapter will examine this issue by redoing the control chart with an unknown value for σ.

One corollary to the preceding analysis should be stated: Because even a single professor experiences wide fluctuations in student evaluations over time, apparent differences among various faculty members can also be deceptive. You should use all such statistics with caution.

You can close the Teach2 workbook now, saving your changes.

Calculating Control Limits When σ Is Unknown

In many instances, the value of σ is not known. You learned in Chapter 6 that the normal distribution does not strictly apply for analysis when σ is unknown and must be estimated. In that chapter, the *t*-distribution was used instead of the standard normal distribution. Because SPC is often implemented on the shop floor by workers who have had little or no formal statistical training (and might not have ready access to Excel), the method for estimating σ is simplified and the normal approximation is used to construct the control chart. The difference is that when σ is unknown, the control limits are estimated using the average range of observations within a subgroup as the measure of the variability of the process. The control limits are

$$\text{LCL} = \bar{\bar{x}} - A_2\bar{R}$$

$$\text{UCL} = \bar{\bar{x}} + A_2\bar{R}$$

$\bar{R}$ represents the average of the subgroup ranges, and $\bar{\bar{x}}$ is the average of the subgroup averages. A_2 is a correction factor that is used in quality-control charts. As you'll see, there are many correction factors for different types of control charts. Table 12-3 displays a list of common correction factors for various subgroup sizes, n.

Table 12-3 QC Correction Factors

n	A_2	d_2	D_1	D_2	D_3	D_4
2	1.88	1.128	0	3.686	0	3.268
3	1.023	1.693	0	4.358	0	2.574
4	0.729	2.059	0	4.698	0	2.282
5	0.577	2.326	0	4.918	0	2.114
6	0.483	2.534	0	5.078	0	2.004
7	0.419	2.704	0.204	5.204	0.076	1.924
8	0.373	2.847	0.388	5.306	0.136	1.864
9	0.337	2.970	0.547	5.393	0.184	1.816
10	0.308	3.078	0.687	5.469	0.223	1.777
11	0.285	3.173	0.811	5.535	0.256	1.744
12	0.266	3.258	0.922	5.594	0.284	1.717
13	0.249	3.336	1.025	5.647	0.308	1.692
14	0.235	3.407	1.118	5.696	0.329	1.671
15	0.223	3.472	1.203	5.741	0.348	1.652
16	0.212	3.532	1.282	5.782	0.363	1.637
17	0.203	3.588	1.356	5.820	0.378	1.622
18	0.194	3.640	1.424	5.856	0.391	1.608
19	0.187	3.689	1.487	5.891	0.403	1.597
20	0.180	3.735	1.549	5.921	0.415	1.585
21	0.173	3.778	1.605	5.951	0.425	1.575
22	0.167	3.819	1.659	5.979	0.434	1.566
23	0.162	3.858	1.710	6.006	0.443	1.557
24	0.157	3.895	1.759	6.031	0.451	1.548
25	0.153	3.931	1.806	6.056	0.459	1.541

Source: Adapted from "1950 ASTM Manual on Quality Control of Materials," American Society for Testing and Materials, in J. M. Juran, ed., *Quality Control Handbook* (New York: McGraw-Hill, 1974), Appendix II, p. 39. Reprinted with permission of McGraw-Hill.

A_2 accounts for both the factor of 3 from the earlier equations (used when σ was known) and for the fact that the average range represents a proxy for the common-cause variation. (There are other alternative methods for calculating control limits when σ is unknown.) As you can see from the table, A_2 depends only on the number of observations in each subgroup. Furthermore, the control limits become tighter when the subgroup sample size increases. The most typical sample size is 5 because this usually ensures normality of sample means. You will learn to use the control factors in the table later in the chapter.

$\bar{x}$-Chart Example: A Coating Process

The data in the workbook COATS.XLS comes from a manufacturing firm that sprays one of its metal products with a special coating to prevent corrosion. Because this company has just begun to implement SPC, σ is unknown for the coating process.

To open COATS.XLS:

1 Open **COATS.XLS** from the folder containing your student files.

2 Save the workbook as **COATS2.XLS**.

Figure 12-8 shows the contents of the workbook.

Figure 12-8
The Coats workbook

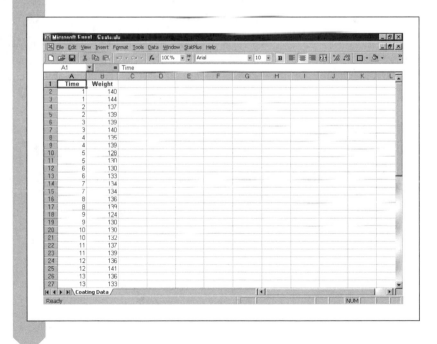

The weight of the spray in milligrams is recorded, with two observations taken at each of 28 times each day. Note that the data is arranged differently, with the Time column indicating the subgroup number. The range names have been defined for the workbook in Table 12-4.

Table 12-4 The Coats Workbook

Range Name	Range	Description
Time	A2:A57	The order of the evaluation (also the subgroup number)
Weight	B2:B57	The weight of the spray in milligrams

As before, you can use StatPlus to create the control chart. Note that because $n = 2$ (there are two observations per subgroup), $A_2 = 1.880$.

To create a control chart of the weight values:

1 Click **StatPlus > QC Charts > Xbar Chart**.

2 Click the **Data Values** button and select the **Weight** range name. Click **OK**.

3 Click the **Subgroups** button and select **Time** from the range names list. Click **OK**.

4 Click the **Output** button and send the control chart to a new chart sheet named **XBar Chart**. Click **OK** twice. See Figure 12-9.

**Figure 12-9
Control chart of the weight values**

value is not within control limits

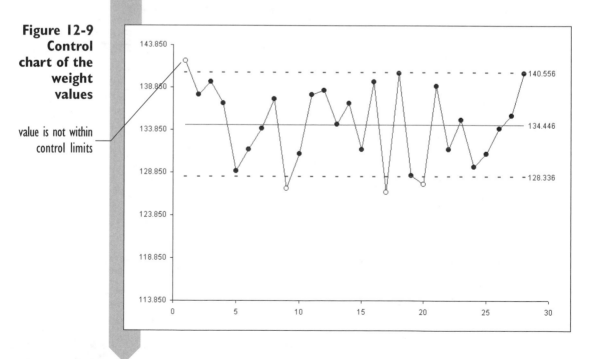

The lower control limit is 128.336, the average of the subgroup averages is 134.446, and the upper control limit is 140.556. Note that although most of the points in the mean chart lie between the control limits, four points (observations 1, 9, 17, and 20) lie outside the limits. This process is not in control.

Because the process is out of control, you should attempt to identify the special causes associated with each out-of-control point. Observation 1, for example, has too much coating. Perhaps the coating mechanism became stuck for an instant while applying the spray to that item. The other three observations indicate too little coating on the associated products. In talking with the operator, you might learn that he had not added coating material to the sprayer on schedule, so there was insufficient material to spray.

It is common practice in SPC to note the special causes either on the front of the control chart (if there is room) or on the back. This is a convenient way of keeping records of special causes.

In many instances, proper investigation leads to identification of the special causes underlying out-of-control processes. However, there might be out-of-control points whose special causes cannot be identified.

The Range Chart

The $\bar{x}$-chart provides information about the variation around the average value for each subgroup. It is also important to know whether the range of values is stable from group to group. In the coating example, if some observations exhibit very large ranges and others very small ranges, you might conclude that the sprayer is not functioning consistently over time. To test this, you can create a control chart of the average subgroup ranges, called a **range chart**. As with the $\bar{x}$-chart, the width of the control limits depends on the variability within each subgroup. If σ is known, the control limits for the Range chart are

$$LCL = D_1\sigma$$

$$\text{Center Line} = d_2\sigma$$

$$UCL = D_2\sigma$$

and if σ is not known, the control limits are

$$LCL = D_3\bar{R}$$

$$UCL = D_4\bar{R}$$

Where d_2, D_1, D_2, D_3, and D_4 are the correction factors from Table 12-3, and $\bar{R}$ is the average subgroup range. It's important to note that the $\bar{x}$-chart is valid only when the range is in control. For this reason the range chart is usually drawn alongside the $\bar{x}$-chart.

Use the information in the Coats workbook to determine whether the range of coating weights is in control.

To create a range chart of the weight values:

1 Return to the Coating Data worksheet.

2 Click **StatPlus > QC Charts > Range Chart**.

3 Select **Weight** as your Data Values variable and **Time** as the Subgroup variable.

4 Verify that the Sigma Known checkbox is *unselected*.

5 Direct the output to a new chart sheet named **Range Chart**. Click **OK** twice.

**Figure 12-10
The range
control
chart**

value is not within
control limits

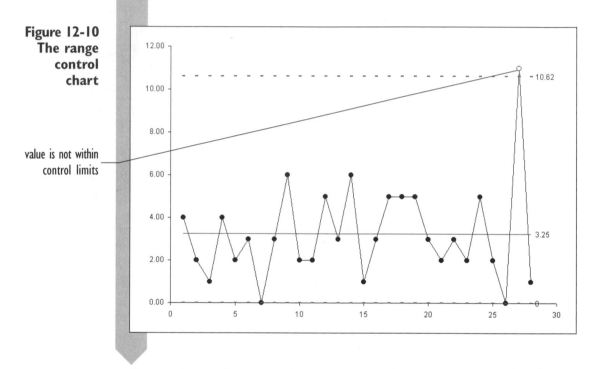

Each point on the range chart represents the range within each subgroup The average subgroup range is 3.25, with the control limits going from 0 to 10.62. According to the range chart shown in Figure 12-10, only the 27th observation has an out-of-control value. The special cause should be identified if possible. However, in discussing the problem with the operator, sometimes you might not be able to determine a special cause. This does not necessarily mean that no special cause exists; it could mean instead that you are unable to determine what the cause is in this instance. It is also possible that there really is no special cause. However, because you are constructing control charts with the width of about 3 standard deviations, an out-of-control value is unlikely unless there is something wrong with the process.

You might have noticed that the range chart identifies as out of control a point that was apparently in control in the x̄-chart but does not identify any of the four observations that are out of control in the x̄ chart. This is a common occurrence. For this reason, the x̄-chart and range charts are often used in conjunction to determine whether a process is in control. In practice, the x̄-chart and range chart often appear on the same page because viewing both charts simultaneously improves the overall picture of the process. In this example, you would judge that the process is out of control with both charts but based on different observations.

You can close the Coats workbook now, saving your changes.

The C-Chart

Both the x̄-chart and the range chart measure the values of a particular variable. Now let's look at an attribute chart that measures an attribute of the process. Some processes can be described by counting a certain feature, such as the number of flaws in a standardized section of continuous sheet metal or the number of defects in a production lot. The number of accidents in a plant might also be counted in this manner. A **C-chart** displays control limits for the counts attribute. The lower and upper control limits are

$$LCL = \bar{c} - 3\sqrt{\bar{c}}$$
$$UCL = \bar{c} + 3\sqrt{\bar{c}}$$

where $\bar{c}$ is the average number of counts in each subgroup. If the LCL is less than zero, by convention it will set to equal zero, because a negative count is impossible.

C-Chart Example: Factory Accidents

The ACCID.XLS workbook contains the number of accidents that occurred each month during a period of a few years at a production site. Let's create control charts of the number of accidents per month to determine whether the process is in control.

To open the ACCID workbook:

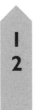

1 Open **ACCID.XLS** from the folder containing your student files.

2 Save the workbook as **ACCID2.XL**. See Figure 12-11.

Figure 12-11 The Accid workbook

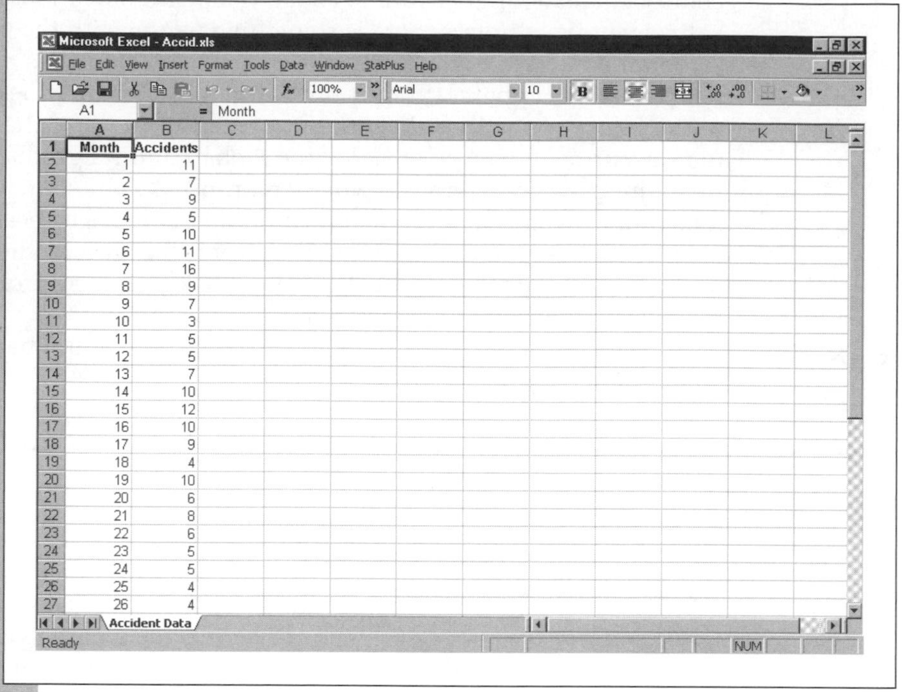

Month	Accidents
1	11
2	7
3	9
4	5
5	10
6	11
7	16
8	9
9	7
10	3
11	5
12	5
13	7
14	10
15	12
16	10
17	9
18	4
19	10
20	6
21	8
22	6
23	5
24	5
25	4
26	4

The range names have been defined for the workbook in Table 12-5.

Table 12-5 The Accid Workbook

Range Name	Range	Description
Month	A2:A45	The month
Accidents	B2:B45	The number of accidents that month

To create a C-chart for accidents at this firm:

1. Click **StatPlus > QC Charts > C-Chart**.
2. Select **Accidents** as the Data Values variable.
3. Direct the output to a new chart sheet named **C-Chart**.
4. Click **OK**.

Figure 12-12
C-chart of
the number
of accidents
per month

number of
accidents exceeded
control limits

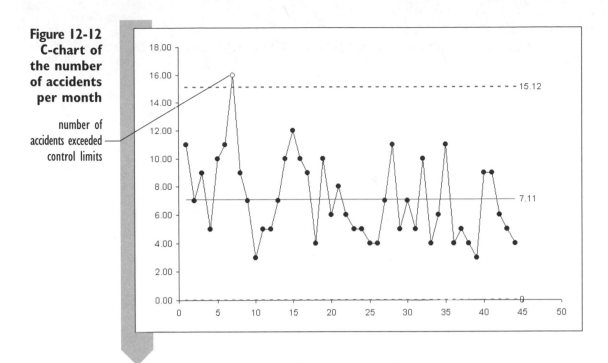

Each point on the C-chart in Figure 12-12 represents the number of accidents per month. The average number of accidents per month was 7.11. Only in the seventh month did the number of accidents exceed the upper control limit of 15.12 with 16 accidents. Since then, the process appears to have been in control. Of course, it is appropriate to determine the special causes associated with the large number of accidents in the seventh month. In the case of this firm, the workload was particularly heavy during that month, and a substantial amount of overtime was required. Because employees put in longer shifts than they were accustomed to working, fatigue is likely to have been the source of the extra accidents.

You can close the Accid workbook, saving your results.

The P-Chart

Closely related to the C-chart is the **P-chart**, which depicts the proportion of items with a particular attribute, such as defects. The P-chart is often used to analyze the proportion of defects in each subgroup.

Let $\bar{p}$ denote the average proportion of the sample that is defective. The distribution of the proportions can be approximated by the normal distribution, provided that $n\bar{p}$ and $n(1 - \bar{p})$ are both at least 5. If $\bar{p}$ is very close to 0 or 1, a very large subgroup size might be required for the approximation to be legitimate.

The lower and upper control limits are

$$LCL = \overline{p} - 3\sqrt{\frac{\overline{p}(1-\overline{p})}{n}}$$

$$UCL = \overline{p} + 3\sqrt{\frac{\overline{p}(1-\overline{p})}{n}}$$

P-Chart Example: Steel Rod Defects

A manufacturer of steel rods regularly tests whether the rods will withstand 50% more pressure than the company claims them to be capable of withstanding. A rod that fails this test is defective. Twenty samples of 200 rods each were obtained over a period of time, and the number and fraction of defects were recorded in the STEEL.XLS workbook.

To open the STEEL workbook:

1   Open **STEEL.XLS** from the folder containing your student files.

2   Save your workbook as **STEEL2.XLS**. See Figure 12-13.

**Figure 12-13
The Steel
workbook**

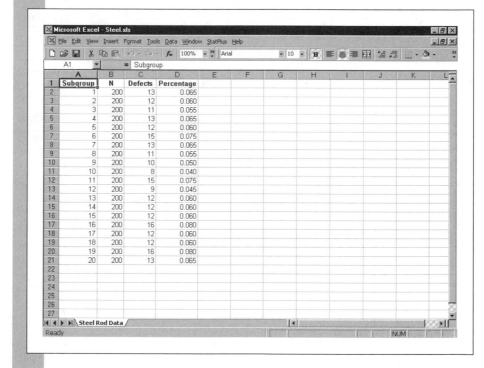

The range names have been defined for the workbook in Table 12-6.

Table 12-6 The Steel Workbook

Range Name	Range	Description
Subgroup	A2:A21	The subgroup number
N	B2:B21	The size of the subgroup
Defects	C2:C21	The number of defects in the subgroup
Percentage	D2:D21	The fraction of defects in the subgroup

To create a P-chart for the percentage of steel rod defects:

1 Click **StatPlus > QC Charts > P-Chart**.

2 Click the **Proportions** button and select **Percentage** from the list of range names. Click **OK**.

3 Type **200** in the Sample Size box, because each subgroup has the same sample size.

4 Send the output to a new chart sheet named **P-Chart**.

5 Click **OK**.

Figure 12-14 P-chart of the percentage of steel rod defects

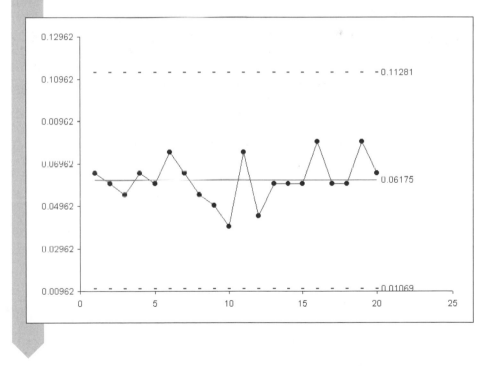

As shown in Figure 12-14, the lower control limit is 0.01069, or a defect percentage of about 1%. The upper control limit is 0.11281, or about 11%. The average defect percentage is 0.06175—about 6%. The control chart clearly demonstrates that no point is anywhere near the three-s limits.

Note that not all out-of-control points indicate the existence of a problem. For example, suppose that another sample of 200 rods was taken and that only one rod failed the stress test. In other words, only one-half of 1% of the sample was defective. In this case, the proportion is 0.005, which falls below the lower control limit, so technically it is out of control. Yet you would not be concerned about the process being out of control in this case, because the proportion of defects is so low. Still, you might be inclined to investigate, just to see whether you could locate the source of your good fortune and then duplicate it!

You can save and close the Steel workbook now.

Control Charts for Individual Observations

Up to now, we've been creating control charts for processes that can be neatly divided into subgroups. Sometimes it's not possible to group your data into subgroups. This could occur when each measurement represents a single batch in a process or when the measurements are widely spaced in time. With a subgroup size of 1, it's not possible to calculate subgroup ranges. This makes many of the regular formulas impractical to apply.

Instead, the recommended method is to create a "subgroup" consisting of each consecutive observation and then calculate the moving average of the data. Thus the subgroup variation is determined by the variation from one observation to another, and that variation will be used to determine the control limits for the variation between subgroups. Because we are setting up our subgroups differently, the formulas for the lower and upper control limits are different as well. The LCL and UCL are

$$LCL = \overline{x} - 3\frac{\overline{R}}{d_2}$$

$$UCL = \overline{x} + 3\frac{\overline{R}}{d_2}$$

Here $\overline{x}$ is the sample average of all of the observations, $\overline{R}$ is the average range of consecutive values in the data set, and d_2 is the control limit factor shown earlier in Table 12-3. We are using a moving average of size 2, so this will be equal to 1.128. Control charts based on these limits are called **individuals charts**.

We can also create a **moving range chart** of the moving range values; that is, the range between consecutive values. In this case, the lower and upper control limits match the ones used earlier for the range chart:

$$\text{LCL} = D_3\overline{R}$$

$$\text{UCL} = D_4\overline{R}$$

Let's apply these formulas to a workbook recording the tensile strength of 25 steel samples. The values are stored in the workbook STRENGTH.XLS.

To open the STRENGTH workbook:

1 Open **STRENGTH.XLS** from the folder containing your student files.

2 Save your workbook as **STRENGTH2.XLS**. See Figure 12-15.

Figure 12-15
The
Strength
workbook

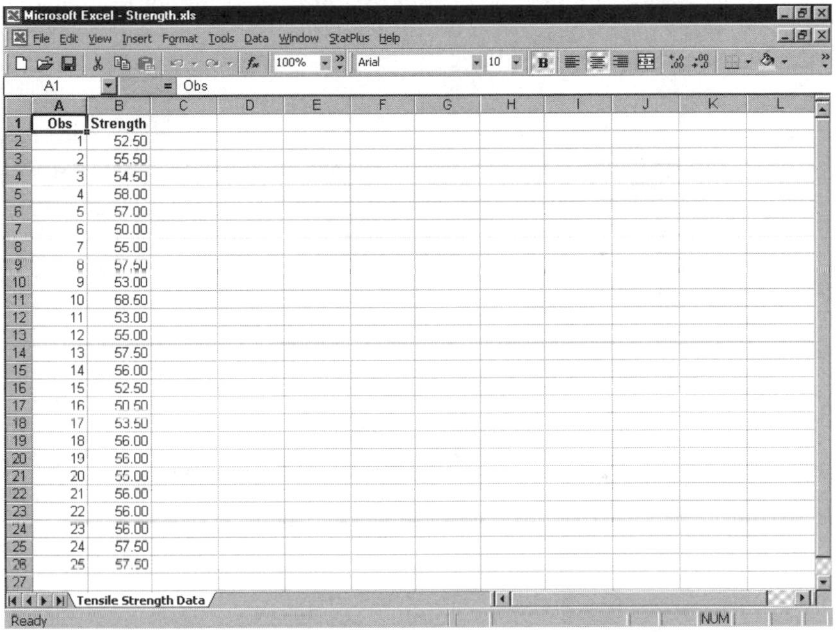

	A	B
1	Obs	Strength
2	1	52.50
3	2	55.50
4	3	54.50
5	4	58.00
6	5	57.00
7	6	50.00
8	7	55.00
9	8	57.50
10	9	53.00
11	10	68.50
12	11	53.00
13	12	55.00
14	13	57.50
15	14	56.00
16	15	52.50
17	16	50.50
18	17	53.50
19	18	56.00
20	19	56.00
21	20	55.00
22	21	56.00
23	22	56.00
24	23	56.00
25	24	57.50
26	25	57.50

Tensile Strength Data

The range names are shown Table 12-7.

Table 12-7 The Strength Workbook

Range Name	Range	Description
Obs	A2:A26	The observation number
Strength	B2:B26	The tensile strength of the sample, measured to the nearest 500 pounds in 1,000-pound units.

To create an individuals chart for the steel samples:

1 Click **StatPlus > QC Charts > Individuals Chart**.

2 Select **Strength** for the Data Values variable.

3 Send the output to a new chart sheet named **I-Chart**.

4 Click **OK**.

Figure 12-16 The individuals chart for the tensile strength samples

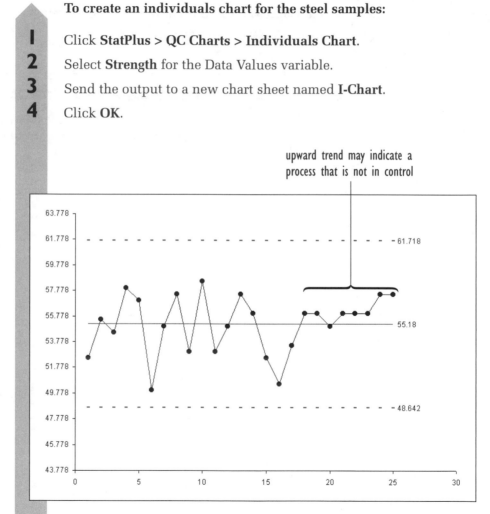

The chart shown in Figure 12-16 gives the values of the individual observations (not the moving averages) plotted alongside the upper and lower control limits. No values fall outside the control limits, which would lead us to conclude that the process is in control. However, the last eight observations were all either above or near the center line, which might indicate a process going out of control toward the end of the process. This is something that should be investigated further.

We should also plot the moving range chart, to see whether there is any evidence in that plot of an out-of-control process.

To create a moving range chart for the steel samples:

1 Click **StatPlus > QC Charts > Moving Range Chart**.

2 Select **Strength** for the Data Values variable.

3 Send the output to a new chart sheet named **I-Chart**.

4 Click **OK**.

**Figure 12-17
The moving range chart for the tensile strength samples**

trend in the moving range indicates a process not in control

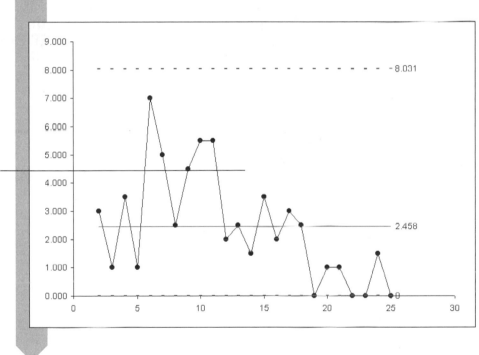

The chart in Figure 12-17 shows additional indications of a process that is not in control. The last seven values all fall below the center line, and there appears to be a generally downward trend to the ranges from the sixth observation on. We would conclude that there is sufficient evidence to warrant further investigation and analysis.

You can save and close the Strength workbook now.

The Pareto Chart

After you have determined that your process is resulting in an unusual number of problems, such as defects or accidents, the next natural step is to determine what component in the process is causing the problem. This investigation can be aided by a **Pareto chart**, which creates a bar chart of the causes of the problem in order from most to least frequent so that you can focus attention on the most important elements. The chart also includes the cumulative percentage of these components so that you can determine what combination of factors causes a certain percentage of the problems.

The workbook POWDER.XLS contains data from a company that manufactures baby powder. Part of the process involves a machine called a filler, which pours the powder into bottles to a specified limit. The quantity of powder placed in the bottle varies because of uncontrolled variation, but the final weight of the bottle filled with powder cannot be less than 368.6 grams. Any bottle weighing less than this amount is rejected and must be refilled manually (at a considerable cost in terms of time and labor). Bottles are filled from a filler that has 24 valve heads so that 24 bottles can be filled at one time. Sometimes a head is clogged with powder, and this causes the bottles being filled on that head to receive less than the minimum amount of powder. To gauge whether the machine is operating within limits, random samples of 24 bottles (one from each head) are selected at about 1-minute intervals over the nighttime shift at the factory. You've been asked to examine the data and determine which part of the filler is most responsible for defective fills.

To open POWDER.XLS:

1   Open the **POWDER** workbook from the folder containing your student files.

2   Save the files as POWDER2.XLS. See Figure 12-18.

Figure 12-18 The Powder workbook

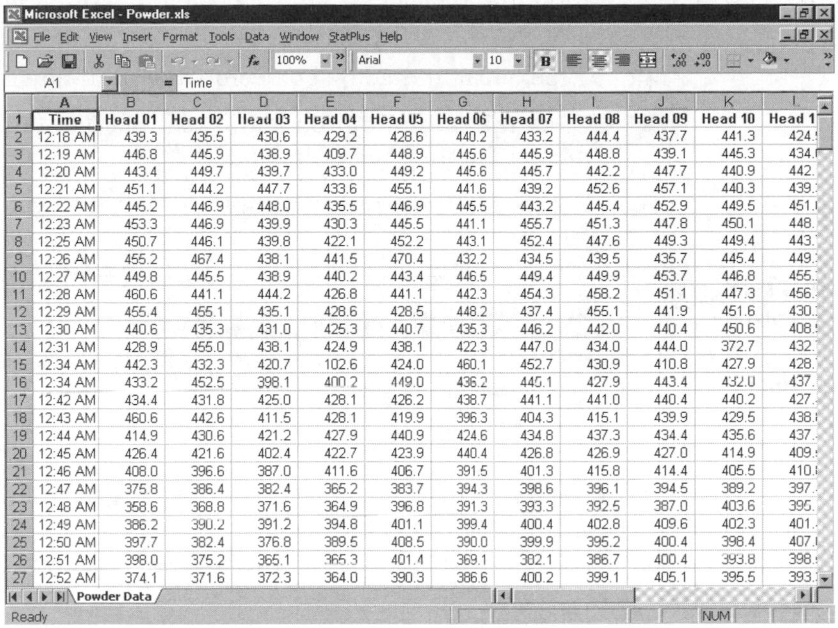

	A	B	C	D	E	F	G	H	I	J	K	L
1	Time	Head 01	Head 02	Head 03	Head 04	Head 05	Head 06	Head 07	Head 08	Head 09	Head 10	Head 1
2	12:18 AM	439.3	435.5	430.6	429.2	428.6	440.2	433.2	444.4	437.7	441.3	424.
3	12:19 AM	446.8	445.9	438.9	409.7	448.9	445.6	445.9	448.8	439.1	445.3	434.
4	12:20 AM	443.4	449.7	439.7	433.0	449.2	445.6	445.7	442.2	447.7	440.9	442.
5	12:21 AM	451.1	444.2	447.7	433.6	455.1	441.6	439.2	452.6	457.1	440.3	439.
6	12:22 AM	445.2	446.9	448.0	435.5	446.9	445.5	443.2	445.4	452.9	449.5	451.
7	12:23 AM	453.3	446.9	439.9	430.3	445.5	441.1	455.7	451.3	447.8	450.1	448.
8	12:25 AM	450.7	446.1	439.8	422.1	452.2	443.1	452.4	447.6	449.3	449.4	443.
9	12:26 AM	455.2	467.4	438.1	441.5	470.4	432.2	434.5	439.5	435.7	445.4	449.
10	12:27 AM	449.8	445.5	438.9	440.2	443.4	446.5	449.4	449.9	453.7	446.8	455.
11	12:28 AM	460.6	441.1	444.2	426.8	441.1	442.3	454.3	458.2	451.1	447.3	456.
12	12:29 AM	455.4	455.1	435.1	428.6	428.5	448.2	437.4	455.1	441.9	451.6	430.
13	12:30 AM	440.6	435.3	431.0	425.3	440.7	435.3	446.2	442.0	440.4	450.6	408.
14	12:31 AM	428.9	455.0	438.1	424.9	438.1	422.3	447.0	434.0	444.0	372.7	432.
15	12:34 AM	442.3	432.3	420.7	102.6	424.0	460.1	452.7	430.9	410.8	427.9	428.
16	12:34 AM	433.2	452.5	398.1	400.2	449.0	436.2	445.1	427.9	443.4	432.0	437.
17	12:42 AM	434.4	431.8	425.0	428.1	426.2	438.7	441.1	441.0	440.4	440.2	427.
18	12:43 AM	460.6	442.6	411.5	428.1	419.9	396.3	404.3	415.1	439.9	429.5	438.
19	12:44 AM	414.9	430.6	421.2	427.9	440.9	424.6	434.8	437.3	434.4	435.6	437.
20	12:45 AM	426.4	421.6	402.4	422.7	423.9	440.4	426.8	426.9	427.0	414.9	409.
21	12:46 AM	408.0	396.6	387.0	411.6	406.7	391.5	401.3	415.8	414.4	405.5	410.
22	12:47 AM	375.8	386.4	382.4	365.2	383.7	394.3	398.6	396.1	394.5	389.2	397.
23	12:48 AM	358.6	368.8	371.6	364.9	396.8	391.3	393.3	392.5	387.0	403.6	395.
24	12:49 AM	386.2	390.2	391.2	394.8	401.1	399.4	400.4	402.8	409.6	402.3	401.
25	12:50 AM	397.7	382.4	376.8	389.5	408.5	390.0	399.9	395.2	400.4	398.4	407.
26	12:51 AM	398.0	375.2	365.1	365.3	401.4	369.1	382.1	386.7	400.4	393.8	398.
27	12:52 AM	374.1	371.6	372.3	364.0	390.3	386.6	400.2	399.1	405.1	395.5	393.

The following range names have been defined for the workbook in Figure 12-18:

Table 12-8 The Powder Workbook

Range Name	Range	Description
Time	A2:A352	The time of the sample
Head_01	B2:B352	Quantity of powder from head 1
Head_02	C2:C352	Quantity of powder from head 2
...	...	...
Head_24	Y2:Y352	Quantity of powder from head 24

Now generate the Pareto chart using StatPlus.

To create the Pareto chart:

1. Click **StatPlus > QC Charts > Pareto Chart**.
2. Click the **Values in separate columns** option button.
3. Click the **Data Values** button and then select the range names from **Head 01** to **Head 24** in the range names list (do *not* select the Time variable). Click **OK**.

4 Click the **Defects occur when the data value is** drop-down list box and select **Less than**.

5 Type **368.6** in the text box below the drop-down list box.

6 Click the **Output** button and direct the output to a new chart sheet named **Pareto Chart**. Click **OK**.

Figure 12-19 shows the completed dialog box.

**Figure 12-19
The Create a
Pareto
Chart dialog
box**

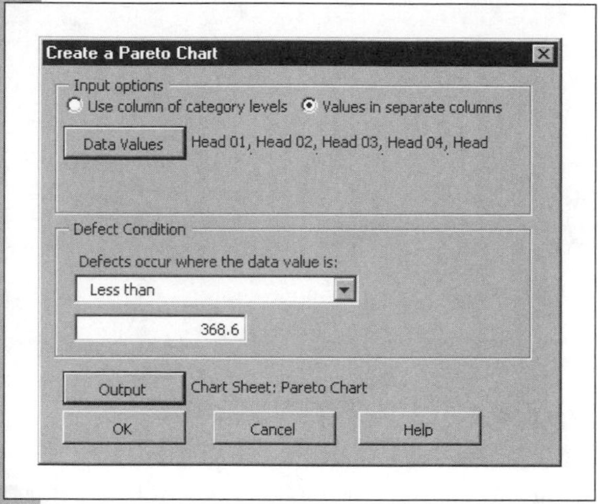

```
Create a Pareto Chart                                    [X]
 ┌─ Input options ────────────────────────────────────────┐
 │  ○ Use column of category levels  ● Values in separate columns
 │  ┌─────────────┐                                        │
 │  │ Data Values │  Head 01, Head 02, Head 03, Head 04, Head
 │  └─────────────┘                                        │
 └─────────────────────────────────────────────────────────┘

 ┌─ Defect Condition ──────────────────────────────────────┐
 │  Defects occur where the data value is:                 │
 │  ┌──────────────────────────────────────┐  ▼            │
 │  │ Less than                            │               │
 │  └──────────────────────────────────────┘               │
 │  ┌──────────────────────────────────────┐               │
 │  │                  368.6               │               │
 │  └──────────────────────────────────────┘               │
 └─────────────────────────────────────────────────────────┘

    ┌────────┐
    │ Output │   Chart Sheet: Pareto Chart
    └────────┘
    ┌────────┐   ┌────────┐      ┌────────┐
    │   OK   │   │ Cancel │      │  Help  │
    └────────┘   └────────┘      └────────┘
```

7 Click **OK**.

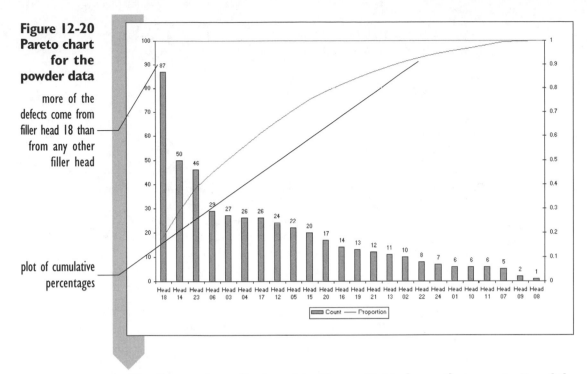

**Figure 12-20
Pareto chart
for the
powder data**

more of the
defects come from
filler head 18 than
from any other
filler head

plot of cumulative
percentages

The Pareto chart displayed in Figure 12-20 shows that a majority of the rejects come from a few heads. Filler head 18 accounts for 87 of the defects, and the first three heads in the chart (18, 14, and 23) account for almost 40% of all of the defects. There might be something physically wrong with the heads that made them more liable to clogging up with powder. If rejects were being produced randomly from the filler heads, you would expect that each filler head would produce $\frac{1}{24}$, or about 4%, of the total rejects. Using the information from the Pareto chart shown in Figure 12-18, you might want to repair or replace those three heads in order to reduce clogging.

You can close the Powder workbook now, saving your changes.

Exercises

1. *True or false (and why):* The purpose of statistical process control is to eliminate all variation from a process.

2. *True or false (and why):* As long as the process values lie between the control limits, the process is in control.

3. Calculate the control limits for an $\bar{x}$-chart where:
   a. $n = 9$, $\mu = 50$, $\sigma = 5$.
   b. $n = 9$, $\bar{\bar{x}} = 50$, $\bar{R} = 8$, σ is unknown.

4. Calculate the control limits for a range chart where:
   a. $n = 4$, $\bar{R} = 10$, $\sigma = 4$ (What is the value of the center line?)
   b. $n = 4$, $\bar{R} = 10$, σ is unknown.

5. Calculate the control limits for a C-chart where:
   a. $\bar{c} = 16$
   b. $\bar{c} = 22$

6. Calculate the control limits for a P-chart where:
   a. $n = 25$, $\bar{p} = 0.5$
   b. $n = 25$, $\bar{p} = 0.2$

7. Reopen the TEACH.XLS workbook from this chapter and perform the following analysis:
   a. Redo the $\bar{x}$-chart; this time do not assume a value for σ.
   b. Create a range chart of the data; once again, do not assume a value for σ.
   c. Examine your control charts. Is there evidence that the teacher's grades are not in control? Report your conclusions and save the workbook as E12TEACH.XLS.

8. A can manufacturing company must be careful to keep the width of its cans consistent. One associated problem is that the metalworking tools tend to wear down during the day. To compensate, the pressure behind the tools is increased as the blades become worn. In the CANS.XLS workbook, the width of 39 cans is measured at four randomly selected points. Open the workbook and do the following:
   a. Use range and $\bar{x}$-charts to determine whether the process is in control. Does the pressure-compensation scheme seem to correct properly for the tool wear? If not, suggest some special causes that seem still to be present in the process.
   b. Report your results, saving your workbook as E12CANS.XLS.

9. Just-in-time inventory management is becoming increasingly important. The ONTIME.XLS workbook contains data regarding the proportion of on-time deliveries during each month over a 2-year period for each of several paperboard products (cartons, sheets, and total). The Total column includes cartons, sheets, and other products. Because sheets were not produced for the entire 2-year period, a few data points are missing for that variable. Assume that 1,044 deliveries occurred during each month. Open the workbook and do the following:
   a. For each of these products (Cartons, Sheets, and Total), use a P-chart to determine whether the delivery process is in control. If not, suggest some special causes that might exist.
   b. Report your results and save your workbook as E12ONTIME.XLS.

10. A steel sheet manufacturer is concerned about the number of defects, such as scratches and dents, that occur as the sheet is made. In order to track defects, 10-foot lengths of sheet are examined at regular intervals.

For each length, the number of defects is counted. Data have been recorded in the SHEET.XLS workbook.

    a. Open the workbook and determine whether the process is in control. If it is not, suggest some special causes that might exist.

    b. Save your workbook as E12SHEET.XLS.

11. A firm is concerned about safety in its workplace. This company does not consider all accidents to be identical. Instead, it calculates a safety index, which assigns more importance to more serious accidents.

    a. Open the workbook and construct a C-chart for the SAFETY.XLS workbook to determine whether safety is in control at this firm.

    b. Report your conclusions. Save your workbook as E12SAFE.XLS.

12. A manufacturer subjects its steel bars to stress tests to be sure they are up to standard. Three bars were tested in each of 23 subgroups. The amount of stress applied before the bar breaks is recorded in the STRESS.XLS workbook.

    a. Open the workbook and create a range chart and a $\bar{x}$-chart to determine whether the production process is in control. If it is not, what factors might be contributing to the lack of control?

    b. Report your results, saving your workbook as E12STRESS.XLS.

13. A steel rod manufacturer has contracted to supply rods 180 millimeters in length to one of its customers. Because the cutting process varies somewhat, not all rods are exactly the desired length. Five rods were measured from each of 33 subgroups during a week.

    a. Open the workbook and create range and $\bar{x}$-charts to determine whether the cutting process is in statistical control.

    b. Report your conclusion. Save your workbook as E12ROD.XLS.

14. SPC can also be applied to services. An amusement park sampled customers leaving the park over an 18-day period. The total number of customers and the number of customers who indicated they were satisfied with their experience in the park were recorded in the SATISFY.XLS workbook.

    a. Open the workbook and determine whether the percentage of people satisfied is in statistical control.

    b. What factors, if any, might have contributed to the lack of control?

    c. Save your workbook as E12SATISFY.XLS.

15. The number of flaws on the surfaces of a particular model of automobile leaving the plant was recorded in the AUTOS.XLS workbook for each of 40 automobiles during a 1-week period.

    a. Open the workbook and determine whether this attribute is in control. If so, is the process perfect?

    b. Save your workbook as E12AUTOS.XLS.

16. You've learned in this chapter that filler head 18 is a major factor in the number of defective fills. To investigate further, you decide to look at the head

18 values from the data set to determine at what points in time the head was out of statistical control.

   a. Open the POWDER.XLS workbook and create an individuals chart and a moving range chart of the Head 18 values. At what times are the head values beyond the control limits?

   b. Repeat part a for filler heads 14 and 23.

   c. Interpret your findings in light of the fact that a new shift comes in at midnight. Does this fact affect the filler process?

   d. Save your workbook as E12POWDER.XLS.

17. Weather can be considered a process with process variables such as temperature and precipitation and attribute variables such as the number of hurricanes and tornadoes in a given season. One theory of meteorology holds that climatic changes in this process take place over long periods of time, whereas over short periods of time, the process should be stable. On the other hand, concerns have been raised about the effect of CO_2 emissions on the atmosphere, which may lead to major changes in the weather. You've been given the yearly temperature values for northern Illinois from 1895 to 1998, saved in the workbook TEMP100.XLS.

   a. Open the workbook, and create an individuals chart and a moving range chart of the average yearly temperature.

   b. What is the average yearly temperature? What are the lower and upper control limits? Do the temperature values appear to be in statistical control?

   c. Create a moving range chart of the average yearly temperature. Does this chart show any violations of process control?

   d. Summarize your results, saving your workbook as E12TEMP100.XLS.

18. The RAIN100.XLS workbook contains the total precipitation for northern Illinois from 1895 to 1998.

   a. Open the workbook and create an individuals chart of the total precipitation.

   b. Create a moving range chart of the total precipitation.

   c. Does the process appear to be in statistical control? Report your conclusions and save the workbook as E12RAIN.XLS.

19. The TORNADO.XLS workbook records the number of tornadoes of various levels of severity in Kansas from 1950 to 1999. Tornadoes are rated on the Fujita Tornado Scale, which ranges from "minor" tornadoes rated at F0 to major tornadoes rated at F5. You've been asked to determine whether the number of tornadoes has changed over this period of time.

   a. Open the workbook and create a C-chart of the number of tornadoes each year for each severity level and then for all types of tornadoes.

   b. Which classes of tornadoes show signs of being out of statistical control? Describe the nature of the problem.

   c. Techniques in recording and counting tornadoes have improved in the last few decades, especially for minor tornadoes. Explain how this fact may be related to the results you noted in part b.

   d. Report your results and save the workbook as E12TORNADO.XLS.

Excel Reference

Contents

The Excel Reference contains the following:

- Summary of Data Sets

- Excel's Data Analysis ToolPak

- Excel's Math and Statistical Functions

- StatPlus™ Commands

- StatPlus™ Math and Statistical Functions

- Bibliography

Summary of Data Sets

ACCID.XLS has the average number of accidents each month for several years (used in Chapter 12).

ALUM.XLS has mass and volume for eight chunks of aluminum, as measured in a high school chemistry class (used in Chapters 2, 3, and 8).

AUTOS.XLS has data on surface flaws for 40 cars (used in Chapter 12).

BASE.XLS is part of a data set collected by Lorraine Denby of AT&T Bell Labs for a graphics exposition at the 1988 Joint Statistical Meetings. BASE.XLS has batting and salary data for 263 major league baseball players (used in Chapters 2, 3, 4, 5, and 10).

BASE26.XLS gives batting data for the 26 major league teams on Tuesday, July 29, 1992, as reported the next day in the Bloomington, Illinois, *Pantagraph* (used in Chapter 9).

BASEINFD.XLS is a subset of BASE.XLS with only the infielders (used in Chapter 10).

BBAVER.XLS has the leading major league batting average for each of the years 1901–1991 (used in Chapter 11).

BCANCER.XLS has data analyzing the relationship between mean annual temperature in 16 regions and the region's mortality index for neoplasms of the female breast. Source: Lea (1965), *British Medical Journal* (used in Chapter 8).

BEER.XLS gives monthly United States beer production for 1980–1991, as given by the *Brewers Almanac* 1992, published by the Beer Institute in Washington, D.C (used in Chapter 11).

BIGTEN.XLS has data from the *Chronicle of Higher Education*, July 5, 1996, pages A41-A42 and *ARCO: The Right College* 1990, New York: ARCO Publishing (used in Chapters 2, 3, and 6).

BOOTH.XLS has data on the largest banks in the U.S.A. in the early 1970s, taken from Booth (1985) (used in Chapter 8).

BREWER.TXT has data on American beer consumption for the major brands. The data are from the *Beverage Industry Annual Manual* 91/92, page 37 (used in Chapter 2).

CALC.XLS has the final grade in first-semester calculus and several predictor variables for 154 students (used in Chapters 3, 8, and 9).

CALCFM.XLS contains a subset of the data from CALC.XLS: just the calculus scores for the males and females, broken into two columns, one for each gender (used in Chapter 6).

CANCERRATE.XLS displays data on cigarette use and various cancer rates by state. The data are expressed as per capita numbers of cigarettes smoked (sold) for 43 states and the District of Columbia in 1960, together with death rates per thousand population from various forms of cancer (used in Chapter 4).

CANS.XLS has the width of 39 cans (used in Chapter 12).

CARS.XLS has data on 392 car models. It is based on material collected by David Donoho and Ernesto Ramos (1982) and has been used by statisticians to compare the abilities of different statistics packages (used in Chapters 2, 3, 7, and 9).

COATS.XLS has data on coating weight for sprayed coatings on metals (used in Chapter 12).

COLA.XLS has data on foam volume, type (diet or regular), and brand for 48 cans of cola (used in Chapter 10).

COLD.XLS contains an analysis of the effect of ascorbic acid on colds. The data are taken from Pauling, L. (1971) "The significance of the evidence about ascorbic acid and the common cold." *Proceedings of the National Academy of Sciences,* 68, 2678–2681 (used in Chapter 7 exercises).

DISCRIM.XLS has a subset of the data from JRCOL.XLS, reorganized to facilitate using the Analysis ToolPak (used in Chapter 9).

DOW.XLS has the closing Dow Jones average for each month, 1981–1990 (used in Chapter 11).

DRAFT.XLS has data on draft numbers from the 1970 draft lottery. Birth dates were placed into a rotating drum and drawn out one at a time. The first date drawn received the number 1, and so forth (used in Chapters 4, 6, 8, and 11).

ECON.XLS has data related to the economy of the United States from 1947 to 1962, as given by Longley (1967). The Deflator variable is a measure of the inflation of the dollar, arbitrarily set to 100 for 1954. Total contains total employment. Population shows us population each year (used in Chapter 2 and the Chapter 2 exercises).

ELECTRIC.XLS has monthly data on U.S. electric power production, 1978–1990. The source is the 1992 *CRB Commodity Year Book*, published by the Commodity Research Bureau in New York (used in Chapter 11).

EMERALD.XLS has data on Emerald Health Network, Inc. claim costs from 1986 to 1992. For comparison with the unadjusted consumer price index for all urban customers, along with the medical care component of the CPI. *The Columbus Dispatch,* June 8, 1993 (used in Chapter 8).

EXPSOLVE.XLS has the monthly percent change in the Dow Jones average from DOW.XLS, set up to facilitate use by the Solver (used in Chapter 11).

FIDELITY.XLS has figures from 1989, 1990, and 1991 for 35 Fidelity sector funds. The source is the *Morningstar Mutual Fund Sourcebook 1992*, Equity Mutual Funds (used in Chapter 8).

FOURYR.XLS has information on 24 colleges from the 1992 edition of *U.S. News America's Best Colleges*. They list 140 national liberal arts colleges, 35 in each of four quartiles. A random sample of six colleges was taken from each quartile. An additional variable on computer availability is from the *Computer Industry Almanac 1991* (used in Chapter 10).

GENDEROP.XLS has data on male and female responses to different issues, taken from the U.S. Census Bureau (used in Chapter 7).

HEART.XLS has the results of an experiment conducted at Ohio State University to estimate the effects of rate of step climbing (0 = 14 steps/min, 1 = 21 steps/min, 2 = 28 steps/min) and step height (0 = 5.75 in., 1 = 11.5 in.) on heart rate (used in Chapter 10).

HOMEDATA.XLS contains data on home prices in Albuquerque, New Mexico. The data are a random sample of records of home sales from 2/15/1993 to 4/30/1993, from the files maintained by the Albuquerque Board of Realtors.

HONDA12.XLS contains a subset of the HONDA25.XLS workbook in which the age variable is categorical with the values 1–3, 4–5, and 6+. Some observations of HONDA25.XLS have been removed to balance the data (used in Chapter 10).

HONDA25.XLS has price, age, and transmission for 25 used Hondas advertised in the *San Francisco Examiner-Chronicle*, November 25, 1990 (used in Chapter 10).

HONDACIV.XLS includes price, age, and miles for used Honda Civics advertised in the *San Francisco Examiner-Chronicle*, November 25, 1990 (used in Chapter 9).

HOTEL.XLS has data on 32 hotels, eight chosen randomly from each of four cities, as given in the 1992 *Mobil Travel Guide to Major Cities* (used in Chapter 10).

HOTELTWOWAY.XLS is taken from the same source as the HOTEL.XLS data. However, the random sample has only two hotels in each of twelve cells, which are defined by four levels of city and three levels of stars (used in Chapter 10).

HOUSE.XLS has Bureau of Labor Statistics data for the consumer price index for housing from 1967 to 1996. Collected on a monthly basis (used in Chapter 11).

JOBS.XLS has Bureau of Labor Statistics data on youth unemployment (16–19) rates from 1981 to 1996 (used in Chapter 11).

JRCOL.XLS has employment data for 81 faculty members at a junior college (used in Chapters 3, 6, 7, and 9).

LONGLEY.XLS has seven variables related to the economy of the United States in 1947–1962, as given by Longley (1967) (used in Chapter 3).

MARRIAGE.XLS contains data from a marriage–height study, taken from G.U. Yule, 1900, "On the association of attributes in statistics: with illustration from the material of the childhood society," *Philos. Trans. Roy. Soc. Ser.* 194 (used in Chapter 7).

MATH.XLS has data from students in a low-level math course (used in Chapter 6).

MORTGAGE.XLS contains data on refusal rates from 20 lending institutions broken down by race and income status. The data was presented to a joint congressional hearing on discrimination in lending (used in Chapters 4 and 6).

MUSTANG.XLS includes prices and ages of used Mustangs from the *San Francisco Examiner-Chronicle*, November 25, 1990 (used in Chapter 8).

NFP.XLS contains daily temperature data taken for 239 consecutive days for a woman in her twenties as one of the measures used in natural family planning (NFP), in which a woman uses her monthly cycle in combination with a number of biological signs to determine the onset of ovulation. The file has four columns: Observation, Period (the menstrual period), Day (the day of the menstrual period; day 1 is the first day of menstruation), and waking temperature (used in Chapter 11).

NOISE.XLS contains data from testimony given to the Senate Public Works Committee on June 26, 1973, on the silencing properties of a special brand of filters (used in Chapter 10).

NURSEHOME.XLS displays data collected by the Department of Health and Social Services of the state of New Mexico, covering 52 of the 60 licensed nursing facilities in New Mexico in 1988 (used in Chapter 6).

OIL.XLS has data on the monthly production of crude cottonseed oil (in thousands of pounds) from 1992 to 1995. Data taken from the U.S. Census Bureau (used in Chapter 11).

ONTIME.XLS has the percentage of on-time deliveries each month for a two-year period (used in Chapter 12).

PARK.XLS contains monthly public-use data (the number of visitors) for Kenai Fjords National Park in Alaska in 1993. Data courtesy of Maria Gillett, chief of interpretation, and Glenn Hart, park ranger (used in Chapter 1).

PCINFO.XLS contains data resulting from tests of 191 prominent 486 PCs, published in *PC Magazine*, Vol. 12, No. 21, pp. 148-149. Used by permission of Ziff-Davis Publishing (used in Chapter 4).

PCSURV.XLS has survey results on 35 makes of personal computers, taken from *PC Magazine*, February 9, 1993. There were 17,000 questionnaires sent to randomly chosen subscribers, and the data are based on 8,176 responses. There are five variables. Used by permission of Ziff-Davis Publishing (used in Chapters 3 and 9).

POLU.XLS gives the number of unhealthful days in each of 14 cities, for six years from the 1980s. The figures are from "Environmental Protection Agency National Air Quality and Emissions Trends Report, 1989," as quoted in the *Universal Almanac 1992*, p. 534 (used in Chapters 2, 4, and 6).

POWDER.XLS contains data from a company that manufactures baby powder, with random samples of bottle weights selected at one-minute intervals over the nighttime shift. Data courtesy of Kemp Wills, Johnson & Johnson (used in Chapter 12).

RACE.XLS has data on the reaction times and race results from the first-round heats of the 1996 Summer Olympics in Atlanta (used in Chapters 3, 5, and 10).

RACEOP.XLS has data on opinions from individuals of different races from the National Election Studies at the Center for Political Studies at the University of Michigan's Institute for Social Research (used in Chapter 7).

RACEPAIR.XLS has data on the reaction times and race results from the first three rounds of the 100-meter race at the 1996 Summer Olympics in Atlanta (used in Chapters 6 and 10).

RAIN100.XLS contains rainfall data for Northern Illinois from 1895 to 1998 (used in the Chapter 12 exercises).

REACT.XLS has data on the reaction times from first-round heats at the 1996 Summer Olympics in Atlanta (used in the Chapters 4 and 10).

ROD.XLS has data on the lengths of steel rods (used in Chapter 12).

SAFETY.XLS has values for the seriousness of accidents at a firm (used in Chapter 12).

SALARY.XLS has salary data recorded by the Ohio State University to compare faculty salaries at 49 different universities (used in Chapter 4).

SATISFY.XLS has satisfaction data for customers leaving an amusement park (used in Chapter 12).

SHEET.XLS has data on defects in steel sheets (used in Chapter 12).

SPACE.XLS has measurements on the electrical resistance at the calf (to get an idea of the loss of blood from the legs to the upper part of the body) over a period of 24 days, for seven men and eight women. The study was financed by NASA (used in Chapter 6).

STATE.XLS has the death rate from cardiovascular problems and the death rate from pulmonary problems, for each of the 50 states. The data are from the 1986 *Metropolitan Area Data Book of the U.S. Census Bureau* (used in Chapters 2 and 8).

STEEL.XLS has the proportion of defects in each of 20 samples of steel rods (used in Chapter 12).

STRENGTH.XLS records the tensile strength of 25 steel samples (used in Chapter 12).

STRESS.XLS has strength data for steel bars (used in Chapter 12).

SURVEY.XLS has data from 390 professors who responded to a questionnaire designed to help plan this book. There are 16 variables (used in Chapter 7).

TEACH.XLS has five evaluations for each of 20 instructors (used in Chapter 12).

TEACHER.XLS contains the average public teacher pay and spending on public schools per pupil in 1985 for 50 states and the District of Columbia as reported by the *Albuquerque Tribune* (used in Chapters 4, 6, and 8).

TEMP.XLS has the average January temperature for several U.S. cities (used in Chapter 9).

TEMP100.XLS contains temperature data for Northern Illinois from 1895 to 1998 (used in Chapter 12).

TORNADO.XLS contains data on the number of tornadoes of various degrees of severity in Kansas from 1950 to 1999. Tornadoes are rated on the Fujita Tornado Scale, which ranges from "minor" tornadoes rated at F0 to major tornadoes rated at F5 (used in Chapter 12).

UNEMP.XLS has unemployment levels from 1950 to 1959 as related to industrial production (used in Chapter 9).

VISIT.XLS contains monthly visitation data for two sites at the Kenai Fjords National Park in Alaska, from January 1990 through June 1994. Data courtesy of Maria Gillett, chief of interpretation, and Glenn Hart, park ranger (used in Chapter 11).

VOTING.XLS has data on the percent of the vote given to the Democratic presidential candidate in the 1980 and 1994 elections (used in Chapters 3 and 6).

WASTE.XLS contains data from a clothing manufacturer on the percent waste relative to what can be achieved via computer layouts. Data is collected for five different plants (used in Chapter 10).

WBUS.XLS contains data from the Wisconsin State Department of Development on the top 50 women-owned businesses (used in Chapter 4).

WHEAT.TXT and WHEAT.XLS have nutrition information on ten wheat products, as given on the labels (used in Chapters 2, 3, 8, and 9).

WHEATDAN.XLS is the same as WHEAT.XLS with an additional case, the Apple Danish from McDonald's (used in Chapter 9).

WLABOR.XLS has data on labor force participation in 19 U.S. cities from 1968 to 1972, from the U.S. Department of Labor statistical record (used in Chapters 4 and 6).

WORK.XLS displays the number of work stoppages of 1,000 workers or more for each month from 1981 to 1996, from the Bureau of Labor Statistics (used in Chapter 11).

The Analysis ToolPak add-ins that come with Excel enable you to perform basic statistical analysis. None of the output from the Analysis ToolPak is updated for changing data, so if the source data changes, you will have to rerun the command. To use the Analysis ToolPak, you must first verify that it is available to your workbook.

To check whether the Analysis ToolPak is available:

1 Click **Tools** from the menu. If the menu option Data Analysis appears, the Analysis ToolPak commands are available to you.

2 If the Data Analysis menu command does not appear, click **Tools > Add-Ins** from the menu and then click **Analysis ToolPak** from the list of add-ins and click **OK**.

3 If the Analysis ToolPak files are installed on your computer or network, the add-in will be ready to use; otherwise, Excel will prompt you to insert the Installation CD and install the files from the CD onto your hard drive.

The rest of this section documents each Analysis ToolPak command, showing each corresponding dialog box and describing the features of the command.

Output Options

All the dialog boxes that produce output share the following output storage options:

Output Range

Click to send output to a cell in the current worksheet, and then type the cell; Excel uses that cell as the upper-left corner of the range.

New Worksheet Ply

Click to send output to a new worksheet; then type the name of the worksheet.

New Workbook

Click to send output to a new workbook.

Anova: Single-Factor

The **Anova: Single-Factor** command calculates the one-way analysis of variance, testing whether means from several samples are equal.

Input Range

Enter the range of worksheet data you want to analyze. The range must be contiguous.

Grouped By

Indicate whether the range of samples is grouped by columns or by rows.

Labels in First Row/Column

Indicate whether the first row (or column) includes header information.

Alpha

Enter the alpha level used to determine the critical value for the F-statistic.
See "Output Options" at the beginning of this section for information on the output storage options.

Anova: Two-Factor with Replication

The **Anova: Two-Factor with Replication** command calculates the two-way analysis of variance with multiple observations for each combination of the two factors. An analysis of variance table is created that tests for the significance of the two factors and the significance of an interaction between the two factors.

Input Range

Enter the range of worksheet data you want to analyze. The range must be rectangular, the columns representing the first factor and the rows representing the second factor. An equal number of rows are required for each level of the second factor.

Rows per Sample

Enter the number of repeated values for each combination of the two factors.

Alpha

Enter the alpha level used to determine the critical value for the F-statistic.
See "Output Options" at the beginning of this section for information on the output storage options.

Anova: Two-Factor without Replication

The **Anova: Two-Factor without Replication** command calculates the two-way analysis of variance with one observation for each combination of the two factors. An analysis of variance table is created that tests for the significance of the two factors.

Input Range

Enter the range of worksheet data you want to analyze. The range must be contiguous, with each row and column representing a combination of the two factors.

Labels

Indicate whether the first row (or column) includes header information.

Alpha

Enter the alpha level used to determine the critical value for the F-statistic.

See "Output Options" at the beginning of this section for information on the output storage options.

Correlation

The **Correlation** command creates a table of the Pearson correlation coefficient for values in rows or columns on the worksheet.

Input Range

Enter the range of worksheet data you want to analyze. The range must be contiguous.

Grouped By

Indicate whether the range of samples is grouped by columns or by rows.

Labels in First Row/Column

Indicate whether the first row (or column) includes header information.

See "Output Options" at the beginning of this section for information on the output storage options.

Covariance

The **Covariance** command creates a table of the covariance for values in rows or columns on the worksheet.

Input Range

Enter the range of worksheet data you want to analyze. The range must be contiguous.

Grouped By

Indicate whether the range of samples is grouped by columns or by rows.

Labels in First Row/Column

Indicate whether the first row (or column) includes header information.
See "Output Options" at the beginning of this section for information on the output storage options.

Descriptive Statistics

The **Descriptive Statistics** command creates a table of univariate descriptive statistics for values in rows or columns on the worksheet.

Descriptive Statistics

Input
Input Range:
Grouped By: ● Columns ○ Rows
□ Labels in First Row
□ Confidence Level for Mean: 95 %
□ Kth Largest: 1
□ Kth Smallest: 1

Output options
○ Output Range:
● New Worksheet Ply:
○ New Workbook
□ Summary statistics

OK
Cancel
Help

Input Range

Enter the range of worksheet data you want to analyze. The range must be contiguous.

Grouped By

Indicate whether the range of samples is grouped by columns or by rows.

Labels in First Row/Column

Indicate whether the first row (or column) includes header information.

Confidence Level for Mean

Click to print the specified confidence level for the mean in each row or column of the input range.

Kth Largest

Click to print the kth largest value for each row or column of the input range; enter k in the corresponding box.

Kth Smallest

Click to print the *k*th smallest value for each row or column of the input range; enter k in the corresponding box.

Summary Statistics

Click to print the following statistics in the output range: Mean, Standard Error (of the mean), Median, Mode, Standard Deviation, Variance, Kurtosis, Skewness, Range, Minimum, Maximum, Sum, Count, Largest (#), Smallest (#), and Confidence Level.

See "Output Options" at the beginning of this section for information on the output storage options.

Exponential Smoothing

The **Exponential Smoothing** command creates a column of smoothed averages using simple one-parameter exponential smoothing.

Input Range

Enter the range of worksheet data you want to analyze. The range must be a single row or a single column.

Damping Factor

Enter the value of the smoothing constant. The value 0.3 is used as a default if nothing is entered.

Labels

Indicate whether the first row (or column) includes header information.

Chart Output

Click to create a chart of observed and forecasted values.

Standard Errors

Click to create a column of standard errors to the right of the forecasted column.

Output Options

You can send output from this command only to a cell on the current worksheet.

F-Test: Two-Sample for Variances

The **F-Test: Two-Sample for Variances** command performs an *F*-test to determine whether the population variances of two samples are equal.

Variable 1 Range

Enter the range of the first sample, either a single row or a single column.

Variable 2 Range

Enter the range of the second sample, either a single row or a single column.

Labels

Indicate whether the first row (or column) includes header information.

Alpha

Enter the alpha level used to determine the critical value for the F-statistic.

See "Output Options" at the beginning of this section for information on the output storage options.

Histogram

The **Histogram** command creates a frequency table for data values located in a row, column, or list. The frequency table can be based on default or customized bin widths. Additional output options include calculating the cumulative percentage, creating a histogram, and creating a histogram sorted in descending order of frequency (also known as a Pareto chart).

Input Range

Enter the range of worksheet data you want to analyze. The range must be a row, column, or rectangular region.

Bin Range

Enter an optional range of values that define the boundaries of the bins.

Labels

Indicate whether the first row (or column) includes header information.

Pareto (sorted histogram)

Click to create a Pareto chart sorted by descending order of frequency.

Cumulative Percentage

Click to calculate the cumulative percents.

Chart Output

Click to create a histogram of frequency versus bin values.

See "Output Options" at the beginning of this section for information on the output storage options.

Moving Average

The **Moving Average** command creates a column of moving averages over the preceding observations for an interval specified by the user.

Input Range

Enter the range of worksheet data for which you want to calculate the moving average. The range must be a single row or a single column containing four or more cells of data.

Labels in First Row

Indicate whether the first row (or column) includes header information.

Interval

Enter the number of cells you want to include in the moving average. The default value is three.

Chart Output

Click to create a chart of observed and forecasted values.

Standard Errors

Click to create a column of standard errors to the right of the forecasted column.

Output Options

You can only send output from this command to a cell on the current worksheet.

Random Number Generation

The **Random Number Generation** command creates columns of random numbers following a user-specified distribution.

Number of Variables

Enter the number of columns of random variables you want to generate. If no value is entered, Excel fills up all available columns.

Number of Random Numbers

Enter the number of rows in each column of random variables you want to generate. If no value is entered, Excel fills up all available columns. This command is not available for the patterned distribution (see below).

Distribution

Click the down arrow to open a list of seven distributions from which you can choose to generate random numbers and then specify the parameters of that distribution.

Random Seed

Enter an optional value used as a starting point, called a random seed, for generating a string of random numbers. You need not enter a random seed, but using the same random seed ensures that the same string of random numbers will be generated. This box is not available for patterned or discrete random data.

See "Output Options" at the beginning of this section for information on the output storage options.

Rank and Percentile

The **Rank and Percentile** command produces a table with ordinal and percentile values for each cell in the input range.

Input Range

Enter the range of worksheet data you want to analyze. The range must be contiguous.

Grouped By

Indicate whether the range of samples is grouped by columns or by rows.

Labels in First Row/Column

Indicate whether the first row (or column) includes header information.
    See "Output Options" at the beginning of this section for information on the output storage options.

Regression

The **Regression** command performs multiple linear regression for a variable in an input column based on up to 16 predictor variables. The user has the option of calculating residuals and standardized residuals and producing line fit plots, residuals plots, and normal probability plots.

Input Y Range

Enter a single column of values that will be the response variable in the linear regression.

Input X Range

Enter up to 16 contiguous columns of values that will be the predictor variables in the regression.

Labels

Indicate whether the first row of the Y range and that of the X range include header information.

Constant is Zero

Click to include an intercept term in the linear regression or to assume that the intercept term is zero.

Confidence Level

Click to indicate a confidence interval for linear regression parameter estimates. A 95% confidence interval is automatically included; enter a different one in the corresponding box.

Residuals

Click to create a column of residuals (observed – predicted) values.

Residual Plots

Click to create a plot of residuals versus each of the predictor variables in the model.

Standardized Residuals

Click to create a column of residuals divided by the standard error of the regression's analysis of variance table.

Line Fit Plots

Click to create a plot of observed and predicted values against each of the predictor variables.

Normal Probability Plots

Click to create a normal probability plot of the Y variable in the Input Y Range.

See "Output Options" at the beginning of this section for information on the output storage options.

Sampling

The **Sampling** command creates a sample of an input range. The sample can be either random or periodic (sampling values a fixed number of cells apart). The sample generated is placed into a single column.

Input Range

Enter the range of worksheet data you want to sample. The range must be contiguous.

Labels

Indicate whether the first row of the Y range and that of the X range include header information.

Sampling Method

Click the sampling method you want.

Periodic

Click to sample values from the input range period cells apart; enter a value for period in the corresponding box.

Random

Click to create a random sample the size of which you enter in the corresponding box.

See "Output Options" at the beginning of this section for information on the output storage options.

t-Test: Paired Two-Sample for Means

The *t*-Test: **Paired Two-Sample for Means** command calculates the paired two-sample Student's *t*-test. The output includes both the one-tail and the two-tail critical values.

Variable 1 Range

Enter the input range of the first sample; it must be a single row or column.

Variable 2 Range

Enter the input range of the second sample; it must be a single row or column.

Hypothesized Mean Difference

Enter a mean difference value with which to calculate the *t*-test. If no value is entered, a mean difference of zero is assumed.

Labels

Indicate whether the first row of the Y range and that of the X range include header information.

Alpha

Enter an alpha value used to calculate the critical values of the t shown in the output.

See "Output Options" at the beginning of this section for information on the output storage options.

t-Test: Two-Sample Assuming Equal Variances

The **t-Test: Two-Sample Assuming Equal Variances** command calculates the unpaired two-sample Student's t-test. The test assumes that the variances in the two groups are equal. The output includes both the one-tail and the two-tail critical values.

Variable 1 Range

Enter an input range for the first sample; it must be a single row or column.

Variable 2 Range

Enter an input range for the second sample; it must be a single row or column.

Hypothesized Mean Difference

Enter a mean difference with which to calculate the t-test. If no value is entered, a mean difference of zero is assumed.

Labels

Indicate whether the first row of the Y range and that of the X range include header information.

Alpha

Enter an alpha value used to calculate the critical values of the t shown in the output.

See "Output Options" at the beginning of this section for information on the output storage options.

t-Test: Two-Sample Assuming Unequal Variances

The **t-Test: Two-Sample Assuming Unequal Variances** command calculates the unpaired two-sample Student's t-test. The test allows the variances in the two groups to be unequal. The output includes both the one-tail and the two-tail critical values.

Variable 1 Range

Enter the input range of the first sample; it must be a single row or column.

Variable 2 Range

Enter the input range of the second sample; it must be a single row or column.

Hypothesized Mean Difference

Enter a mean difference with which to calculate the *t*-test. If no value is entered, a mean difference of zero is assumed.

Labels

Indicate whether the first row of the Y range and that of the X range include header information.

Alpha

Enter an alpha value used to calculate the critical values of the *t* shown in the output.

See "Output Options" at the beginning of this section for information on the output storage options.

---

z-Test: Two-Sample for Means

The **z-Test: Two-Sample for Means** command calculates the unpaired two-sample *z*-test. The test assumes that the variances in the two groups are known (though not necessarily equal to each other). The output includes both the one-tail and the two-tail critical values.

Variable 1 Range

Enter an input range of the first sample; it must be a single row or column.

Variable 2 Range

Enter an input range of the second sample; it must be a single row or column.

Hypothesized Mean Difference

Enter a mean difference with which to calculate the t-test. If no value is entered, a mean difference of zero is assumed.

Variable 1 Variance (known)

Enter the known variance s_1^2 of the first sample.

Variable 2 Variance (known)

Enter the known variance s_2^2 of the second sample.

Labels

Indicate whether the first row of the Y range and that of the X range include header information.

Alpha

Enter an alpha value used to calculate the critical values of the t shown in the output.

See "Output Options" at the beginning of this section for information on the output storage options.

This section documents all the functions provided with Excel that are relevant to statistics. So that you can more easily find the function you need, similar functions are grouped together in six categories: Descriptive Statistics for One Variable, Descriptive Statistics for Two or More Variables, Distributions, Mathematical Formulas, Statistical Analysis, and Trigonometric Formulas.

Descriptive Statistics for One Variable

Function Name	Description
AVEDEV	AVEDEV(*number1*, *number2*, . . .) returns the average of the (absolute) deviations of the points from their mean.
AVERAGE	AVERAGE(*number1*, *number2*, . . .) returns the average of the *numbers* (up to 30).
CONFIDENCE	CONFIDENCE(*alpha*, *standarddev*, *n*) returns a confidence interval for the mean.
COUNT	COUNT(*value1*, *value2*, . . .) returns how many numbers are in the *value(s)*.
COUNTA	COUNTA(*value1*, *value2*, . . .) returns the count of non-blank values in the list of arguments.
COUNTBLANK	COUNTBLANK(*range*) returns the count of blank cells in the *range*.
COUNTIF	COUNTIF(*range*, *criteria*) returns the count of non-blank cells in the *range* that meet the *criteria*.
DEVSQ	DEVSQ(*number1*, *number2*, . . .) returns the sum of squared deviations from the mean of the *numbers*.
FREQUENCY	FREQUENCY(*data-array*, *bins-array*) returns the frequency distribution of *data-array* as a vertical array, based on *bins-array*).
GEOMEAN	GEOMEAN(*number1*, *number2*, . . .) returns the geometric mean of up to 30 *numbers*.
HARMEAN	HARMEAN(*number1*, *number2*, . . .) returns the harmonic mean of up to 30 *numbers*.
KURT	KURT(*number1*, *number2*, . . .) returns the kurtosis of up to 30 *numbers*.
LARGE	LARGE(*array*, *n*) returns the *n*th-largest value in *array*.
MAX	MAX(*number1*, *number2*, . . .) returns the largest of up to 30 *numbers*.

MEDIAN	MEDIAN(*number1, number2, . . .*) returns the median of up to 30 *numbers*.
MIN	MIN(*number1, number2, . . .*) returns the smallest of up to 30 *numbers*.
MODE	MODE(*number1, number2, . . .*) returns the value most frequently occurring in up to 30 *numbers* or in a specified array or reference.
PERCENTILE	PERCENTILE(*array, n*) returns the *n*th percentile of the values in *array*.
PERCENTRANK	PERCENTRANK(*array, value, significant-digits*) returns the percent rank of the *value* in the *array*, with the specified number of *significant digits* (optional).
PRODUCT	PRODUCT(*number1, number2, . . .*) returns the product of up to 30 *numbers*.
RANK	RANK(*number, range, order*) returns the rank of the *number* in the *range*. If *order* = 0, then the range is ranked from largest to smallest; if *order* = 1, then the range is ranked from smallest to largest.
QUOTIENT	QUOTIENT(*dividend, divisor*) returns the quotient of the numbers, truncated to integers.
SKEW	SKEW(*number1, number2, . . .*) returns the skewness of up to 30 *numbers* (or a reference to numbers).
SMALL	SMALL(*array, n*) returns the *n*th-smallest number in *array*.
STANDARDIZE	STANDARDIZE(*x, mean, standard-deviation*) normalizes a distribution and returns the *z*-score of *x*.
STDEV	STDEV(*number1, number2, . . .*) returns the sample standard deviation of up to 30 *numbers*, or of an array of numbers.
STDEVP	STDEVP(*number1, number2, . . .*) returns the population standard deviation of up to 30 *numbers* or of an array of numbers.
SUM	SUM(*number1, number2, . . .*) returns the sum of up to 30 *numbers* or of an array of numbers.
SUMIF	SUMIF(*range, criteria, sum-range*) returns the sum of the numbers in *range* (optionally in *sum-range*) according to *criteria*.
SUMSQ	SUMSQ(*number1, number2, . . .*) returns the sum of the squares of up to 30 *numbers* or of an array of numbers.
TRIMMEAN	TRIMMEAN(*array, percent*) returns the mean of a set of values in an *array*, excluding *percent* of the values, half from the top and half from the bottom.

VAR	VAR(*number1*, *number2*, . . .) returns the sample variance of up to 30 *numbers* (or an array or reference).
VARP	VARP(*number1*, *number2*, . . .) returns the population variance of up to 30 *numbers* (or an array or reference).

Descriptive Statistics for Two or More Variables

Function Name	Description
CORREL	CORREL(*array1*, *array2*) returns the coefficient of correlation between *array1* and *array2*.
COVAR	COVAR(*array1*, *array2*) returns the covariance of *array1* and *array2*.
PEARSON	PEARSON(*array1*, *array2*) returns the Pearson correlation coefficient between *array1* and *array2*.
RSQ	RSQ(*known-y's*, *known-x's*) returns the square of Pearson's product moment correlation coefficient.
SUMPRODUCT	SUMPRODUCT(*array1*, *array2*, . . .) returns the sum of the products of corresponding entries in up to 30 *arrays*.
SUMX2MY2	SUMX2MY2(*array1*, *array2*) returns the sum of the differences of squares of corresponding entries in two *arrays*.
SUMX2PY2	SUMX2PY2(*array1*, *array2*) returns the sum of the sums of squares of corresponding entries in two *arrays*.
SUMXMY2	SUMXMY2(*array1*, *array2*) returns the sum of the squares of differences of corresponding entries in two arrays.

Distributions

Function Name	Description
BETADIST	BETADIST(*x*, *alpha*, *beta*, *a*, *b*) returns the value of the cumulative beta probability density function.
BETAINV	BETAINV(*p*, *alpha*, *beta*, *a*, *b*) returns the value of the inverse of the cumulative beta probability density function.

BINOMDIST	BINOMDIST(*successes*, *trials*, *p*, *type*) returns the probability for the binomial distribution (*type* is TRUE for cumulative distribution function, FALSE for probability mass function).
CHIDIST	CHIDIST(*x*, *df*) returns the probability for the chi-squared distribution.
CHIINV	CHIINV(*p*, *df*) returns the inverse of the chi-squared distribution.
CRITBINOM	CRITBINOM(*trials*, *p*, *alpha*) returns the smallest value so that the cumulative binomial distribution is greater than or equal to the criterion value, *alpha*.
EXPONDIST	EXPONDIST(*x*, *lambda*, *type*) returns the probability for the exponential distribution (*type* is TRUE for cumulative distribution function, FALSE for probability density function).
FDIST	FDIST(*x*, *df1*, *df2*) returns the probability for the *F*-distribution.
FINV	FINV(*p*, *df1*, *df2*) returns the inverse of the *F*-distribution.
GAMMADIST	GAMMADIST(*x*, *alpha*, *beta*, *type*) returns the probability for the gamma distribution with parameters *alpha* and *beta* (*type* is TRUE for cumulative distribution function, FALSE for probability mass function).
GAMMAINV	GAMMAINV(*p*, *alpha*, *beta*) returns the inverse of the gamma distribution.
GAMMALN	GAMMALN(*x*) returns the natural log of the gamma function evaluated at *x*.
HYPGEOMDIST	HYPGEOMDIST(*sample-successes*, *sample-size*, *population-successes*, *population_size*) returns the probability for the hypergeometric distribution.
LOGINV	LOGINV(*p*, *mean*, *sd*) returns the inverse of the lognormal distribution, where the natural logarithm of the distribution is normally distributed with mean *mean* and standard deviation *sd*.
LOGNORMDIST	LOGNORMDIST(*x*, *mean*, *sd*) returns the probability for the lognormal distribution, where the natural logarithm of the distribution is normally distributed with mean *mean* and standard deviation *sd*.
NEGBINOMDIST	NEGBINOMDIST(*failures*, *threshold-successes*, *probability*) returns the probability for the negative binomial distribution.
NORMDIST	NORMDIST(*x*, *mean*, *sd*, *type*) returns the probability for the normal distribution with mean *mean* and standard deviation *sd* (*type* is TRUE for the cumulative distribution function, FALSE for the probability mass function).

NORMINV	NORMINV(p, *mean*, *sd*) returns the inverse of the normal distribution with mean *mean* and standard deviation *sd*.
NORMSDIST	NORMSDIST(*number*) returns the probability for the standard normal distribution.
NORMSINV	NORMSINV(*probability*) returns the inverse of the standard normal distribution.
POISSON	POISSON(x, *mean*, *type*) returns the probability for the Poisson distribution (*type* is TRUE for the cumulative distribution, FALSE for probability mass function).
TDIST	TDIST(x, *df*, *number-of-tails*) returns the probability for the t-distribution.
TINV	TINV(p, *df*) returns the inverse of the t-distribution.
WEIBULL	WEIBULL(x, *alpha*, *beta*, *type*) returns the probability for the Weibull distribution (*type* is TRUE for cumulative distribution function, FALSE for probability mass function).

Mathematical Formulas

Function Name	Description
ABS	ABS(*number*) returns the absolute value of *number* to the point specified.
COMBIN	COMBIN(x, n) returns the number of combinations of x objects taken n at a time.
EVEN	EVEN(*number*) returns *number* rounded up to the nearest even integer.
EXP	EXP(*number*) returns the exponential function of *number* with base e.
FACT	FACT(*number*) returns the factorial of *number*.
FACTDOUBLE	FACTDOUBLE(*number*) returns the double factorial of *number*.
FLOOR	FLOOR(*number*, *significance*) returns *number* rounded down to the nearest multiple of the *significance* value.
GCD	GCD(*number1*, *number2*, . . .) returns the greatest common divisor of up to 29 *numbers*.
GESTEP	GESTEP(*number*, *step*) returns 1 if *number* is greater than or equal to *step*, 0 if not.
LCM	LCM(*number1*, *number2*, . . .) returns the least common multiple of up to 29 *numbers*.
INT	INT(*number*) truncates *number* to the units place.
LN	LN(*number*) returns the natural logarithm of *number*.

LOG	LOG(*number*, *base*) returns the logarithm of *number*, with the specified (optional, default is 10) *base*.
LOG10	LOG10(*number*) returns the common logarithm of *number*.
MOD	MOD(*number*, *divisor*) returns the remainder of the division of *number* by *divisor*.
MULTINOMIAL	MULTINOMIAL(*number1*, *number2*, . . .) returns the quotient of the factorial of the sum of *numbers* and the product of the factorials of *numbers*.
ODD	ODD(*number*) returns *number* rounded up to the nearest odd integer.
POWER	POWER(*number*, *power*) returns *number* raised to the *power*.
PERMUT	PERMUT(*x*, *n*) returns the number of permutations of *x* items taken *n* at a time.
RAND	RAND() returns a randomly chosen number from 0 to but not including 1.
ROUND	ROUND(*number*, *places*) rounds *number* to a certain number of decimal *places* (if *places* is positive), or to an integer (if *places* is 0), or to the left of the decimal point (if *places* is negative).
ROUNDDOWN	ROUNDDOWN(*number*, *places*) rounds like ROUND, except always toward 0.
ROUNDUP	ROUNDUP(*number*, *places*) rounds like ROUND, except always away from 0.
SERIESSUM	SERIESSUM(*x*, *n*, *m*, *coefficients*) returns the sum of the power series: $a_1 x^n + a_2 x^{n+m} + \ldots + a_i x^{n+(i-1)m}$, where $a_1, a_2, \ldots a_i$ are the *coefficients*.
SIGN	SIGN(*number*) returns 0, 1, or −1, the sign of *number*.
SQRT	SQRT(*number*) returns the square root of *number*.
SQRTPI	SQRTPI(*number*) returns the square root of *number**π.
TRIM	TRIM(*text*) returns *text* with spaces removed, except for single spaces between words.
TRUNC	TRUNC(*number*, *digits*) truncates *number* to an integer (optionally, to a number of digits).

Statistical Analysis

Function Name	Description
CHITEST	CHITEST(*observed, expected*) calculates the Pearson chi-square for observed and expected counts.
GROWTH	GROWTH(*known-y's*, *known-x's*, *new-x's*, *constant*) returns the predicted (*y*) values for the *new-x's*, based on exponential regression of the *known-y's* on the *known-x's*.
FISHER	FISHER(*x*) returns the value of the Fisher transformation evaluated at *x*.
FISHERINV	FISHERINV(*y*) returns the value of the inverse Fisher transformation evaluated at *y*.
FORECAST	FORECAST(*x*, *known-y's*, *known-x's*) returns a predicted (*y*) value for *x*, based on linear regression of the *known-y's* on the *known-x's*.
FTEST	FTEST(*array1*, *array2*) returns the *p*-value of the one-tailed *F*-statistic, based on the hypothesis that the variances *array1* and *array2* are not significantly different (which is rejected for low *p*-values).
INTERCEPT	INTERCEPT(*known-y's*, *known-x's*) returns the *y*-intercept of the linear regression of *known-y's* on *known-x's*.
LINEST	LINEST(*known-y's*, *known-x's*, *constant*, *stats*) returns coefficients in linear regression of *known-y's* on *known-x's* (*constant* is TRUE if the intercept is forced to be 0, and *stats* is TRUE if regression statistics are desired).
LOGEST	LOGEST(*known-y's*, *known-x's*, *constant*, *stats*) returns the exponential regression of *known-y's* on *known-x's* (*constant* is TRUE if the leading coefficient is forced to be 0, and *stats* is true if regression statistics are desired).
PROB	PROB(*x-values*, *probabilities*, *value*) returns the *probability* associated with *value*, given the *probabilities* of a range of values.
PROB	PROB(*x-values*, *probabilities*, *lower-limit*, *upper-limit*) returns the *probability* associated with values between *lower-limit* and *upper-limit*.
SLOPE	SLOPE(*known-y's*, *known-x's*) returns the slope of a linear regression line.
STEYX	STEYX(*known-y's*, *known-x's*) returns the standard error of the linear regression.
TREND	TREND(*known-y's*, *known-x's*, *new-x's*, *constant*) returns the *y*-values of given input values (*new-x's*)

	based on regression of *known-y's* on *known-x's*. If constant = FALSE, the constant value is zero.
TTEST	TTEST(*array1, array2, number-of-tails, type*) returns the *p*-value of a *t*-test, of *type* paired (1), two-sample equal variance (2), or two-sample unequal variance (3).
ZTEST	ZTEST(*array, x, sigma*) returns the *p*-value of a two-tailed z-test, where *x* is the value to test and *sigma* is the population standard deviation.

Trigonometric Formulas

Function Name	Description
ACOS	ACOS(*number*) returns the arccosine (inverse cosine) of *number*.
ACOSH	ACOSH(*number*) returns the inverse hyperbolic cosine of number.
ASIN	ASIN(*number*) returns the arcsine (inverse sine) of *number*.
ASINH	ASINH(*number*) returns the inverse hyperbolic sine of *number*.
ATAN	ATAN(*number*) returns the arctangent (inverse tangent) of *number*.
ATAN2	ATAN2(*x,y*) returns the arctangent (inverse tangent) of the angle from the positive *x*-axis.
ATANH	ATANH(*number*) returns the inverse hyperbolic tangent of *number*.
COS	COS(*angle*) returns the cosine of *angle*.
COSH	COSH(*number*) returns the hyperbolic cosine of *number*.
DEGREES	DEGREES(*angle*) returns the degree measure of an *angle* given in radians.
PI	PI() returns π accurate to 15 digits.
RADIANS	RADIANS(*angle*) returns the radian measure of an *angle* given in degrees.
SIN	SIN(*angle*) returns the sine of *angle*.
SINH	SINH(*number*) returns the hyperbolic sine of *number*.
TAN	TAN(*angle*) returns the tangent of *angle*.
TANH	TANH(*number*) returns the hyperbolic tangent of *number*.

StatPlus™ is supplied with the textbook *Data Analysis with Microsoft Excel* to perform basic statistical analysis not covered by Excel or the Analysis ToolPak. To use StatPlus you must first verify that it is available to your workbook.

To check whether StatPlus is available:

1 If the StatPlus menu option appears on Excel's menu bar, StatPlus is loaded.

2 If the menu command does not appear, click **Tools > Add-Ins** from the menu. If the StatPlus option is listed in the Add-Ins list box, click the checkbox. StatPlus is now available to you.

3 If StatPlus is not listed in the Add-Ins list box, you will have to install it from your instructor's disk. See Chapter 1 for more information.

The rest of this section documents each StatPlus Add-In command, showing each corresponding dialog box, and describes the command's options and output.

Creating Data

Bivariate Normal Data

The **StatPlus>Create Data>Bivariate Normal** command creates two columns of random normal data where the standard deviation, σ, of the data in the first column and the correlation between the columns are specified by the user. The standard deviation of the second column of data is a function of the standard deviation of the data in the first column.

Patterned Data

The **StatPlus>Create Data>Patterned Data** command generates a column of data following a specified pattern. The pattern can be created on the basis of a sequence of numbers or taken from a number sequence entered in a data column already existing in the workbook. The user can specify how often each number in pattern is repeated and how many times the entire sequence is repeated.

Random Numbers

The **StatPlus>Create Data>Random Numbers** command generates columns of random numbers for a specified probability distribution. The user specifies the number of samples (columns) of random numbers and the sample size (rows) of each sample.

Manipulating Columns

Indicator Columns

The **StatPlus>Manipulate Columns>Create Indicator Columns** command takes a column of category levels and creates columns of indicator variables, one for each category level in the input range. An indicator variable for a particular category = 1 if the row comes from an observation belonging to that category and 0 otherwise.

```
┌─────────────────────────────────────────────────────┐
│ Make Indicator Variables                          [X]│
│  ┌─ Input ───────────────────────────────────────┐  │
│  │                                                 │  │
│  │  [  Categories  ]   <None Selected>            │  │
│  │                                                 │  │
│  │  [   Output    ]    <None selected>            │  │
│  │   [   OK   ]     [  Cancel  ]     [  Help  ]   │  │
│  └─────────────────────────────────────────────────┘│
└─────────────────────────────────────────────────────┘
```

Two-Way Table

The **StatPlus>Manipulate Columns>Create Two-Way Table** command takes data arranged into three columns—a column of values, a column of category levels for one factor, and a second column of category levels for a second factor—and arranges the data into a two-way table. The columns of the table consist of the different levels of the first factor; the rows of the table consist of different levels of the second factor. Multiple values for each combination of the two factors show up in different rows within the table. Output from this command can be used in the Analysis ToolPak's ANOVA commands. The numbers of rows in the three columns must be equal. The user can choose whether or not to sort the row and column headers of the table.

```
┌─────────────────────────────────────────────────────┐
│ Make a Two-way Table                              [X]│
│  ┌─ Input ───────────────────────────────────────┐  │
│  │  [ Data Values  ]   <None Selected>            │  │
│  │  [ Column Levels]   <None Selected>            │  │
│  │  [  Row Levels  ]   <None Selected>            │  │
│  │  [✓] Sort the Column Levels  [✓] Sort theRow Levels│
│  │                                                 │  │
│  │  [   Output    ]    <None selected>            │  │
│  │   [   OK   ]     [  Cancel  ]     [  Help  ]   │  │
│  └─────────────────────────────────────────────────┘│
└─────────────────────────────────────────────────────┘
```

Unstack Column

The **StatPlus>Manipulate Column>Unstack Column** takes data found in two columns—a column of data values and a column of categories—and outputs to values into different columns, one column for each category level. The length of the values column and the length of the category column must be equal. The user can choose whether or not to sort the columns in ascending order of the category variable.

Stack Columns

The **StatPlus>Manipulate Column>Stack Columns** command takes data that lie in separate columns and stacks the values into two columns. The column to the left contains the values; the column to the right is a category column. Values for the category are found from the header rows in the input columns, or, if there are no header rows, the categories are labeled as Level 1, Level 2, and so forth. The input range need not be contiguous.

Sampling Data

Conditional Sampling

The **StatPlus>Sampling>Conditional Sample** command extracts data values from a collection of columns corresponding to a specified condition.

```
┌─────────────────────────────────────────────────────┐
│ Generate Conditional Sample                      [X] │
│ ┌─ Input ────────────────────────────────────────┐  │
│ │  [   Data Values   ]   <None Selected>          │  │
│ │                                                 │  │
│ │  [ Conditional Variable ]  <None Selected>      │  │
│ │  Extract rows where the conditional variable is │  │
│ │  [                                      ▼]      │  │
│ │                                                 │  │
│ │  [                      ]                       │  │
│ └─────────────────────────────────────────────────┘ │
│  [   Output   ]   <None selected>                    │
│  [   OK   ]      [  Cancel  ]      [  Help  ]        │
└─────────────────────────────────────────────────────┘
```

Periodic Sampling

The **StatPlus>Sampling>Periodic Sample** command samples data values from a collection of columns starting at a specified row and then extracting every ith row, where i is specified by the user.

```
┌─────────────────────────────────────────────────────┐
│ Generate Periodic Sample                         [X] │
│ ┌─ Input ────────────────────────────────────────┐  │
│ │  [   Data Values   ]   <None Selected>          │  │
│ │                                                 │  │
│ │  Extract every  [        ]  th value            │  │
│ │  Starting with the [      ]  th row             │  │
│ └─────────────────────────────────────────────────┘ │
│  [   Output   ]   <None selected>                    │
│  [   OK   ]      [  Cancel  ]      [  Help  ]        │
└─────────────────────────────────────────────────────┘
```

Random Sampling

The **StatPlus>Sampling>Random Sample** command extracts a random sample of a given size from a collection of columns. The user can choose whether to sample with replacement or without replacement.

Frequency Tables

The **StatPlus>Frequency Tables** command creates a table containing frequency, cumulative frequency, percentage, and cumulative percentage. The frequency table can be displayed either by discrete values in the data column or by bin values. If bin values are used, the user can specify how the data is counted relative to the placement of the bins. The frequency table can also be broken down by the values of a By variable.

Standardizing Data

The **StatPlus>Standardize** command standardizes values in a collection of data columns. The user can choose one of five different standardization methods.

```
┌─ Standardize Data Values ──────────────────────── ✕ ┐
│  ┌─ Input ─────────────────────────────────────────┐ │
│  │  [ Data Values ]   <None Selected>               │ │
│  │                                                   │ │
│  │   ○ Subtract the mean and divide by the standard deviation │ │
│  │   ○ Subtract the mean                             │ │
│  │   ○ Divide by the standard deviation              │ │
│  │   ○ Subtract [          ] and divide by [        ]│ │
│  │   ○ Range from [    0   ] to [      1 ]            │ │
│  └───────────────────────────────────────────────────┘ │
│  [ Output ]   <None selected>                        │
│  [ OK ]        [ Cancel ]        [ Help ]            │
└─────────────────────────────────────────────────────┘
```

Table Statistics

The **StatPlus>Table Statistics** command creates a table of descriptive statistics for a two-way cross-classification table. The first column of the table contains the titles of the descriptive statistics, the second column shows their values, the third column indicates the degrees of freedom, and the fourth column shows the *p*-value or asymptotic standard error. The user must select the range containing the two-way table, *excluding* the row and column totals but *including* the row and column headers.

```
┌─ Analyze a Two Way Table ──────────────────────── ✕ ┐
│  ┌─ Input ─────────────────────────────────────────┐ │
│  │  Input Columns:                                   │ │
│  │  [                              ] [ - ]           │ │
│  │  Select the range containing the table labels and │ │
│  │  cell counts, but NOT the row or column totals    │ │
│  └───────────────────────────────────────────────────┘ │
│  [ Output ]   <None selected>                        │
│  [ OK ]        [ Cancel ]        [ Help ]            │
└─────────────────────────────────────────────────────┘
```

Univariate Statistics

The **StatPlus>Univariate Statistics** command creates a table of univariate statistics. The user can choose from a selection of 33 different statistics, either by selecting the statistics individually or by selecting entire groups of statistics. Statistics can be displayed in different columns or in different rows. The table can be broken down using a By variable.

Single-Variable Charts

Boxplots

The **StatPlus>Single Variable Charts>Boxplots** command creates a boxplot. The data values can be arranged either as separate columns or in one column with a category variable. Users can choose to add a dotted line for the sample average and to connect the medians between the boxes. The boxplot can be sent to an embedded chart on a worksheet or to its own chart sheet.

Fast Scatterplot

The **StatPlus>Single Variable Charts>Fast Scatterplot** command creates a quick scatterplot bypassing many of the prompts and queries of Excel's Chart wizard. The scatterplot can be sent to an embedded chart on a worksheet or to its own chart sheet.

Histograms

The **StatPlus>Single Variable Charts>Histograms** command creates a histogram. The user can specify a frequency, cumulative frequency, percentage, or cumulative percentage chart. Also, the histogram can be broken down into the different levels of a categorical variable. If a categorical variable is used, the histogram bars can be 1) stacked, 2) displayed side by side, or 3) displayed in 3-D. The user can choose to add a normal curve to the histogram, as well as display the corresponding frequency table. The histogram can be sent to an embedded chart on a worksheet or to its own chart sheet.

Stem and Leaf Plots

The **StatPlus>Single Variable Charts>Stem and Leaf** command creates a stem and leaf plot. The data values can be arranged either as separate columns or in one column with a category variable. If more than one stem and leaf plot is generated, the user can choose to apply the same stem values to each of the plots and to add a summary stem and leaf plot. The user can also choose to truncate outliers of either moderate or major size. The stem and leaf plot appears as values within a worksheet.

Normal Probability Plot

The **StatPlus>Single Variable Charts>Normal P-Plots** command creates a normal probability plot with a table of normal scores for data in a single column. The normal probability plot can be sent to an embedded chart on a worksheet or to its own chart sheet.

Multi-variable Charts

Multiple Histograms

The **StatPlus>Multi-variable Charts>Multiple Histograms** command creates stacked histogram charts. The source data can be arranged in separate columns or within a single column along with a column of category values. The user can choose to display frequencies, cumulative frequencies, percentages, or cumulative percentages. A normal curve can also be added to each of the histograms. The histograms have common bin values and are shown in the same vertical-axis scale. The histogram charts are sent to embedded charts on a worksheet.

Scatterplot Matrix

The **StatPlus>Multi-variable Charts>Scatterplot Matrix** command creates a matrix of scatterplots. The scatterplots are sent to embedded charts on a worksheet.

Quality-Control Charts

C-Charts

The **StatPlus>QC Charts>C-Chart** command creates a C-chart (count chart) of quality-control data for a single column of counts (for example, the number of defects in an assembly line). The quality control chart includes a mean line and lower and upper control limits. The C-chart can be sent to an embedded chart on a worksheet or to its own chart sheet.

Individuals Charts

The **StatPlus>QC Charts>Individuals Chart** command creates an Individuals chart of quality-control data for a single column of quality-control values, where there is no subgroup available. The Individuals chart can be sent to an embedded chart on a worksheet or to its own chart sheet.

Pareto Charts

The **StatPlus>QC Charts>Pareto Chart** command creates a Pareto chart of quality-control data. Data values can be arranged in separate columns or within a single column along with a column of category values. The user specifies conditions for a defective value. The Pareto chart can be sent to an embedded chart on a worksheet or to its own chart sheet.

P-Charts

The **StatPlus>QC Charts>P-Chart** command creates a P-chart (proportion chart) of quality-control data. Proportion values are placed in a single column. The user can specify a single sample size for all proportion values or can use a column of sample-size values. The P-chart can be sent to an embedded chart on a worksheet or to its own chart sheet.

Range Charts

The **StatPlus>QC Charts>Range Chart** command creates a Range chart of quality-control data. The subgroups can be arranged in rows across separate columns or within a single column of data values alongside a column of subgroup levels. The user can use a known value of σ or create the Range chart with an unknown σ-value. The Range chart can be sent to an embedded chart on a worksheet or to its own chart sheet.

Moving Range Charts

The **StatPlus>QC Charts>Moving Range Chart** command creates a Moving Range chart of quality-control data where there is no subgroup available. The quality-control values must be placed in a single column. The Moving Range chart can be sent to an embedded chart on a worksheet or to its own chart sheet.

XBAR Charts

The **StatPlus>QC Charts>XBAR Chart** command creates an xbar chart of quality-control data. The subgroups can be arranged in rows across separate columns or within a single column of data values alongside a column of subgroup levels. The user can use known values of μ and σ or create the xbar chart with unknown μ- and σ-values. The xbar chart can be sent to an embedded chart on a worksheet or to its own chart sheet.

Multivariate Analyses

Correlation Matrix

The **StatPlus>Multivariate>Correlation Matrix** command creates a correlation matrix for data arranged in different columns. The correlation matrix can use either the Pearson correlation coefficient or the nonparametric Spearman rank correlation coefficient. You can also output a matrix of *p*-values for the correlation matrix.

Means Matrix

The **StatPlus>Multivariate>Means Matrix** command creates a matrix of pairwise mean differences for data. The data values can be arranged in separate columns or within a single column alongside a column of category levels. The output includes a matrix of *p*-values with an option to adjust the *p*-value for the number of comparisons using the Bonferroni correction factor.

One-Sample Tests

One-Sample *t*-test

The **StatPlus>One Sample Test>1 Sample t-test** command performs a one sample *t*-test and calculates a confidence interval. The data values can be arranged either as a single column or as two columns (in which case the command will analyze the paired difference between the columns). If two columns are used, the columns must have the same number of rows. Users can specify the null and alternative hypotheses as well as the size of the confidence interval. The output can be broken down by the levels of a By variable.

One-Sample z-test

The **StatPlus>One Sample Test>1 Sample z-test** command performs a one-sample z-test and calculates a confidence interval for data with a known standard deviation. The data values can be arranged either as a single column or as two columns (in which case the command will analyze the paired difference between the columns). If two columns are used, the columns must have the same number of rows. Users can specify the null and alternative hypotheses as well as the size of the confidence interval. The output can be broken down by the levels of a By variable. Users must specify the value of the standard deviation.

One-Sample Sign Test

The **StatPlus>One Sample Test>1 Sample Sign test** command performs a one-sample Sign test and calculates a confidence interval. The data values can be arranged either as a single column or as two columns (in which case the command will analyze the paired difference between the columns). If two columns are used, the columns must have the same number of rows. Users can specify the null and alternative hypotheses as well as the size of the confidence interval. For confidence intervals, the user specifies that the calculated interval be approximately, at least, or at most the size of the specified interval. The output can be broken down by the levels of a By variable.

One-Sample Wilcoxon Signed Rank Test

The **StatPlus>One Sample Test>1 Sample Wilcoxon Signed Rank test** command performs a one-sample Wilcoxon Signed Rank test and calculates a confidence interval. The data values can be arranged either as a single column or as two columns (in which case the command will analyze the paired difference between the columns). If two columns are used, the columns must have the same number of rows. Users can specify the null and alternative hypotheses as well as the size of the confidence interval. The output can be broken down by the levels of a By variable.

Two-Sample Tests

Two-Sample *t*-test

The **StatPlus>Two Sample Tests>2 Sample t-test** command performs a two-sample *t*-test for data values, arranged either in two separate columns or within a single column alongside a column of category levels. Users can specify the null and alternative hypotheses as well as the size of the confidence interval. The test can use either a pooled or an unpooled variance estimate. The output can be broken down by the levels of a By variable.

Two-Sample z-test

The **StatPlus>Two Sample Tests>2 Sample z-test** command performs a two-sample *t*-test for data values, arranged either in two separate columns or within a single column alongside a column of category levels. Users can specify the null and alternative hypotheses as well as the size of the confidence interval. Users must enter the standard deviation for each sample. The output can be broken down by the levels of a By variable.

```
┌─────────────────────────────────────────────────────┐
│ Perform Two Sample or Unpaired z-test          [X]   │
│ ┌─Type────────────────────────────────────────────┐ │
│ │ (•) Use column of category values ( ) Values in 2│ │
│ │                                    separate columns│ │
│ │  [ Data Values ]   <None selected>               │ │
│ │  [ Categories  ]   <None selected>               │ │
│ └──────────────────────────────────────────────────┘ │
│ ┌─Analysis─────────────────────────────────────────┐ │
│ │ Ho: Mean Diff. =      [    0 ] Confidence Level   │ │
│ │                                    [ 0.95 ]       │ │
│ │ Ha:     [ Not equal to ]                          │ │
│ └──────────────────────────────────────────────────┘ │
│ ┌─Sigma────────────────────────────────────────────┐ │
│ │ Enter the population standard deviation           │ │
│ │ for group 1                        [          ]   │ │
│ │ Enter the population standard deviation           │ │
│ │ for group 2                        [          ]   │ │
│ └──────────────────────────────────────────────────┘ │
│  [ Output ]   <None selected>                        │
│  [  By   ]    <None selected>                        │
│  [  OK  ]      [ Cancel ]      [ Help ]              │
└─────────────────────────────────────────────────────┘
```

Two-Sample Mann-Whitney Rank test

The **StatPlus>Two Sample Tests>2 Sample Mann-Whitney Rank test** command performs a two-sample Mann-Whitney Rank test for data values, arranged either in two separate columns or within a single column alongside a column of category levels. Users can specify the null and alternative hypotheses as well as the size of the confidence interval. The output can be broken down by the levels of a By variable.

Time Series Analyses

ACF Plot

The **StatPlus>Time Series>ACF Plot** command creates a table of the autocorrelation function and a chart of the autocorrelation function, for time series data arranged in a single column. The first column in the output table contains the lag values up to a number specified by the user, the second column contains the autocorrelation, the third column of the table contains the lower 95% confidence boundary, and the fourth column contains the upper 95% confidence boundary. Autocorrelation values that lie outside the 95% confidence interval are shown in red. The chart shows the autocorrelations and the 95% confidence width.

Exponential Smoothing

The **StatPlus>Time Series>Exponential Smoothing** command calculates one-, two-, or three-parameter exponential smoothing models for a single column of time series data. You can forecast future values of the time series based on the smoothing model for a specified number of units and include a confidence interval of size specified by the user. The output includes a table of observed and forecasted values, future forecasted values, and a table of descriptive statistics including the mean square error and final values of the smoothing factors. A plot of the seasonal indices (for three-parameter exponential smoothing) is included. The exponential smoothing output is not dynamic and will not update if the source data in the input range changes.

Runs Test

The **StatPlus>Time Series>Runs Test** command performs a Runs test on time series data. The test displays the number of runs, the expected number of runs, and the statistical significance. The cut point can either be the sample mean or be specified by the user.

Seasonal Adjustment

The **StatPlus>Time Series>Seasonal Adjustment** command creates a column of seasonally adjusted values for time series data that show periodicity and creates a plot of unadjusted and adjusted values. A plot of the seasonal indices is included in the output (multiplicative seasonality is assumed). The seasonal adjustment output is not dynamic and will not update if the source data in the input range changes.

Unload Modules

The **StatPlus>Unload Modules** command unloads StatPlus modules. Select the individual modules to unload from the list of loaded modules.

Chart Commands

Label Chart Points

Right-click the chart series in a scatterplot and click the **Label points** command. You can link the labels to a cell range in the workbook and copy the cell format. You can also replace the points in the scatterplot with the labels. Individual points (rather than the entire chart series) can also be labeled.

Display Chart Series by Category

Right-click the chart series in a scatterplot, and click **Display by category**. The command divides the chart series into several different series based on the levels of the category variable. Note that you cannot undo this command. Once the chart series is broken down, it cannot be joined again.

Select Row from Chart Series

Right-click an individual point from a scatterplot and click **Select Row**. The command selects the row in the worksheet corresponding to the point you selected.

The following functions are available in Excel when StatPlus™ is loaded.

Descriptive Statistics for One Variable

Function Name	Description
COUNTBETW	COUNTBETW(*range*, *lower*, *upper*, *boundary*) returns the count of non-blank cells in the *range* that lie between the *lower* and *upper* values. The *boundary* variable determines how the end points are used. If *boundary* = 1, the interval is > the lower value and < the upper value. If *boundary* = 2, the interval is ≥ the lower value and < the upper value. If *boundary* = 3, the interval is > the lower value and ≤ the upper value. If *boundary* = 4, the interval is ≥ lower value and ≤ the upper value.
IQR	IQR(*range*) calculates the interquartile range for the data in *range*.
MODEVALUE	MODEVALUE(*range*) calculates the mode of the data in *range*. The data is assumed to be in one column.
NSCORE	NSCORE(*number*, *range*) returns the normal score of *number* (or cell reference to *number*) from a *range* of values.
RANGEVALUE	RANGEVALUE(*range*) calculates the difference between the maximum and minimum values from a *range* of values.
RANKTIED	RANKTIED(*number*, *range*, *order*) returns the rank of the *number* in *range*, adjusting the rank for ties. If *order* = 0, then the range is ranked from largest to smallest; if *order* = 1, the range is ranked from smallest to largest.
RUNS	RUNS(*range*, [*center*]) returns the number of runs in the data column *range*. The center = 0 unless a *center* value is entered.
SE	SE(*range*) calculates the standard error of the values in *range*.
SIGNRANK	SIGNRANK(*number*, *range*) returns the sign rank of the *number* in *range*, adjusting the rank for ties. Values of zero receive a sign rank of 0. If *order* = 0, then the range is ranked from largest to smallest

in absolute value; if *order* = 1, the range is ranked from smallest to largest in absolute value.

Descriptive Statistics for Two or More Variables

Function Name	Description
CORRELP	CORRELP(*range1*, *range2*) returns the *p*-value for the Pearson coefficient of correlation between *range1* and *range2*. **NOTE: Range values must be in two columns.**
MWMedian	MWMedian(*range*, *range2*) calculates the median of the Walsh averages between two columns of data.
MWMedian2	MWMedian2(*range*, *range2*) calculates the median of the Walsh averages for data values in one column (*range*) with category levels in a second column (*range2*). There can be only two levels in the categories column.
PEARSONCHISQ	PEARSONCHISQ(*range*) returns the Pearson chi-square test statistic for data in *range*.
PEARSONP	PEARSONP(*range*) returns the *p*-value for the Pearson chi-square test statistic for data in *range*.
SPEARMAN	SPEARMAN(*range*) returns the Spearman nonparametric rank correlation for values in *range*. **NOTE: Range values must be in one column only.**
SPEARMANP	SPEARMANP(*range*) returns the *p*-value for the Spearman nonparametric rank correlation for values in *range*. **NOTE: Range values must be in one column only.**

Distributions

Function Name	Description
NORMBETW	NORMBETW(*lower, upper, mean, stdev*) calculates the area under the curve between the *lower* and *upper* limits for a normal distribution with μ = *mean* and σ = *stdev*.
TDF	TDF(*number, df, cumulative*) calculates the area under the curve to the left of *number* for a *t*-distribution with degrees of freedom *df*, if *cumulative* = TRUE. If *cumulative* = FALSE, this function calculates the probability density function for *number*.

Mathematical Formulas

Function Name	Description
IFFUNC	IFFUNC(*Fname, IFRange, IFValue,* [*Arg1, Arg2, ...*]) calculates the value of the Excel function *Fname*, for rows in a data set where the values of *IFRange* are equal to *IFValue*. Parameters of the *Fname* function can be inserted as *Arg1, Arg2,* and so forth.
IF2FUNC	IF2FUNC(*Fname, IFRange1, IFValue1, IFRange2, IFValue2, RangeAnd,* [*Arg1, Arg2,...*]) calculates the value of the Excel function *Fname*, for rows in a data set where the values of *IFRange1* are equal to *IFValue1* and the values of *IFRange2* are equal to *IFValue2*. Parameters of the *Fname* function can be inserted as *Arg1, Arg2,* and so forth. If *RangeAnd* = TRUE, an "AND" clause is assumed between the two values. If *RangeAnd* = FALSE, an "OR" clause is assumed.
RANDBETA	RANDBETA(*alpha, beta,* [*a*], [*b*]) returns a random number from the Beta distribution with parameters *alpha, beta,* and (optionally) *a* and *b* where *a* and *b* are the endpoints of the distribution.
RANDBERNOULLI	RANDBERNOULLI(*prob*) returns a random number from the Bernoulli distribution with probability = *prob*.
RANDBINOMIAL	RANDBINOMIAL(*prob, trials*) returns a random number from the binomial distribution with probability = *prob* and number of trials = *trial*.
RANDCHISQ	RANDCHISQ(*df*) returns a random number from the chi-square distribution with degrees of freedom *df*.
RANDDISCRETE	RANDDISCRETE(*range, prob*) returns a random number from a discrete distribution where the values of the distribution are found in the cell range *range*, and the associated probabilities are found in the cell range *prob*.
RANDEXP	RANDEXP(*lambda*) returns a random number from the exponential distribution where λ = *lambda*.
RANDF	RANDF(*df1, df2*) returns a random number from the *F*-distribution with numerator degrees of freedom *df1* and denominator degrees of freedom *df2*.
RANDGAMMA	RANDGAMMA(*alpha, beta*) returns a random number from the gamma distribution with parameters *alpha* and *beta*.
RANDINTEGER	RANDINTEGER(*lower, upper*) returns a random integer from a discrete uniform distribution with the lower boundary = *lower* and the upper boundary = *upper*.

Function Name	Description
RANDLOG	RANDLOG(*mean, stdev*) returns a random number from the log normal distribution with $\mu = mean$ and $\sigma = stdev$.
RANDNORM	RANDNORM(*mean, stdev*) returns a random number from the normal distribution with $\mu = mean$ and $\sigma = stdev$.
RANDPOISSON	RANDPOISSON(*lambda*) returns a random number from the Poisson distribution where $\lambda = lambda$.
RANDT	RANDT(*df*) returns a random number from the *t*-distribution with degrees of freedom *df*.
RANDUNI	RANDUNI(*lower, upper*) returns a random number from the uniform distribution where the lower boundary = *lower* and the upper boundary = *upper*.

Statistical Analysis

Function Name	Description
ACF	ACF(*range, lag*) calculates the autocorrelation function for values in *range* for lag = *lag*. Note: Range values must lie within one column.
MannW	MannW(*range, range2,* [*median*]) calculates the Mann-Whitney test statistic for data values in two columns. The median difference is assumed to be 0, unless a *median* value is specified.
MannWp	MannWp(*range, range2,* [*median*], [*Alt*]) calculates the *p*-value of the Mann-Whitney test statistic for data values in two columns. The median difference is assumed to be 0, unless a *median* value is specified. The *p*-value is for a two-sided alternative hypothesis unless *Alt* = 1, in which case a one-sided test is performed.
MannW2	MannW2(*range, range2,* [*median*]) calculates the Mann-Whitney test statistic for data values in one column (*range*) and category values in a second column (*range2*). There can be only two levels in the categories column. The median difference is assumed to be 0, unless a *median* value is specified.
MannWp2	MannWp2(*range, range2,* [*median*]) calculates the *p*-value of the Mann-Whitney test statistic for data values in one column (*range*) and category values in a second column (*range2*). There can be only two levels in the categories column. The median difference is assumed to be 0, unless a *median* value is specified. The *p*-value is for a two-sided alternative hypothesis unless *Alt* = 1, in which case a one-sided test is performed.

RUNSP	RUNSP(*range*, [*center*]) calculates the *p*-value of the Runs test for values in the data column *range*. Center = 0 unless a *center* value is entered.
TSTAT	TSTAT(*range*, [*mean*]) calculates the one-sample t-test statistic for values in the data column *range*. The mean value under the null hypothesis is assumed to be 0, unless a *mean* value is specified.
TSTATP	TSTATP(*range*, [*mean*], [*Alt*]) calculates the *p*-value for the one-sample t-test statistic for values in the data column *range*. The mean value under the null hypothesis is assumed to be 0, unless a *mean* value is specified. A two-sided alternative hypothesis is assumed unless *Alt* = −1, in which case the "less than" alternative hypothesis is assumed, or Alt=1, in which case the "greater than" alternative hypothesis is assumed.
WILCOXON	WILCOXON(*range*, [*median*]) calculates the Wilcoxon Signed Rank statistic for values in the data column *range*. The median value under the null hypothesis is assumed to be 0, unless a *median* value is specified.
WILCOXONP	WILCOXONP(*range*, [*median*], [*Alt*]) calculates the *p*-value of the Wilcoxon Signed Rank statistic for values in the data column *range*. The median value under the null hypothesis is assumed to be 0, unless a *median* value is specified. A two-sided alternative hypothesis is assumed unless *Alt* = −1, in which case the "less than" alternative hypothesis is assumed, or *Alt* = 1, in which case the "greater than" alternative hypothesis is assumed.
ZSTAT	ZSTAT(*range, sigma*, [*mean*]) calculates the z-test statistic for values in the data column *range* with a standard deviation *sigma*. The mean value under the null hypothesis is assumed to be 0, unless a *mean* value is specified.
ZSTATP	ZSTATP(*range, sigma*, [*mean*], [*Alt*]) calculates the *p*-value for the z-test statistic for values in the data column *range* with a standard deviation *sigma*. The mean value under the null hypothesis is assumed to be 0, unless a *mean* value is specified. A two-sided alternative hypothesis is assumed unless *Alt* = −1, in which case the "less than" alternative hypothesis is assumed, or *Alt* = 1, in which case the "greater than" alternative hypothesis is assumed.

Bibliography

Bliss, C. I. (1964). *Statistics in Biology*. New York: McGraw-Hill.

Booth, D. E. (1985). Regression methods and problem banks. *Umap Modules: Tools for Teaching 1985*. Arlington, MA: Consortium for Mathematics and Its Applications, pp.179–216.

Bowerman, B.L., and O'Connell, R.T. (1987). *Forecasting and Time Series, An Applied Approach*. Pacific Grove, CA: Duxbury Press.

Cushny, A. R., and Peebles, A. R. (1905). The action of optical isomers, II: Hyoscines. *Journal of Physiology* 32: 501–510.

D'Agostino, R. B., Chase, W., and Belanger A. (1988). The appropriateness of some common procedures for testing the equality of two independent binomial populations. *The American Statistician* 42: 198–202.

Deming, W. E. (1982). *Quality, Productivity, and Competitive Position*. Cambridge, MA: M.I.T. Center for Advanced Engineering Study.

Deming, W. E. (1982). *Out of the Crisis*. Cambridge, MA: M.I.T. Center for Advanced Engineering Study.

Donoho, D., and Ramos, E. (1982). PRIMDATA: Data Sets for Use with PRIM-H (DRAFT). FTP stat library at Carnegie Mellon University.

Edge, O. P., and Friedberg, S. H. (1984). Factors affecting achievement in the first course in calculus. *Journal of Experimental Education* 52: 136–140.

Fosback, N. G. (1987) *Stock Market Logic*. Fort Lauderdale, FL: Institute for Econometric Research.

Halio, M. P. (1990). Student writing: Can the machine maim the message? *Academic Computing*, January 1990, 16–19, 45.

Juran, J. M., ed.(1974) *Quality Control Handbook*. New York: McGraw-Hill.

Longley, J. W. (1967). An appraisal of least squares programs for the electronic computer from the point of view of the user. *Journal of the American Statistical Association* 62: 819–831.

Milliken, G., and Johnson, D. (1984). *Analysis of Messy Data*, Volume 1: Designed Experiments, Princeton, NJ: Van Nostrand.

Neave, H. R. (1990). *The Deming Dimension*. Knoxville, TN: SPC Press.

Rosner, B., and Woods, C. (1988). Autoregressive modeling of baseball performance and salary data. *1988 Proceedings of the Statistical Graphics Section*, American Statistical Association, pp. 132–137.

Shewhart, W. A. (1931). *Economic Control of Quality of Manufactured Product*. Princeton, NJ: Van Nostrand.

Tukey, J. W. (1977). *Exploratory Data Analysis*. Reading, MA: Addison-Wesley.

Weisberg, S. (1985). *Applied Linear Regression*, 2nd ed. New York: Wiley.

Index

ANOVA one-way and, 387–391
command, 527–529
equation, 300–301
fitted regression line, 300, 301–302
functions in Excel, 302–303
model, checking, 314–320
parameter estimates and statistics, 312–313
plotting data, 304–308
residuals, predicted values and, 300, 313–314
residuals, testing, 316–320
simple linear, 300–303
statistics, calculating, 308–311
straight-line assumption, testing, 315–316
Regression, multiple
coefficients and prediction equation, 345–346
example using, 355–367
F-distribution, 338–340
multiple correlation, 344–345
output, interpreting, 343–344
prediction using, 340–343
t-tests for coefficients, 346–347
Regression assumptions, testing
normal errors and plot, 353–354
observed versus predicted values, 347–349
plotting residuals versus predicted values, 350–351
plotting residuals versus predictor variables, 351–353
Rejection region, 219
Relative frequency, 173–174
Relative reference, 51
Replicates, 391
Residuals, 300, 313–314
predicted values versus plotting, 350–351
predictor variables versus plotting, 351–353
testing for constant variance in, 317–318
testing for independence of, 318–320
testing for normal distribution of, 316–317
Robustness 229–235
Row headings, 15
R^2 value, 308
Runs test, 319–320, 566–567

S

Sample, 180
Sampling command, 529–530
Sampling data commands, 548–549
Sampling distributions
creating, 196–200
defined, 196
standard deviation/error, 200–201
Saving work, 30–32
Scatterplots
adding moving average to, 431–432
breaking into categories, 106–108
commands, 552, 555
components of, 83–85
lagged values and, 423–424
matrix (SPLOM), creating, 328–330, 356–357, 555
regression data plotting and use of, 304–308
variables, plotting, 109–110
Scroll bars, 15
vertical, 21
Seasonal/cyclical autocorrelation, 428, 429
Seasonality
additive, 451
adjusting for, 458–459
autocorrelation function and, 457–458
boxplots and, 454–455
command, 567
example of, 452–459
line plots and, 455–458
multiplicative, 450
Shewhart, Walter A., 475
Sign test, 238–240, 562–563
Significance level, 219
Single-Factor command, 515
Skewness, 132, 133, 152
Slope, 300
correlation and, 322
Smoothing factor/constant, 433–434
Solver, 465–468
Somers' D, 283, 289
Sorting data, 54–55, 69–73
custom, 290–292
Spearman's rank correlation coefficient, 323
Special causes, 476

Licensing and Warranty Agreement

Notice to Users:

Do not install or use the CD-ROM until you have read and agreed to this agreement. You will be bound by the terms of this agreement if you install or use the CD-ROM or otherwise signify acceptance of this agreement. If you do not agree to the terms contained in this agreement, do not install or use any portion of this CD-ROM.

License:

The material in the CD-ROM (the "Software") is copyrighted and is protected by United States copyright laws and international treaty provisions. All rights are reserved to the respective copyright holders. No part of the Software may be reproduced, stored in a retrieval system, distributed (including but not limited to over the www/Internet), decompiled, reverse engineered, reconfigured, transmitted, or transcribed, in any form or by any means — electronic, mechanical, photocopying, recording, or otherwise — without the prior written permission of Brooks/Cole (the "Publisher"). Adopters of Berk & Carey's *Data Analysis with Microsoft® Excel* may place the Software on the adopting school's network during the specific period of adoption for classroom purposes only in support of that text. The Software may not, under any circumstances, be reproduced and/or downloaded for sale. For further permission and information, contact Brooks/Cole, 511 Forest Lodge Road, Pacific Grove, California 93950.

U.S. Government Restricted Rights:

The enclosed Software and associated documentation are provided with RESTRICTED RIGHTS. Use, duplication, or disclosure by the Government is subject to restrictions as set forth in subdivision(c)(1)(ii) of the Rights in Technical Data and Computer Software clause at DFARS 252.277.7013 for DoD contracts, paragraphs(c)(1) and (2) of the Commercial Computer Software-Restricted Rights clause in the FAR (48 CFR 52.227-19) for civilian agencies, or in other comparable agency clauses. The proprietor of the enclosed software and associated documentation is Brooks/Cole, 511 Forest Lodge Road, Pacific Grove, CA 93950.

Limited Warranty:

The warranty for the media on which the Software is provided is for ninety (90) days from the original purchase and valid only if the packaging for the Software was purchased unopened. If, during that time, you find defects in the workmanship or material, the Publisher will replace the defective media. The Publisher provides no other warranties, expressed or implied, including the implied warranties of merchantability or fitness for a particular purpose, and shall not be liable for any damages, including direct, special, indirect, incidental, consequential, or otherwise.

For Technical Support:

Voice: 1-800-423-0563
Fax: 1-606-647-5045
E-mail: support@kdc.com

Worksheets, 17
    cells, 22–25
    hidden, 36–37

X

x-axis, 84, 92–93
x-charts (x-bar charts), 480
    calculating, when standard deviation is
        known, 481–482
    calculating, when standard deviation is
        not known, 485–487
    command, 558–559

examples of, 482–485, 487–489
false-alarm rate, 481
X^2-distribution, 278–281

Y

y-axis, 84, 93–94

Z

z-test commands, 534–535, 561, 564
z-test statistic and z-values, 211–214